Student Study Guide with Solutions for

PRECALCULUS

SECOND EDITION

J. Douglas Faires
Youngstown State University

James DeFranza
St. Lawrence University

Brooks/Cole
Thomson Learning™

Pacific Grove • Albany • Belmont • Boston • Cincinnati • Johannesburg • London • Madrid • Melbourne
Mexico City • New York • Scottsdale • Singapore • Tokyo • Toronto

Assistant Editor: Seema Atwal
Marketing Manager: Caroline Croley
Marketing Assistant: Debra Johnston
Editorial Assistant: Carol Ann Benedict
Production Coordinator: Stephanie Andersen
Print Buyer: Micky Lawler
Printing and Binding: Webcom Limited

For more information, contact:
BROOKS/COLE
511 Forest Lodge Road
Pacific Grove, CA 93950 USA
www.brookscole.com

Printed in Canada

5   4   3

ISBN: 0-534-37353-4

PREFACE

This Student Study Guide for PreCalculus was written to accompany the Second Edition of PreCalculus by Faires and DeFranza. The textbook weaves the algebra review material into the sections as it is needed, rather than present it as a block of material at the beginning of the book. This Guide has been prepared to help students who would like to have access to more review material than the textbook provides. While it has been written to accompany our PreCalculus book, the material would be valuable to any student studying precalculus or calculus.

In this Guide you will find extensive algebra review in the early sections, and additional algebra, geometry, and trigonometry topics throughout. There are numerous additional examples and supplemental explanations in the Guide, as well as detailed solutions to all the odd exercises in the Second Edition of Faires and DeFranza PreCalculus.

The topics presented in the Guide should give you an indication of the material in the book that must be mastered if you are to succeed in Precalculus and in Calculus. We recommend that when you are studying calculus you keep this Guide readily available to help you with the background material that will be expected in that course.

In the Appendix to this Guide you will find two copies of a placement examination of the type that has been shown to be effective at Youngstown State University. The answers to the placement examinations are also included, and both can also be downloaded for the WWW site

http://www.as.ysu.edu/~faires/PreCalculus2/Placement

We suggest that you work one of these examinations before you take your Precalculus course and the other after you complete the course. If you score 16 or higher on the 40 question examination you should be sufficiently prepared to take a Precalculus course based on this book. A score of 28 indicates that you are well-prepared for a University Calculus sequence, but you should review the concepts associated with the problems that you missed.

We hope this Guide helps in your study of Precalculus and the Calculus courses you will be taking. If you have any suggestions for improvements that can be incorporated into future editions of the book or into this supplement, we would be most grateful for your comments. We can be most easily contacted by electronic mail at the addresses listed below.

Finally, we would like thank Laurie Marinelli, who managed the production of this Guide, and Tom Wakefield who generated much of the art. Your hard work is much appreciated.

St. Lawrence University

Youngstown State University

June 10, 1999

James DeFranza
defranza@vm.stlawu.edu

J. Douglas Faires
faires@math.ysu.edu

Table of Contents

CHAPTER 1
FUNCTIONS

1.1 Introduction

All branches of mathematics are layered vertically, with the most basic material at the bottom of the stack and the most sophisticated at the top. In order to master the most difficult material you must first become versed in the most basic and work up slowly toward the top. This is certainly the case with Algebra, PreCalculus, and Calculus. In this chapter we review topics from elementary algebra that will be used freely in the material that follows as part of PreCalculus. The first topic is the most basic of all, the number system that is the foundation for most of our work in PreCalculus.

1.2 The Real Line

REAL NUMBERS

The collection of *real numbers* $\mathbb{R}$ consists of several different kinds of numbers. The *natural numbers* $\mathbb{N}$ consist of all positive whole numbers, 1, 2, 3, 4, ..., and the *integers* $\mathbb{Z}$ consist of the positive and negative whole numbers and 0, that is, $\ldots -3, -2, -1, 0, 1, 2, 3, \ldots$. The *rational numbers* $\mathbb{Q}$ include all fractions of the form $\frac{p}{q}$, where p and q are integers with $q \neq 0$. Finally, the *irrational numbers* consist of all numbers that cannot be represented as a rational number, for example, $\sqrt{2}$ or π.

An important way to distinguish between rational and irrational numbers is through the decimal expansions of these numbers. Rational numbers have decimal expansions that have a repeating pattern, and irrational numbers are those with no repeating pattern.

EXAMPLE 1.2.1   Classify each number as a natural, integer, rational, or irrational number.

(a) 192.237   (b) $-2.132132132...$ (c) $2.01001000100001...$ (d) -3

SOLUTION:

(a) It is rational since $192.237 = 192\frac{237}{1000} = \frac{192237}{1000}$.

(b) It is rational since the decimal expansion has the pattern 132 repeated indefinitely. It also isn't hard to find the rational number that is represented by $-2.132\overline{132}$. The trick is to move the decimal point to the right one repeated block. We first set

$$x = 2.132\overline{132}.$$

Then

$$1000x = 2132.132\overline{132}$$

$$1000x - x = 2130.000\overline{000}$$

$$999x = 2130$$

and

$$x = \frac{2130}{999}.$$

So

$$-2.132\overline{132} = -\frac{2130}{999}.$$

(c) This is irrational since the number can not have any finite sequence of digits that repeats indefinitely. Each successive block of zeros contains one additional zero.

(d) This can be described as an integer or a rational number, but not a natural number.

◇

A letter used to represent an arbitrary real number is called a *variable*. A letter used to represent a fixed value, which does not change, is called a *constant*. An *algebraic expression*, is any combination of variables and constants formed using a finite number of the operations of addition, subtraction, multiplication, division, raising to a power, or taking roots. Examples are,

$$ax^2, \quad \frac{x^3 + \sqrt{2x - 1}}{x^5 + x^3 - 1}, \quad \text{and} \quad x^2yz^4 - 3xyz^2 + 2x^3y - 6.$$

The first and third examples are called polynomials and will be considered in more detail later.

The basic rules for combining real numbers are simple but important to recognize when manipulating algebraic expressions.

Commutative Laws: $x + y = y + x$ and $xy = yx$

Associative Laws: $(x + y) + z = x + (y + z)$ and $(xy)z = x(yz)$

Distributive Law: $x(y + z) = xy + xz$

EXAMPLE 1.2.2 Simplify the algebraic expression using the properties for combining real numbers.

(a) $2(x - 3) - 3(2 + 5x)$ (b) $(2x^2 + 3xy)(x - 2xy^2)$

SOLUTION:

(a) First use the Distributive Law to multiply the 2 and the -3 through the parentheses, being careful to multiply both terms in the second parentheses by -3. Then, combining like terms gives

$$2(x-3) - 3(2+5x) = 2x - 6 - 6 - 15x = -13x - 12.$$

(b) Similarly,

$$(2x^2 + 3xy)(x - 2xy^2) = (2x^2 + 3xy)x - (2x^2 + 3xy)2xy^2$$
$$= 2x^3 + 3x^2y - 4x^3y^2 - 6x^2y^3.$$

We used the Distributive Law twice, with the end result that each term in the first parentheses is multiplied by each term in the second.   ◇

SET AND INTERVAL NOTATION

A *set* is a collection of objects called *elements*. Sets are denoted using capital letters, and elements are usually denoted using lower case letters. There are three standard ways of representing a set:

(i) Listing all the elements in brackets { };

(ii) Listing a few of the elements that give the pattern followed by ...;

(iii) Describing a property or properties that all elements satisfy in the form, $\{x \mid x$ satisfies the property $P\}$.

The set with no elements, called the **empty set**, is denoted ϕ. The two basic operations on sets are **union**($\cup$) and **intersection** ($\cap$). If A and B are sets, then $A \cup B$ is the set of all elements from either A or B, and the intersection $A \cap B$ is the set of all elements common to both sets.

EXAMPLE 1.2.3   Determine the intersection and union of the given sets.

(a) $A = \{-5, -2, 0, 1, 2, 3, 6, 8, 13\}, \quad B = \{-3, -2, -1, 5, 7, 9, 11, 13, 15\}$

(b) $A = \{x \mid -2 < x < 10\}, \quad B = \{x \mid x \geq 3\}$

SOLUTION:

(a) The union of the two sets is,

$$A \cup B = \{-5, -3, -2, -1, 0, 1, 2, 3, 5, 6, 7, 8, 9, 11, 13, 15\}.$$

Simply list all the unique elements that are members of A or B.

The intersection is,

$$A \cap B = \{-2, 13\}$$

which consists of the elements common to both sets.

(b) The union is the set of all real numbers to the right of -2, not including -2, so $A \cup B = \{x \mid x > -2\}$. Since 3 lies between -2 and 10, the intersection is $A \cap B = \{x \mid 3 \leq x < 10\}$.

◇

An **interval** is a set of real numbers between two given real numbers where the endpoints may or may not be included.

EXAMPLE 1.2.4   Rewrite the sets using interval notation. Sketch the intervals on a real line.

(a) $\{x \mid -1 < x < 2\}$ (b) $\{x \mid 2 \leq x < 5\}$ (c) $\{x \mid -2 < x \leq 0\}$

(d) $\{x \mid -5 \leq x \leq -2\}$ (e) $\{x \mid x < 3\}$ (f) $\{x \mid x \leq -1\}$

(g) $\{x \mid x > -2\}$ (h) $\{x \mid x \geq 0\}$

SOLUTION:

(a) $\{x \mid -1 < x < 2\} = (-1, 2)$

Figure 1.1

(b) $\{x \mid 2 \le x < 5\} = [2, 5)$

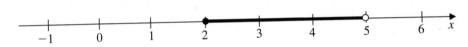

Figure 1.2

(c) $\{x \mid -2 < x \le 0\} = (-2, 0]$

Figure 1.3

(d) $\{x \mid -5 \le x \le -2\} = [-5, -2]$

Figure 1.4

(e) $\{x \mid x < 3\} = (-\infty, 3)$

Figure 1.5

(f) $\{x \mid x \le -1\} = (-\infty, -1]$

Figure 1.6

(g) $\{x \mid x > -2\} = (-2, \infty)$

Figure 1.7

(h) $\{x \mid x \ge 0\} = [0, \infty)$

Figure 1.8

◇

Notice when an endpoint is not included in the interval, a round parenthesis, (or), is used. When the endpoint is included, a square bracket, [or], is used. Since ∞ is only a symbol, not a real number, only round parenthesis are placed next to ∞ and $-\infty$.

The length of an interval is the distance between two points on the real line. The *absolute value* of a real number is the distance from the number to the origin, 0. The absolute value of a number x is defined by

$$|x| = \begin{cases} x, & \text{if } x \geq 0 \\ -x, & \text{if } x < 0. \end{cases}$$

EXAMPLE 1.2.5   Determine the exact value of the absolute value.

(a) $|3.5|$ (b) $|-3|$ (c) $\left|\sqrt{2} - 1\right|$ (d) $\left|1 - \sqrt{3}\right|$

SOLUTION:

(a) $|3.5| = 3.5$

(b) $|-3| = -(-3) = 3$

(c) $\left|\sqrt{2} - 1\right| = \sqrt{2} - 1$, since $\sqrt{2} > 1$.

(d) $\left|1 - \sqrt{3}\right| = -(1 - \sqrt{3}) = \sqrt{3} - 1$, since $\sqrt{3} > 1$. So we have $1 - \sqrt{3} < 0$.   ◇

The distance between two real numbers a and b is $|a - b| = |b - a|$.

EXAMPLE 1.2.6   Determine the distance between the two real numbers.

(a) 2 and 12.7   (b) -3 and 5

SOLUTION:

(a) $|2 - 12.7| = |-10.7| = 10.7$

(b) $8 = |8| = |5 - (-3)| = |-3 - 5| = |-8| = 8$ ◇

Notice that the order of subtraction does not matter as long as the computations are done inside the absolute value.

Exponents and Radicals

Listed are the basic definitions and properties.

$$x^n = \underbrace{x \cdot x \cdot x \cdots x}_{n\text{-times}} \qquad\qquad x^{-n} = \frac{1}{x^n} \qquad\qquad x^{1/n} = \sqrt[n]{x}$$

$$x^m x^n = x^{m+n} \qquad\qquad (xy)^n = x^n y^n \qquad\qquad (x^m)^n = x^{mn}$$

$$\frac{x^m}{x^n} = x^{m-n} \qquad\qquad \left(\frac{x}{y}\right)^n = \frac{x^n}{y^n}$$

$$\sqrt[n]{xy} = \sqrt[n]{x}\,\sqrt[n]{y} \qquad\qquad \sqrt[n]{\frac{x}{y}} = \frac{\sqrt[n]{x}}{\sqrt[n]{y}} \qquad\qquad x^{m/n} = \sqrt[n]{x^m} = \left(\sqrt[n]{x}\right)^m$$

EXAMPLE 1.2.7 Simplify each of the expressions. Do not use negative exponents in the final answer.

(a) $x^3 x^4$ (b) $x^2 x^{-9}$ (c) $\dfrac{x^5}{x^2}$ (d) $\dfrac{x^2}{x^7}$

SOLUTION:

(a) $x^3 x^4 = x^{3+4} = x^7$

(b) $x^2 x^{-9} = x^{2-9} = x^{-7} = \dfrac{1}{x^7}$

(c) $\dfrac{x^5}{x^2} = x^{5-2} = x^3$

(d) $\dfrac{x^2}{x^7} = x^{2-7} = x^{-5} = \dfrac{1}{x^5}$

A shortcut in problems like (c) and (d) is to subtract the smaller exponent from the larger, placing the result in the numerator or denominator depending on which has the term with the larger exponent. ◇

EXAMPLE 1.2.8 Simplify each of the expressions.

(a) $(2x^3 y^2)(x^4 y^3)^3$ (b) $\left(\dfrac{x^2 y}{xy}\right)^3 \left(\dfrac{xy^5}{x^6 y^2}\right)^2$ (c) $\left(\dfrac{2x^{-2} y^3}{3x^4 y}\right)^{-2}$

SOLUTION:

(a) First bring the outer exponent of 3 inside the second parentheses by multi-plying each inside exponent by 3, and then combine the exponents on like terms. This gives

$$(2x^3y^2)(x^4y^3)^3 = (2x^3y^2)(x^{12}y^9) = 2x^{15}y^{11}.$$

(b) $\left(\dfrac{x^2y}{xy}\right)^3 \left(\dfrac{xy^5}{x^6y^2}\right)^2 = (x)^3 \left(\dfrac{y^3}{x^5}\right)^2 = x^3\dfrac{y^6}{x^{10}} = \dfrac{y^6}{x^7}$

(c) $\left(\dfrac{2x^{-2}y^3}{3x^4y}\right)^{-2} = \left(\dfrac{2y^2}{3x^6}\right)^{-2} = \dfrac{2^{-2}y^{-4}}{3^{-2}x^{-12}} = \dfrac{9x^{12}}{4y^4}$

In parts (b) and (c) it is slightly easier to simplify inside the parentheses first and then apply the outside exponent, as we did in the solution. Try to solve the problem another way.   ◇

EXAMPLE 1.2.9   Rationalize each of the fractions involving radicals.

(a) $\dfrac{1}{\sqrt{3}}$ (b) $\sqrt{\dfrac{3}{5}}$ (c) $\dfrac{\sqrt{3}-1}{\sqrt{3}+1}$

SOLUTION: *Rationalizing* a fraction involving radicals refers to eliminating all radicals from the denominator of the expression by multiplying by a suitable fraction equaling 1. This was done originally to simplify calculations that were performed by hand. It is much easier to divide radical approximations by inte-gers, for example, than it is to divide integers by radical approximations. While this is no longer an important application because of the wide-spread availability of calculators, it is still a useful procedure.

(a) $\dfrac{1}{\sqrt{3}} = \dfrac{1}{\sqrt{3}}\dfrac{\sqrt{3}}{\sqrt{3}} = \dfrac{\sqrt{3}}{3}$

(b) $\sqrt{\dfrac{3}{5}} = \dfrac{\sqrt{3}}{\sqrt{5}} = \dfrac{\sqrt{3}}{\sqrt{5}}\dfrac{\sqrt{5}}{\sqrt{5}} = \dfrac{\sqrt{3}\sqrt{5}}{5} = \dfrac{\sqrt{15}}{5}$

(c) $\dfrac{\sqrt{3}-1}{\sqrt{3}+1} = \dfrac{\sqrt{3}-1}{\sqrt{3}+1}\dfrac{\sqrt{3}-1}{\sqrt{3}-1} = \dfrac{3-2\sqrt{3}+1}{3-1} = \dfrac{4-2\sqrt{3}}{2} = 2 - \sqrt{3}$

In (c), simply change the sign between the $\sqrt{3}$ and 1 in the denominator from plus to minus and multiply and divide by $\sqrt{3}-1$. If the sign in between were a minus, then switch it to a plus sign.   ◇

EXAMPLE 1.2.10   Simplify each of the expressions.

(a) $\sqrt[3]{8x^6y^5}$ (b) $\left(\dfrac{9x^{2/3}y^4}{4x^9}\right)^{3/2}$

SOLUTION:

(a) $\sqrt[3]{8x^6y^5} = \left(8x^6y^5\right)^{1/3} = 8^{1/3}\left(x^6\right)^{1/3}\left(y^5\right)^{1/3} = 2x^2y^{5/3}$.

(b) $\left(\dfrac{9x^{2/3}y^4}{4x^9}\right)^{3/2} = \dfrac{(9)^3xy^6}{(\sqrt{4})^3x^{27/2}} = \dfrac{27xy^6}{8x^{27/2}} = \dfrac{27y^6}{8x^{25/2}}$

In part (b), it is a bit more difficult to first simplify inside the parentheses and then apply the rational exponent $3/2$.   ◇

SPECIAL PRODUCTS AND FACTORING

The following formulas are very useful for quickly multiplying certain algebraic expressions which arise frequently, and are especially useful when going in the reverse direction of factoring an algebraic expression.

$$(x+y)^2 = x^2 + 2xy + y^2 \qquad\qquad (x-y)^2 = x^2 - 2xy + y^2$$
$$(x+y)(x-y) = x^2 - y^2$$
$$(x+y)^3 = x^3 + 3x^2y + 3xy^2 + y^3 \qquad (x-y)^3 = x^3 - 3x^2y + 3xy^2 - y^3$$
$$x^3 + y^3 = (x+y)(x^2 - xy + y^2) \qquad x^3 - y^3 = (x-y)(x^2 + xy + y^2)$$

Notice that $(x+y)^2 \neq x^2 + y^2$, $(x-y)^2 \neq x^2 - y^2$, $(x+y)^3 \neq x^3 + y^3$, and $(x-y)^3 \neq x^3 - y^3$.

EXAMPLE 1.2.11   Perform the indicated multiplication.

(a) $(2x - 2y)(2x + 2y)$ (b) $\left(x^2 + y^2\right)^2$ (c) $\left(3x^3 - y\right)^3$ (d) $(x - \sqrt{x})(x + \sqrt{x})$

SOLUTION: In each part we will apply the appropriate special product formula.

(a) $(2x - 2y)(2x + 2y) = (2x)^2 - (2y)^2 = 4x^2 - 4y^2.$

(b) $\left(x^2 + y^2\right)^2 = \left(x^2\right)^2 + 2x^2y^2 + \left(y^2\right)^2 = x^4 + 2x^2y^2 + y^4.$

(c)

$$(3x^3 - y)^3 = (3x^3)^3 - 3(3x^3)^2 y + 3(3x^3)y^2 - y^3$$
$$= 27x^9 - 27x^6 y + 9x^3 y^2 - y^3.$$

(d) $(x - \sqrt{x})(x + \sqrt{x}) = x^2 - (\sqrt{x})^2 = x^2 - x.$

The special product formulas are not needed in computing the products, although it is easier to use them. For example, in part (b) we can write

$$(x^2 + y^2)^2 = (x^2 + y^2)(x^2 + y^2),$$

then multiply each term in the first parenthesis by each term in the second and collect terms to get the desired answer. This would be more difficult in part (c). To convince yourself of this, you might try the problems this alternative way.

◇

EXAMPLE 1.2.12   Factor each expression.

(a) $4x^2 - 9y^2$ (b) $x^2 + 4x + 4$ (c) $4x^3 - 20x$ (d) $x^4 + 2x^2 + 1$

SOLUTION: Each problem fits one of the special formulas given.

(a) To get the expression to fit the formula for the difference of two squares, we note that $4x^2 = (2x)^2$ and $9y^2 = (3y)^2$, so

$$4x^2 - 9y^2 = (2x)^2 - (3y)^2 = (2x + 3y)(2x - 3y).$$

(b) $x^2 + 4x + 4 = (x + 2)^2$.

(c) Always look first for common terms that can be factored out of the expression. This usually simplifies the work. In this case we have

$$4x^3 - 20x = 4x(x^2 - 5) = 4x(x^2 - (\sqrt{5})^2) = 4x(x - \sqrt{5})(x + \sqrt{5}).$$

(d) There are no common factors, but treating the term x^2 as the x in the special formulas and the 1 as the y we have

$$x^4 + 2x^2 + 1 = (x^2)^2 + 2(x^2) + 1 = (x^2 + 1)^2.$$

◇

Usually the more difficult problems to factor are those that do not fit one of the special formulas. Factoring is needed so often that it is very important to feel comfortable with the process. To factor an expression of the form $ax^2 + bx + c$, it is necessary to choose factors of the numbers a and c and their signs in just the right way.

EXAMPLE 1.2.13   Factor each expression.

(a) $x^2 - x - 2$ (b) $6x^2 + 7x + 2$ (c) $2x^3 - 3x^2 - 2x$

Solution:

(a) The only factor of $a = 1$ is 1, and the only factors of $c = 2$ are 1 and 2. So the only choice for a possible factoring is

$$x^2 - x - 2 = (x \pm 1)(x \pm 2).$$

To get the constant term of -2, the signs of 1 and 2 must differ. Since the middle term is $-x$ rather then x, the minus sign must be with the 2. So

$$x^2 - x - 2 = (x + 1)(x - 2).$$

(b) The factors of $a = 6$ are $1, 2, 3, 6$, and the only factors of $c = 2$ are 1 and 2. Since the constant term is positive, and the middle term is also positive, the signs inside the factors will also be both positive. If both signs were negative, we would have a positive constant term but a negative middle term. The possibilities are

$$(6x + 1)(x + 2) = 6x^2 + 13x + 2$$
$$(6x + 2)(x + 1) = 6x^2 + 8x + 2$$
$$(3x + 1)(2x + 2) = 6x^2 + 8x + 2$$
$$(3x + 2)(2x + 1) = 6x^2 + 7x + 2,$$

and we see the last one gives the answer.

(c) $2x^3 - 3x^2 - 2x = x(2x^2 - 3x - 2) = x\,(2x + 1)\,(x - 2)\,.$ ◇

EXAMPLE 1.2.14   Simplify each expression.

(a) $\dfrac{x^2 + 5x + 6}{x^2 - 2x - 8}$ (b) $\dfrac{x^2 + 2x + 1}{x^2 - x - 2} \cdot \dfrac{2x^2 - 5x + 2}{x^2 + 4x + 3}$ (c) $\dfrac{2x^2 + 8x}{x^2 - 3x + 2} \div \dfrac{x^2 + 6x + 8}{x - 1}$

SOLUTION:

(a) When simplifying fractional expressions, always try to first factor as much as possible and then cancel any like terms from the numerator and denominator. We have,

$$\frac{x^2 + 5x + 6}{x^2 - 2x - 8} = \frac{(x+2)(x+3)}{(x+2)(x-4)} = \frac{x+3}{x-4}.$$

(b) The rule

$$\frac{a}{b} \cdot \frac{c}{d} = \frac{ac}{bd}$$

for multiplying two number fractions is also used for algebraic expressions. So,

$$\begin{aligned}
\frac{x^2 + 2x + 1}{x^2 - x - 2} \cdot \frac{2x^2 - 5x + 2}{x^2 + 4x + 3} &= \frac{(x+1)^2}{(x+1)(x-2)} \cdot \frac{(2x-1)(x-2)}{(x+3)(x+1)} \\
&= \frac{(x+1)^2(2x-1)(x-2)}{(x+1)^2(x+3)(x-2)} \\
&= \frac{2x-1}{x+3}.
\end{aligned}$$

(c) The rule

$$\frac{\frac{a}{b}}{\frac{c}{d}} = \frac{a}{b} \cdot \frac{d}{c}$$

for dividing two number fractions is also used for algebraic expressions. So,

$$\begin{aligned}
\frac{2x^2 + 8x}{x^2 - 3x + 2} \div \frac{x^2 + 6x + 8}{x - 1} &= \frac{2x^2 + 8x}{x^2 - 3x + 2} \cdot \frac{x - 1}{x^2 + 6x + 8} \\
&= \frac{2x(x+4)}{(x-1)(x-2)} \cdot \frac{(x-1)}{(x+2)(x+4)} \\
&= \frac{2x}{(x-2)(x+2)} \\
&= \frac{2x}{x^2 - 4}.
\end{aligned}$$

◇

EXAMPLE 1.2.15   Simplify each expression.

(a) $\dfrac{1}{x-2} + \dfrac{4}{x+1}$ (b) $\dfrac{1}{\sqrt{x}} - \dfrac{1}{\sqrt{x+a}}$

SOLUTION:

(a) To add the fractions, we need to find the least common denominator. The rule

$$\frac{a}{b} + \frac{c}{d} = \frac{ad+cb}{bd}$$

for adding number fractions is used. So

$$\frac{1}{x-2} + \frac{4}{x+1} = \frac{(x+1) + 4(x-2)}{(x-2)(x+1)}$$

$$= \frac{5x-7}{x^2 - x - 2}.$$

(b) Taking the common denominator, we have

$$\frac{1}{\sqrt{x}} - \frac{1}{\sqrt{x+a}} = \frac{\sqrt{x+a} - \sqrt{x}}{\sqrt{x}\sqrt{x+a}}.$$

Expressions like these can also be rationalized to eliminate some of the radicals. In this case we eliminate the radical in the numerator. That is,

$$\frac{1}{\sqrt{x}} - \frac{1}{\sqrt{x+a}} = \frac{\sqrt{x+a} - \sqrt{x}}{\sqrt{x}\sqrt{x+a}}$$

$$= \frac{\sqrt{x+a} - \sqrt{x}}{\sqrt{x}\sqrt{x+a}} \cdot \frac{\sqrt{x+a} + \sqrt{x}}{\sqrt{x+a} + \sqrt{x}}$$

$$= \frac{(x+a) - x}{\sqrt{x}\sqrt{x+a}(\sqrt{x+a} + \sqrt{x})}$$

$$= \frac{a}{\sqrt{x}\sqrt{x+a}(\sqrt{x+a} + \sqrt{x})}.$$

◇

SOLVING EQUATIONS AND INEQUALITIES

Solving equations and inequalities arise as subproblems to many problems encountered in a PreCalculus or Calculus course. To solve an equation or inequality means to find all real numbers that satisfy the given condition. Equations and inequalities are solved by using the basic rules of arithmetic.

EXAMPLE 1.2.16   Find all values of x that satisfy the equation.

(a) $5x - 9 = 6$ (b) $\dfrac{2}{3}x + 3 = 2$ (c) $\dfrac{2x}{x+1} + \dfrac{1}{2} = 3$

SOLUTION:

(a) To solve the equation, first add 9 to *both* sides of the equation and then divide by 5. So

$$5x - 9 = 6$$

$$5x - 9 + 9 = 6 + 9$$

$$5x = 15$$

$$\frac{5x}{5} = \frac{15}{5}$$

and

$$x = 3.$$

(b) First subtract 3 from both sides of the equation, and then multiply by 3/2. So

$$\frac{2}{3}x + 3 = 2$$

$$\frac{2}{3}x = -1$$

$$\frac{3}{2}\left(\frac{2}{3}x\right) = \frac{3}{2}(-1)$$

and

$$x = -\frac{3}{2}.$$

(c) First subtract $1/2$ from both sides of the equation, multiply both sides by $(x + 1)$ assuming $x \neq -1$, and then multiply both sides by 2. So

$$\frac{2x}{x + 1} + \frac{1}{2} = 3$$

$$\frac{2x}{x + 1} = \frac{5}{2}$$

$$(x + 1)\frac{2x}{x + 1} = \frac{5}{2}(x + 1)$$

$$2x = \frac{5(x + 1)}{2}$$

$$4x = 5x + 5$$

$$-x = 5$$

and

$$x = -5.$$

◇

EXAMPLE 1.2.17   Solve each equation.

(a) $2x^2 + 5x - 3 = 0$ (b) $x^2 + x - 2 = 4$ (c) $x^4 + x^2 - 2 = 0$

SOLUTION:

(a) To solve the equation for x, we first factor the left side. Therefore,

$$2x^2 + 5x - 3 = 0$$

and

$$(2x - 1)(x + 3) = 0.$$

The product of two real numbers can be 0, only if at least one of the numbers is 0. So

$$2x - 1 = 0 \quad \text{or} \quad x + 3 = 0$$

and

$$x = \frac{1}{2} \quad \text{or} \quad x = -3.$$

(b) First rewrite the equation so 0 is on the right side, then factor. So

$$x^2 + x - 2 = 4$$

$$x^2 + x - 6 = 0$$

$$(x + 3)(x - 2) = 0$$

and

$$x = -3 \quad \text{or} \quad x = 2.$$

(c) If we let $u = x^2$, then $x^4 + x^2 - 2 = u^2 + u - 2 = (u + 2)(u - 1)$, and

$$x^4 + x^2 - 2 = 0$$

$$(x^2 + 2)(x^2 - 1) = 0.$$

Therefore,

$$x^2 + 2 = 0 \quad \text{or} \quad x^2 - 1 = 0.$$

Since $x^2 + 2 = 0$ has no real solutions, the only solutions are when

$$x^2 - 1 = 0$$

$$x^2 = 1$$

and $x = \pm 1.$ ◇

If the solutions to an equation of the form $ax^2 + bx + c = 0$ are rational numbers, with $a \neq 0$, then the expression can be factored and the solutions found as we did above.

In general, the solutions to the **quadratic equation** $ax^2 + bx + c = 0$ with $a \neq 0$, can be found using the **quadratic formula.** The solutions are

$$x = \frac{-b + \sqrt{b^2 - 4ac}}{2a} \quad \text{and} \quad x = \frac{-b - \sqrt{b^2 - 4ac}}{2a},$$

which are often written compactly as

$$x = \frac{-b \pm \sqrt{b^2 - 4ac}}{2a}.$$

EXAMPLE 1.2.18 Find all solutions to $x^2 + 4x + 2 = 0$ and factor the expression $x^2 + 4x + 2$.

SOLUTION: Using the Quadratic formula with $a = 1, b = 4$, and $c = 2$, gives

$$\begin{aligned} x &= \frac{-4 \pm \sqrt{16 - 4(1)(2)}}{2} \\ &= \frac{-4 \pm \sqrt{8}}{2} = \frac{-4 \pm 2\sqrt{2}}{2} \\ &= -2 \pm \sqrt{2}. \end{aligned}$$

Once the solutions (called the **roots**) are known, the expression can be factored completely as

$$x^2 + 4x + 2 = (x - (-2 + \sqrt{2}))(x - (-2 - \sqrt{2})).$$

◇

In the previous example, the solutions are irrational numbers, and we needed to use the Quadratic formula to solve the equation. We cannot find the

factors of the expression using a trial and error approach, as was done when
the solutions were rational numbers.

EXAMPLE 1.2.19   Find all solutions to the equations.

(a) $\sqrt{2x-1} = 3$ (b) $\sqrt{x-2} + x = 2$

SOLUTION:

(a) First square both sides and then isolate x, giving

$$\sqrt{2x-1} = 3$$

$$2x - 1 = 9$$

and

$$x = 5.$$

Because we squared the equation, we need to check the answer to be sure that
we have not introduced an *extraneous* solution. That is, a solution to the final
equation that was not a solution to the original equation. In this case we did
not since $\sqrt{2(5)-1} = \sqrt{9} = 3$.

(b) In problems like this, it is easiest to first isolate the radical term and then
square both sides. So

$$\sqrt{x-2} + x = 2$$

$$\sqrt{x-2} = 2 - x$$

$$x - 2 = (2-x)^2$$

$$x - 2 = 4 - 4x + x^2$$

$$x^2 - 5x + 6 = 0$$

$$(x-2)(x-3) = 0$$

and

$$x = 2 \quad \text{or} \quad x = 3.$$

It is important that you check the answers, since squaring both sides of an equation may introduce extraneous solutions. Checking,

$$\sqrt{3 - 2} + 3 = 4 \neq 2$$

but

$$\sqrt{2 - 2} + 2 = 2$$

and we see $x = 2$ *is* a solution, but $x = 3$ *is not* a solution.   ◇

EXAMPLE 1.2.20   Find all values of x that satisfy the equation $\left| \dfrac{2x - 1}{x + 3} \right| = 4$.

SOLUTION: The exact expression inside the absolute value is unimportant. The solution to an equation of the form

$$|\,\square\,| = a$$

can always be solved by solving the two equations

$$\square = a \quad \text{and} \quad \square = -a,$$

since the absolute value of a quantity and its negative are the same. We have

$$\left| \frac{2x - 1}{x + 3} \right| = 4$$

$$\frac{2x - 1}{x + 3} = 4 \quad \text{or} \quad \frac{2x - 1}{x + 3} = -4.$$

So

$$2x - 1 = 4x + 12 \quad \text{or} \quad 2x - 1 = -4x - 12$$

$$2x = -13 \quad \text{or} \quad 6x = -11$$

and

$$x = -\frac{13}{2} \quad \text{or} \quad x = -\frac{11}{6}.$$

$\diamond$

To solve inequalities, a few basic properties are needed.

Properties of Inequalities

1. If $a < b$ and $b < c$, then $a < c$.
2. If $a < c$, then $a + b < c + b$.
3. If $a < b$, and $c > 0$, then $ac < bc$.
4. If $a < b$, and $c < 0$, then $ac > bc$.

EXAMPLE 1.2.21   Solve each inequality.

(a) $2x - 3 > 2$ (b) $-5x + 2 < 3$ (c) $-1 < 2x - 3 < 7$

SOLUTION:

(a) First apply Property 2, adding 3 to both sides to get,

$$2x - 3 + 3 > 2 + 3$$

so

$$2x > 5.$$

Then use Property 3, and divide both sides by 2, so

$$\frac{2x}{2} > \frac{5}{2}$$

and

$$x > \frac{5}{2}.$$

In interval notation the solution is $\left(\frac{5}{2}, \infty\right)$.

(b) We have

$$-5x + 2 < 3$$

$$-5x + 2 - 2 < 3 - 2$$

$$-5x < 1$$

$$\frac{-5x}{-5} > \frac{1}{-5}$$

and

$$x > -\frac{1}{5}.$$

In interval notation the solution can be written $\left(-\frac{1}{5}, \infty\right)$. In the division step, be very careful to reverse the inequality sign since both sides of the inequality were divided by a negative number (dividing by -5 is the same as multiplying by $-1/5$).

(c) The inequality $-1 < 2x - 3 < 7$ is shorthand for the two inequalities,

$$-1 < 2x - 3 \quad \text{and} \quad 2x - 3 < 7.$$

Solving the pair of inequalities simultaneously gives

$$-1 < 2x - 3 \quad \text{and} \quad 2x - 3 < 7$$

$$2 < 2x \quad \text{and} \quad 2x < 10$$

$$1 < x \quad \text{and} \quad x < 5$$

so

$$1 < x < 5.$$

In interval notation the solution is $(1, 5)$.   ◇

EXAMPLE 1.2.22   Solve each inequality.

(a) $x^2 - x - 2 > 0$ (b) $-x^2 - 2x + 3 > 0$ (c) $\dfrac{x^2 + 3x + 2}{x^2 + 1} < 0$

SOLUTION: To solve a quadratic inequality, first try to factor the quadratic expression. In the case of a fractional expression, as in part (c), try to factor numerator and denominator.

(a) Factoring we have

$$x^2 - x - 2 = (x - 2)(x + 1) > 0.$$

In order for the product of two linear factors to be positive, either both factors must be positive or both must be negative. For example:

$$\text{If } x = 3, \text{ then } (3 - 2)(3 + 1) = (1)(4) = 4 > 0;$$

$$\text{If } x = -2, \text{ then } (-2 - 2)(-2 + 1) = (-4)(-1) = 4 > 0;$$

$$\text{But if } x = 1.5, \text{ then } (1.5 - 2)(1.5 + 1) = (-0.5)(2.5) = -1.25 < 0.$$

The values that make the linear factors 0, in this case at $x = 2$ and $x = -1$, separate the real line into three intervals, $(-\infty, -1)$, $(-1, 2)$, and $(2, \infty)$. To solve the inequality, select test values from each interval, substitute these values into the equation, and the inequality will have the same sign for all values in the interval.

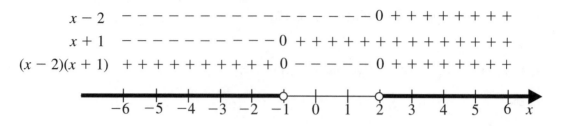

$$
\begin{aligned}
x - 2 \quad &- - - - - - - - - - - - - - - - 0 + + + + + + + + \\
x + 1 \quad &- - - - - - - - - - - 0 + + + + + + + + + + + + + \\
(x - 2)(x + 1) \quad &+ + + + + + + + + + 0 - - - - - 0 + + + + + + + +
\end{aligned}
$$

Figure 1.9

From the chart, the solution to the inequality is all $x < -1$ or all $x > 2$, that is, $(-\infty, -1) \cup (2, \infty)$.

(b) Since

$$-x^2 - 2x + 3 = -(x^2 + 2x - 3) = -(x + 3)(x - 1)$$

we have the chart

$$
\begin{aligned}
x + 3 \quad &- - - - - - 0 + \\
x - 1 \quad &- - - - - - - - - - - - - - - - 0 + + + + + + + + + + \\
(x + 3)(x - 1) \quad &+ + + + + + + 0 - - - - - - - 0 + + + + + + + + + + + \\
-(x + 3)(x - 1) \quad &- - - - - - - 0 + + + + + + + + 0 - - - - - - - - - - -
\end{aligned}
$$

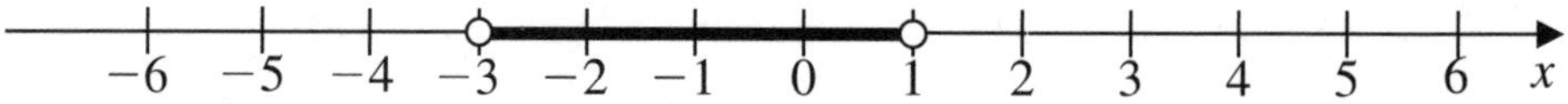

Figure 1.10

The solution is all x in the interval $(-3, 1)$.

(c) Since the denominator can not be factored,

$$\frac{x^2 + 3x + 2}{x^2 + 1} = \frac{(x + 2)(x + 1)}{x^2 + 1}.$$

We could proceed as we did in parts (a) and (b), but in examples like this it is easier to first observe that the denominator is always greater than 0, so

$$\frac{x^2 + 3x + 2}{x^2 + 1} < 0 \text{ precisely when } (x + 2)(x + 1) < 0.$$

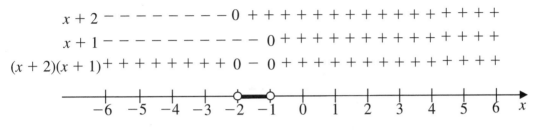

Figure 1.11

The solution is the interval $(-2, -1)$.   ◇

To solve inequalities involving absolute values use the two facts:

$$|x| < a \quad \text{means} \quad -a < x < a$$

$$|x| > a \quad \text{means} \quad x < -a \quad \text{or} \quad x > a.$$

Often in problems like this, an expression involving x is inside the absolute value signs rather than x alone. This is not a concern since the x is just a place holder for any real number. The first statement, for example, could be written as

$$|\square| < a \quad \text{means} \quad -a < \square < a,$$

where $\square$ can be any expression. A similar interpretation can be given to the second statement.

EXAMPLE 1.2.23 Solve each inequality.

(a) $|2x - 1| < 2$ (b) $|3x + 2| > 1$ (c) $\left| \dfrac{2}{x+1} \right| < 3$

SOLUTION:

(a) We have

$$|2x - 1| < 2 \Rightarrow 2 \left| x - \frac{1}{2} \right| < 2 \Rightarrow \left| x - \frac{1}{2} \right| < 1$$

Figure 1.12

and

$$-1 < x - \frac{1}{2} < 1$$

so

$$-\frac{1}{2} < x < \frac{3}{2}.$$

The solution set is the interval $\left(-\frac{1}{2}, \frac{3}{2} \right)$.

(b) We have

$$|3x + 2| > 1 \Rightarrow \left| x + \frac{2}{3} \right| > \frac{1}{3}$$

and

Figure 1.13

$$x + \frac{2}{3} < -\frac{1}{3} \quad \text{or} \quad x + \frac{2}{3} > \frac{1}{3}$$

so

$$x < -1 \quad \text{or} \quad x > -\frac{1}{3}.$$

In interval notation the solution set is $(-\infty, -1) \cup \left(-\frac{1}{3}, \infty\right)$.

(c) For this problem we have

$$\left| \frac{2}{x+1} \right| < 3 \Rightarrow \frac{2}{3} < |x+1|$$

and

$$\frac{2}{3} < x + 1 \quad \text{or} \quad x + 1 < -\frac{2}{3}.$$

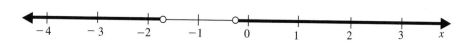

Figure 1.14

So

$$x < -\frac{5}{3} \quad \text{or} \quad x > -\frac{1}{3}$$

and the solution set is $\left(-\infty, -\frac{5}{3}\right) \cup \left(-\frac{1}{3}, \infty\right)$.

Solutions for Exercise Set 1.2

1. $-2 \le x \le 4$

3. $-\sqrt{3} < x \le \sqrt{2}$

5. $x < 3$

7. $x \ge \sqrt{2}$

9. $[-1, 2]$

11. $[2, 5)$

13. $(-\infty, 2)$

15. $[3, \infty)$

17. (a) $|3 - 7| = |-4| = 4$ (b) $\frac{3+7}{2} = 5$

19. (a) $|-3 - 5| = |-8| = 8$ (b) $\frac{-3+5}{2} = 1$

21. $x + 4 < 7 \Leftrightarrow x < 3$

Interval Notation: $(-\infty, 3)$

23. $2x - 2 \ge 8 \Leftrightarrow 2x \ge 10 \Leftrightarrow x \ge 5$

Interval Notation: $[5, \infty)$

25. $-3x + 4 < 5 \Leftrightarrow -3x < 1 \Leftrightarrow x > -\frac{1}{3}$

Interval Notation: $\left(-\frac{1}{3}, \infty\right)$

27. $2x + 9 \le 5 + x \Leftrightarrow x \le -4$

Interval Notation: $(-\infty, -4]$

29. $-2 < 3x - 2 < 5 \Leftrightarrow 0 < 3x < 7 \Leftrightarrow 0 < x < \frac{7}{3}$

Interval Notation: $\left(0, \frac{7}{3}\right)$

31. For $(x + 1)(x - 2) \ge 0$.

$$(x+1) \quad -------- \; 0 +++++++++++++$$
$$(x-2) \quad --------------- \; 0 ++++++$$
$$(x+1)(x-2) \quad ++++++++0 ----- \; 0 ++++++$$

```
◄━━━┿━━┿━━┿━━┿━━●━━┿━━┿━●━━┿━━┿━━┿━━►
           −1   0      2              x
```

The solution set is where the last row is positive or 0, so it is $(-\infty, -1] \cup [2, \infty)$.

33. $x^2 - 3x + 2 \le 0 \Leftrightarrow (x-2)(x-1) \le 0$

$$x - 2 \quad --------------- \; 0 +++++++$$
$$x - 1 \quad ------------- \; 0 +++++++++$$
$$(x-2)(x-1) \quad +++++++++++++ \; 0 - 0 ++++++$$

```
┿━━┿━━┿━━┿━━┿━━┿━━┿━━●━━━●━━┿━━┿━━►
        −5              0  1  2       5   x
```

The solution set is where the last row is negative or 0, so it is $[1, 2]$.

35. For $(x-1)(x-2)(x+1) \le 0$.

$$x - 1 \quad ------------ \; 0 +++++++++$$
$$x - 2 \quad -------------- \; 0 ++++++$$
$$x + 1 \quad -------- \; 0 ++++++++++++++$$
$$(x-1)(x-2)(x+1) \quad -------- \; 0 +++0 - 0 ++++++$$

```
◄━━━┿━━━┿━━━┿━━━●━━┿━━●━━●━━┿━━┿━━►
        −5         −1  0  1  2       5   x
```

The solution set is where the last row is negative or 0, so it is $(-\infty, -1] \cup [1, 2]$.

37. For $x^3 - 3x^2 + 2x \ge 0 \Leftrightarrow x(x^2 - 3x + 2) \ge 0 \Leftrightarrow x(x-2)(x-1) \ge 0.$

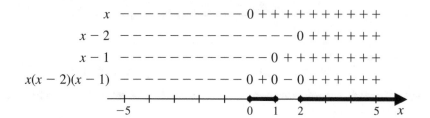

The solution set is where the last row is positive or 0, so it is $[0,1] \cup [2, \infty)$.

39. We have $x^3 - 2x^2 < 0 \Leftrightarrow x^2(x-2) < 0 \Leftrightarrow (x-2) < 0 \Leftrightarrow x < 2$. In addition, x cannot be equal to 0 because that would make the whole expression equal to 0. The solution set is $(-\infty, 0) \cup (0, 2)$.

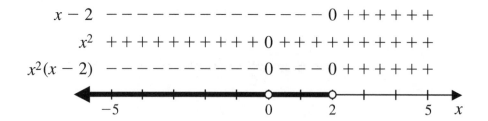

41. For $\dfrac{x+3}{x-1} \geq 0$.

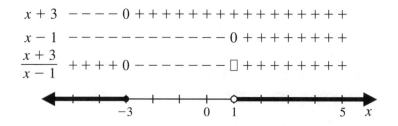

The solution set is where the last row is positive or 0, so it is $(-\infty, -3] \cup (1, \infty)$.

43. For $\dfrac{x(x+2)}{x-2} \leq 0$.

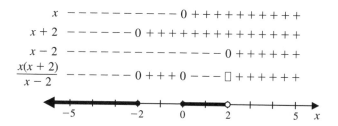

The solution set is where the last row is negative or 0, so it is $(-\infty, -2] \cup$
$[0, 2)$.

45. For $\dfrac{(1 - x)(x + 2)}{x(x + 1)} > 0$.

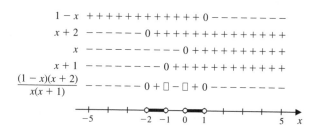

The solution set is where the last row is positive, so it is $(-2, -1) \cup (0, 1)$.

47. $\dfrac{1}{x} \le 5 \Leftrightarrow \dfrac{1}{x} - 5 \le 0 \Leftrightarrow \dfrac{1 - 5x}{x} \le 0$

Notice that we have not multiplied by x in the original equation. If we
had, we would have needed to analyze the situation for $x > 0$ and for
$x < 0$.

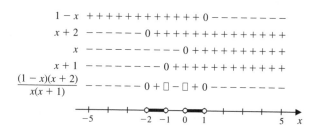

The solution set is where the last row is negative or 0, so it is $(-\infty, 0) \cup$
$\left[\frac{1}{5}, \infty\right)$.

49. $\dfrac{2}{x-1} \geq \dfrac{3}{x+2} \Leftrightarrow \dfrac{2}{x-1} - \dfrac{3}{x+2} \geq 0 \Leftrightarrow \dfrac{2(x+2) - 3(x-1)}{(x-1)(x+2)} \geq 0 \Leftrightarrow$
$\dfrac{7-x}{(x-1)(x+2)} \geq 0$

The solution set is where the last row is positive or 0, so it is $(-\infty, -2) \cup$
$(1, 7]$.

51. For $|5x - 3| = 2 \Leftrightarrow 5x - 3 = 2$ or $5x - 3 = -2 \Leftrightarrow x = 1$ or $x = \frac{1}{5}$.

53. $\left|\dfrac{x-1}{2x+3}\right| = 2 \Leftrightarrow \dfrac{x-1}{2x+3} = 2$ or $\dfrac{x-1}{2x+3} = -2 \Leftrightarrow x - 1 = 4x + 6$ or
$x - 1 = -4x - 6 \Leftrightarrow -7 = 3x$ or $5x = -5 \Leftrightarrow x = -\frac{7}{3}$ or $x = -1$

55. $|x - 4| \leq 1 \Leftrightarrow -1 \leq x - 4 \leq 1 \Leftrightarrow 3 \leq x \leq 5$

In interval notation the solution is $[3, 5]$.

57. $|3 - x| \geq 2 \Leftrightarrow 3 - x \geq 2$ or $3 - x \leq -2 \Leftrightarrow -x \geq -1$ or $-x \leq -5 \Leftrightarrow x \leq 1$
or $x \geq 5$

In interval notation the solution is $(-\infty, 1] \cup [5, \infty)$.

59. $\dfrac{1}{|x+5|} > 2 \Leftrightarrow 0 < |x + 5| < \dfrac{1}{2} \Leftrightarrow -\dfrac{1}{2} < x + 5 < \dfrac{1}{2}$ and $x \neq -5 \Leftrightarrow$
$-\dfrac{11}{2} < x < -\dfrac{9}{2}$ and $x \neq -5$.

In interval notation the solution is $\left(-\frac{11}{2}, -5\right) \cup \left(-5, -\frac{9}{2}\right)$.

61. $|x^2 - 4| > 0 \Leftrightarrow x^2 - 4 \neq 0 \Leftrightarrow x \neq \pm 2$

In interval notation the solution is $(-\infty, -2) \cup (-2, 2) \cup (2, \infty)$.

63. Since, in general, $|x| = \sqrt{x^2}$, we have $0 < a < b \Rightarrow |a| < |b| \Leftrightarrow \sqrt{a^2} < \sqrt{b^2} \Leftrightarrow a^2 < b^2$.

Or, since $0 < a < b$, we have both $0 < a^2 < ab$ and $0 < ab < b^2$. Hence,

$$0 < a^2 < ab < b^2.$$

65. (a) $20 \leq F \leq 50 \Leftrightarrow -12 \leq F - 32 \leq 18 \Leftrightarrow -12\left(\frac{5}{9}\right) \leq \frac{5}{9}(F - 32) \leq 18\left(\frac{5}{9}\right) \Leftrightarrow -\frac{20}{3} \leq C \leq 10$

(b) $20 \leq C \leq 50 \Leftrightarrow 20 \leq \frac{5}{9}(F - 32) \leq 50 \Leftrightarrow \frac{9}{5}(20) \leq F - 32 \leq \frac{9}{5}(50) \Leftrightarrow 68 \leq F \leq 122$.

1.3 The Cartesian Plane

The xy-coordinate system is called the coordinate plane or Cartesian plane. Points are specified using an ordered pair (x, y), with the horizontal or x-coordinate specified first and the vertical or y-coordinate specified second.

DESCRIBING REGIONS IN THE PLANE

EXAMPLE 1.3.1   Describe the set of points in the xy-plane satisfying the inequalities $-2 \leq x \leq 1$.

SOLUTION: Since the y-coordinate can be any real number, and the x-coordinate is restricted to lie between -2 and 1, the inequality describes all points between the vertical lines determined by $\{(x, y) \mid x = -2\}$ and $\{(x, y) \mid x = 1\}$. For example, $(-2, 0), (-2, 2), (-2, -2), (0, 1), (-1, 1)$ and $(1/2, 3)$ are all in the region, whereas $(-3, 1)$ and $(3, -1)$ are not in the region.   ◇

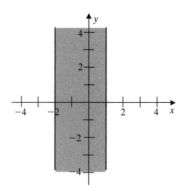

Figure 1.15

EXAMPLE 1.3.2   Describe the set of points in the xy-plane with $|x - 1| < 2$ and $|y + 1| < 3$.

SOLUTION: Rewriting the inequalities we have,

$$|x - 1| < 2$$
$$-2 < x - 1 < 2$$
$$-1 < x < 3$$

and

$$|y + 1| < 3$$
$$-3 < y + 1 < 3$$
$$-4 < y < 2.$$

This means the points in the solution set all have x-coordinates between -1 and 3, not including the values -1 and 3, and y-coordinates between -4 and 2, not including the values -4 and 2. That is, the points lie between, but not on, the vertical lines $x = -1$ and $x = 3$ and between, but not on, the horizontal lines $y = -4$ and $y = 2$, describing a rectangular region as shown in the figure.   ◇

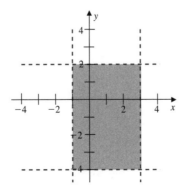

Figure 1.16

THE DISTANCE FORMULA

The **distance** between two points $P\left(x_1, y_1\right)$ and $Q\left(x_2, y_2\right)$ is given by the formula

$$d\left(P, Q\right) = \sqrt{\left(x_2 - x_1\right)^2 + \left(y_2 - y_1\right)^2}.$$

EXAMPLE 1.3.3   Find the distances between the points $(1, 1), (4, 4)$, and $(0, 8)$, and show that they are vertices of a right triangle. At which vertex is the right angle?

SOLUTION: Let $A = (1, 1), B = (4, 4)$, and $C = (0, 8)$. The sketch of the points in the figure indicates the right angle is at B. To verify that the triangle is a right triangle, we determine the lengths a, b, and c and verify that $a^2 + b^2 = c^2$. So,

$$a = d(A, B) = \sqrt{(4 - 1)^2 + (4 - 1)^2} = \sqrt{18}$$
$$b = d(B, C) = \sqrt{(4 - 0)^2 + (4 - 8)^2} = \sqrt{32}$$

and

$$c = d(A, C) = \sqrt{(0 - 1)^2 + (8 - 1)^2} = \sqrt{50}.$$

Hence

$$a^2 + b^2 = 18 + 32 = 50$$

and

$$a^2 + b^2 = c^2.$$

The right angle is at $(4, 4)$.   ◇

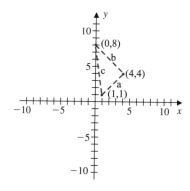

Figure 1.17

THE EQUATION OF A CIRCLE

A **circle** is a set of all points that lie at a fixed distance, called the **radius**, from a fixed point, called the **center**. If the radius is r, and the fixed point in the plane is (h, k), the distance formula implies a point (x, y) will lie on the circle provided

$$\sqrt{(x - h)^2 + (y - k)^2} = r \quad \text{or} \quad (x - h)^2 + (y - k)^2 = r^2.$$

EXAMPLE 1.3.4 Find the equation of the circle with center $(-1, 2)$ and radius 2.

SOLUTION: Substituting into the formula, we have the equation of the circle is

$$(x - (-1))^2 + (y - 2)^2 = 4$$

or

$$(x + 1)^2 + (y - 2)^2 = 4.$$

◇

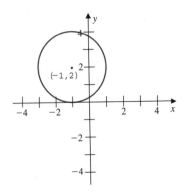

Figure 1.18

EXAMPLE 1.3.5   Find the center and radius of the circle $x^2 - 2x + y^2 + 2y + 1 = 0$.

SOLUTION: To find the center and radius, rewrite the equation in the general form. To do this, complete the square on the x-terms and the y-terms of the equation. To complete the square on $x^2 - 2x$, take half the coefficient of the x-term, square it, and then add and subtract (so the net change is 0) the value. So,

$$x^2 - 2x = x^2 - 2x + \left(\frac{2}{2}\right)^2 - \left(\frac{2}{2}\right)^2$$
$$= x^2 - 2x + 1 - 1$$
$$= (x - 1)^2 - 1.$$

Doing the same process on the y-terms gives

$$x^2 - 2x + y^2 + 2y + 1 = 0$$
$$\left(x^2 - 2x + 1\right) - 1 + \left(y^2 + 2y + 1\right) - 1 + 1 = 0$$
$$(x - 1)^2 + (y + 1)^2 = 1$$

and

$$(x - 1)^2 + (y - (-1))^2 = 1.$$

So the circle has center $(1, -1)$ and radius $\sqrt{1} = 1$.   ◇

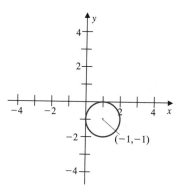

Figure 1.19

EXAMPLE 1.3.6 Find an equation of the circle with center $(-1, 2)$ that passes through $(2, 3)$.

SOLUTION: If $(2, 3)$ is on the circle, then the line segment from the center point $(-1, 2)$ to $(2, 3)$ is a radius, and the length of the radius is

$$d((-1, 2), (2, 3)) = \sqrt{(2 + 1)^2 + (3 - 2)^2} = \sqrt{10}.$$

The equation of the circle is then given by

$$(x + 1)^2 + (y - 2)^2 = 10.$$

◇

EXAMPLE 1.3.7   Find the area of the region that lies outside the circle $x^2 + y^2 = 2$ and inside the circle $x^2 + y^2 = 5$.

SOLUTION: The *annular* region is shown in the figure.

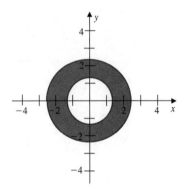

Figure 1.20

The shaded region consists of all points that satisfy $x^2 + y^2 \geq 2$ and $x^2 + y^2 \leq 5$ at the same time.

Since the area of a circle is πr^2, where r is the radius, and the radius of the inner circle is $\sqrt{2}$ and outer circle is $\sqrt{5}$, the area of the annular region is

$$\pi(\sqrt{5})^2 - \pi(\sqrt{2})^2 = 5\pi - 2\pi = 3\pi.$$

◇

Solutions for Exercise Set 1.3

1.

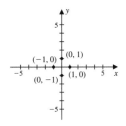

3.

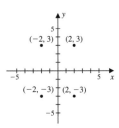

5. (a) Distance: $d = \sqrt{(2 - (-1))^2 + (4 - 3)^2} = \sqrt{9 + 1} = \sqrt{10}$;

(b) midpoint: $\left(\frac{2-1}{2}, \frac{4+3}{2}\right) = \left(\frac{1}{2}, \frac{7}{2}\right)$

7. (a) Distance: $d = \sqrt{(\pi - (-1))^2 + (0 - 2)^2} = \sqrt{(\pi + 1)^2 + 4} = \sqrt{\pi^2 + 2\pi + 5}$;

(b) midpoint: $\left(\frac{\pi-1}{2}, \frac{0+2}{2}\right) = \left(\frac{\pi-1}{2}, 1\right)$

9. For $x = 3$

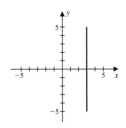

11. For $x > 1$

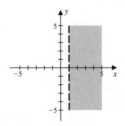

13. For $x \geq 1$ and $y \geq 2$

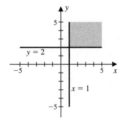

15. For $-3 < y \leq 1$

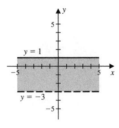

17. For $-1 \leq x \leq 2$ and $2 < y < 3$

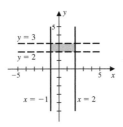

19. $|x - 1| < 3 \Leftrightarrow -3 < x - 1 < 3 \Leftrightarrow -2 < x < 4$ and $|y + 1| < 2 \Leftrightarrow -2 < y + 1 < 2 \Leftrightarrow -3 < y < 1$

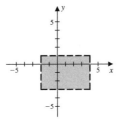

21. $|x| \geq 3 \Leftrightarrow x \geq 3$ or $x \leq -3$.

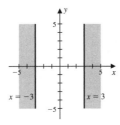

23. $(x - 2)^2 + y^2 = 9$

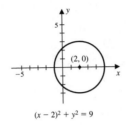

$(x - 2)^2 + y^2 = 9$

25. $(x + 2)^2 + (y - 3)^2 = 4$

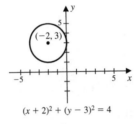

$(x + 2)^2 + (y - 3)^2 = 4$

27. $(x + 1)^2 + (y + 2)^2 = 4$

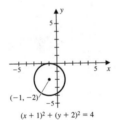

$(x + 1)^2 + (y + 2)^2 = 4$

29. $(x - 2)^2 + (y - 2)^2 = 4$

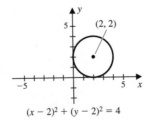

$(x - 2)^2 + (y - 2)^2 = 4$

31. $x^2 + y^2 = 9$ (a)Center: $(0, 0)$ Radius: 3

  (b)

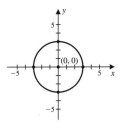

33. $x^2 + (y - 1)^2 = 1$ (a) Center: $(0, 1)$ Radius: 1

  (b)

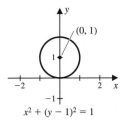

35. $(x - 2)^2 + (y + 1)^2 = 9$ (a) Center: $(2, -1)$ Radius: 3

  (b)

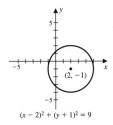

37. Center: $(1, 0)$ Radius: 2

$$x^2 - 2x + y^2 = 3$$
$$x^2 - 2x + 1 - 1 + y^2 = 3$$
$$(x - 1)^2 + y^2 = 4$$

39. Center: $(-1, 2)$ Radius: 1

$$x^2 + 2x + y^2 - 4y = -4$$
$$x^2 + 2x + 1 - 1 + y^2 - 4y + 4 - 4 = -4$$
$$(x + 1)^2 + (y - 2)^2 = 1$$

41. Center: $(2, 1)$ Radius: 3

$$x^2 - 4x + y^2 - 2y - 4 = 0$$
$$x^2 - 4x + 4 - 4 + y^2 - 2y + 1 - 1 = 4$$
$$(x - 2)^2 + (y - 1)^2 = 9$$

43. $x^2 + y^2 \leq 1$

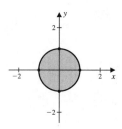

45. $1 < x^2 + y^2 < 4$

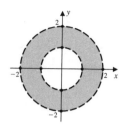

47. $x^2 + y^2 \leq 4$ and $y \geq x$

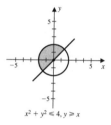

$x^2 + y^2 \leq 4, y \geq x$

49. The point $(6, 3)$ is closer to the origin, since $d((6, 3), (0, 0)) = \sqrt{45}$ and $d((-7, 2), (0, 0)) = \sqrt{53}$.

51. (a) Let $a = d((-3, -4), (2, -1)) = \sqrt{34}$, $b = d((-1, 4), (2, -1)) = \sqrt{34}$, and $c = d((-1, 4), (-3, -4)) = \sqrt{68}$. Then $a^2 + b^2 = c^2$, so the points are the vertices of a right triangle. (b)The right angle is located at vertex $(2, -1)$.

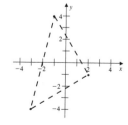

53. The point $(-3, -4)$ is 5 units left and 3 units down from the point $(2, -1)$. The unique point producing a square is the same distance from $(-1, 4)$, that is, the point $(-1 - 5, 4 - 3) = (-6, 1)$.

55. The radius $r = d((0, 0), (2, 3)) = \sqrt{13}$, so the equation is $x^2 + y^2 = 13$.

57. The radius $r = d((3, 7), (0, 7)) = \sqrt{9} = 3$, so the equation is $(x - 3)^2 + (y - 7)^2 = 9$.

59. A circle with radius 3, tangent to both the x- and y-axes, and center in the second quadrant, has center $(-3, 3)$. The equation is $(x + 3)^2 + (y - 3)^2 = 9$.

61. The area of the shaded region in the figure is

$$\pi(3)^2 - \pi(1)^2 = 8\pi.$$

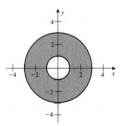

63.

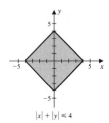

$|x| + |y| \le 4$

65. The area of the shaded region is the area of the circle minus the area of the square. Since the area of the square with side $\sqrt{2}$ is 2, the area of the region is $\pi(1)^2 - 2 \approx 1.142$.

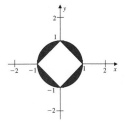

1.4 The Graph of an Equation

The graph of an equation is the set of all points (x, y) that satisfy the equation. The variable x is called the *independent* variable since its values can vary freely over a collection of real numbers. The variable y is called the *dependent* variable since its value depends on the particular value of x selected.

One way to obtain the graph of an equation is to plot several representative points that lie on the graph. The points where the graph crosses the x-axis and the y-axis , called the x- and y-intercepts, are particularly useful. This simple point-plotting approach will work in some cases but usually is not sufficient for our purposes in PreCalculus and Calculus.

PLOTTING POINTS

EXAMPLE 1.4.1   Sketch the graph of the given equation.

(a) $y = x - 1$ (b) $y = x^2 + 1$

SOLUTION:

(a) To sketch the graph, we first make a table of representative points on the graph by selecting particular values of x and then determining the associated y values. The x-intercept is the point with y-coordinate zero. From the table we see this is $(1, 0)$. The y-intercept is obtained by setting x to 0 and is the point $(0, -1)$. The figure shows the graph of the equation is a straight line.   ◇

x	-3	-2	-1	0	1	2	3
$y = x - 1$	-4	-3	-2	-1	0	1	2

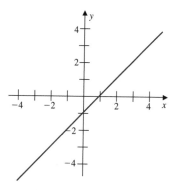

Figure 1.21

(b) Since $x^2 + 1 \geq 1$, the graph does not cross the x-axis and so has no x-intercepts. The table shows that the y-intercept is $(0, 1)$. The graph is the parabola shown in the figure.

x	-2	-1	0	1	2
$y = x^2 + 1$	5	2	1	2	5

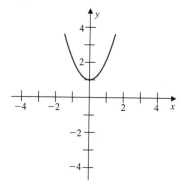

Figure 1.22

SYMMETRY

The left portion of the graph in part (b) of the proceeding example is the reflection through the y-axis of the right portion of the graph. *Symmetries of this type are often useful in sketching graphs.*

<u>Three types of symmetry</u>

1. A curve has *y-axis symmetry* if its equation is unchanged when x is replaced with $-x$. Geometrically this means that whenever the point (x, y) is on the curve, the point $(-x, y)$ also is on the curve. For example, the graph of $y = x^2$.

2. A curve has *x-axis symmetry* if its equation is unchanged when y is replaced with $-y$. Geometrically this means that whenever the point (x, y) is on the curve, the point $(x, -y)$ also is on the curve. For example, the graph of $x = y^2$.

3. A curve has *origin symmetry* if its equation remains unchanged when x is replaced with $-x$ and y is replaced with $-y$. Geometrically this means that whenever the point (x, y) is on the curve, the point $(-x, -y)$ also is on the curve. For example, the graph of $y = x^3$.

EXAMPLE 1.4.2   Test each curve for symmetry, find any x- and y-intercepts and sketch the graph of the equation.

(a) $y = x^2 - 3$ (b) $y = x^3 + 2$

SOLUTION:

(a) <u>y-axis symmetry:</u> Yes, since if we replace x with $-x$,

$$(-x)^2 - 3 = x^2 - 3$$

and the equation remains unchanged.

<u>x-axis symmetry:</u> No, since if y is replaced with $-y$ the equation is changed. For example $(2, 1)$, is on the curve but $(2, -1)$ is not on the curve.

<u>origin symmetry:</u> No, since for example $(2, 1)$, is on the curve but $(-2, -1)$ is not on the curve.

<u>x-intercepts:</u> Solve the equation

$$x^2 - 3 = 0$$
$$x^2 = 3$$
$$x = \pm\sqrt{3}.$$

So the graph crosses the x-axis at the points $(-\sqrt{3}, 0)$ and $(\sqrt{3}, 0)$.

<u>y-intercepts:</u> Set $x = 0$, then $y = -3$ and the y-intercept is the point $(0, -3)$.

Since the graph is symmetric with respect to the y-axis, we can plot the graph for positive values of x and then reflect the graph through the y-axis as shown in the figure.   ◇

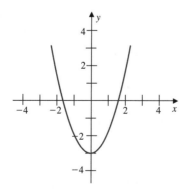

Figure 1.23

(b) If x is replaced with $-x$, the equation becomes

$$y = (-x)^3 + 2 = -x^3 + 2$$

which has altered the original equation, so the graph is not symmetric with respect to the y-axis. Replacing y with $-y$ gives

$$-y = x^3 + 2$$
$$y = -x^3 - 2$$

which gives a new equation, so the graph is not symmetric with respect to the y-axis. Replacing x with $-x$ and y with $-y$ gives

$$-y = (-x)^3 + 2 = -x^3 + 2$$
$$y = x^3 - 2,$$

and the curve is also not symmetric with respect to the origin. The curve is sketched in the figure. Notice that although there is no axis or origin symmetry, this particular curve is symmetric with respect to the point (0,2).

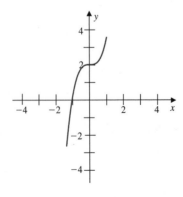

Figure 1.24

EXAMPLE 1.4.3   Sketch the graph of $y = \frac{x^2-x-6}{x-3}$.

SOLUTION: First factor the numerator to get,

$$y = \frac{x^2 - x - 6}{x - 3} = \frac{(x-3)(x+2)}{(x-3)} = x + 2, \quad \text{for} \quad x \neq 3.$$

It is important to realize that the original equation is not defined when x is 3. This is illustrated by removing the point $(3,5)$ from the graph of $y = x + 2$, as shown in the figure.   ◇

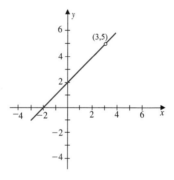

Figure 1.25

EXAMPLE 1.4.4   Find the distance between the points of intersection of the graphs $y = x^2 - 2x + 2$ and $y = x + 2$.

SOLUTION: The points of intersection of the two curves are given by those x values that yield the same y value when substituted in both equations. To find the x values of the points of intersection, set the two equations for y equal and solve for x. So

$$x^2 - 2x + 2 = x + 2$$

$$x^2 - 3x = 0$$

$$x(x - 3) = 0$$

and

$$x = 0 \quad \text{or} \quad x = 3.$$

The y coordinates of the points of intersection are found by substituting the x values into either of the equations for y. We choose to use $y = x + 3$, so

$$(0, 2) \quad \text{and} \quad (3, 5)$$

are the points of intersection.

The distance between the two points is

$$d((0, 2), (3, 5)) = \sqrt{(3 - 0))^2 + (5 - 2)^2}$$

$$= \sqrt{9 + 9} = \sqrt{18}$$

$$= 3\sqrt{2}.$$

The two curves are shown in the figure.　◇

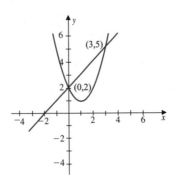

Figure 1.26

Solutions for Exercise Set 1.4

1. y-axis symmetry.

3. x-axis, y-axis, and origin symmetry.

5. Intercepts: $(-1, 0), (0, 1)$; symmetry: none

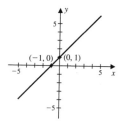

7. $x + y = 1 \Leftrightarrow y = 1 - x$

Intercepts: $(1, 0), (0, 1)$; symmetry: none

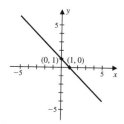

9. Intercepts: $(-1, 0), (1, 0), (0, -1)$; symmetry: y-axis

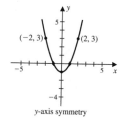

11. Intercepts: $(-1, 0), (1, 0), (0, 1)$; symmetry: y-axis

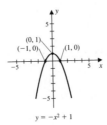

13. Intercepts: $(0, 0)$; symmetry: y-axis

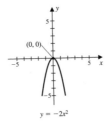

15. Intercepts: $(-1, 0), (0, 1), (0, -1)$; symmetry: x-axis

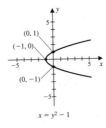

17. Intercepts: $(-1, 0), (0, 1)$; symmetry: none

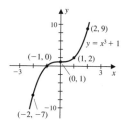

19. Intercepts: $(0, 0)$; symmetry: origin

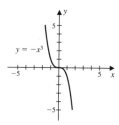

21. Intercepts: $(-2, 0), (0, 2)$; symmetry: none

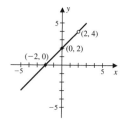

23. $y = \dfrac{x^2 - x - 6}{x + 2} = \dfrac{(x - 3)(x + 2)}{x + 2} = x - 3$, for $x \neq -2$

Intercepts: $(3, 0), (0, -3)$; symmetry: none

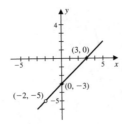

25. Intercepts: $(0, 2)$; symmetry: none

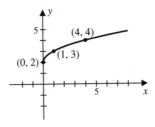

27. Intercepts: $(-2, 0), (2, 0), (0, -2), (0, 2)$; symmetry: x-axis, y-axis, and origin

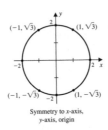

29. Intercepts: $(-3, 0), (3, 0), (0, 3)$; symmetry: y-axis

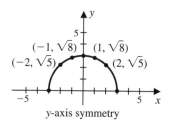

y-axis symmetry

31. Intercepts: $(0,0)$; symmetry: y-axis

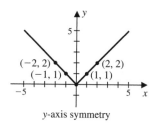

y-axis symmetry

33. Intercepts: $(-1,0), (1,0), (0,-1)$; symmetry: y-axis

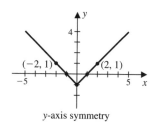

y-axis symmetry

35. $y = x^2 - 1$

Since $(-x)^2 - 1 = x^2 - 1$, the graph is symmetric with respect to the y-axis. Since $-y = -x^2 + 1 \neq x^2 - 1$, the graph is not symmetric with respect to the x-axis. Since $(-x)^2 - 1 = x^2 - 1 \neq -y$, the graph is not symmetric with respect to the origin.

37. $y = x^3 + x^2$

Since $(-x)^3 + (-x)^2 = -x^3 + x^2 \neq x^3 + x^2$, the graph is not symmetric with respect to the y-axis. Since $-y = -x^3 - x^2 \neq x^3 + x^2$, the graph is not symmetric with respect to the x-axis. Since $(-x)^3 + (-x)^2 = -x^3 + x^2 \neq -y$, the graph is not symmetric with respect to the origin.

39. $y = |x|$

Since $|-x| = |x|$, the graph is symmetric with respect to the y-axis. Since $-y = -|x| \neq |x|$, the graph is not symmetric with respect to the x-axis. Since $|-x| = |x| \neq -y$, the graph is not symmetric with respect to the origin.

41. Reflect the graph about the x-axis.

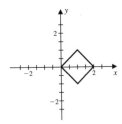

43. Reflect the graph about the origin.

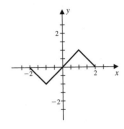

45. From the graph we can see the points of intersection of $y = x^2 + 1$ and $y = 2$ are $(1, 2)$ and $(-1, 2)$. Or

$$x^2 + 1 = 2$$
$$x^2 = 1$$
$$x = \pm 1.$$

The distance between the two points of intersection is

$$d((1, 2), (-1, 2)) = \sqrt{(1 - (-1))^2 + (2 - 2)^2}$$
$$= \sqrt{4} = 2.$$

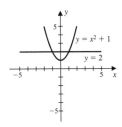

47. Assume the graph has symmetry with respect to both the x- and y-axes. To show the graph is symmetric with respect to the origin, we need to show that if (x, y) is on the graph then $(-x, -y)$ is also on the graph. If (x, y) is on the graph and the graph is symmetric with respect to the x-axis, then $(x, -y)$ is also on the graph. But if $(x, -y)$ is on the graph and the graph is symmetric with respect to the y-axis, then $(-x, -y)$ is also on the graph. Hence the graph is symmetric with respect to the origin.

It is also true that if the graph has symmetry with respect to the origin and to one of the axes, then it has symmetry with respect to the other axis. Consider, for example, the situation of symmetry with respect to the origin and the x-axis. To show the graph is symmetric with respect to the y-axis, we need to show that if (x, y) is on the graph then $(-x, y)$ is also on the graph. If (x, y) is on the graph and the graph is symmetric with respect to the x-axis, then $(x, -y)$ is also on the graph. But if $(x, -y)$ is on the graph and the graph is symmetric with respect to the origin, then $(-x, -(-y)) = (-x, y)$ is also on the graph. Hence the graph is symmetric with respect to the y-axis.

1.5 Using Technology to Graph Equations

Graphing devices sketch curves by plotting many points which are connected by very small line segments. The *viewing rectangle* of a graphing device is the rectangular portion of the plane in which the plot is displayed. Selecting the appropriate viewing rectangle when using a graphing device is very important. A viewing rectangle specified as $[a, b] \times [c, d]$, defines the rectangular region in the plane with (x, y) restricted by $a \leq x \leq b$ and $c \leq y \leq d$.

EXAMPLE 1.5.1   Use a graphing device to sketch a graph of $y = x^3 - 12x + 20$ with the following viewing rectangles, and determine which gives the best representation for the graph of the equation.

(a) $[-1, 1] \times [-1, 1]$ (b) $[-5, 5] \times [-5, 5]$

(c) $[-10, 10] \times [-50, 50]$ (d) $[-100, 100] \times [-200, 200]$

SOLUTION: The viewing rectangle in (a) is too small to show any portion of the graph. We expect the graph to be a smooth curve, so we also reject the graph in part (b). The viewing rectangle in part (c) gives a nice smooth curve which we accept with confidence as a good representative of the graph. The viewing rectangle used in part (d) is too large in both the x and y directions and clearly has distorted the graph. It does illustrate, however, that the curve in part (c) shows the important features of the graph.   ◇

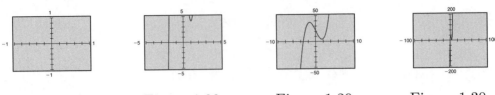

Figure 1.27   Figure 1.28   Figure 1.29   Figure 1.30

EXAMPLE 1.5.2   Determine an appropriate viewing rectangle for the graph of the equation $y = \dfrac{x+6}{x^2 - 1}$.

SOLUTION: We start by selecting a viewing rectangle of $[-5, 5] \times [-5, 5]$. The graph in part (a) appears to be a reasonable representation of the graph. However, we need to be careful when using a graphing device to be certain we have selected a viewing rectangle that shows all the important features of the graph. In this case, if we set $x = 0$, then the y-value of the point on the curve is -6, which we do not see on the graph in part (a). In part (b) we have used a viewing rectangle of $[-5, 5] \times [-10, 5]$, and see we indeed missed a great deal of the curve in part (a).   ◇

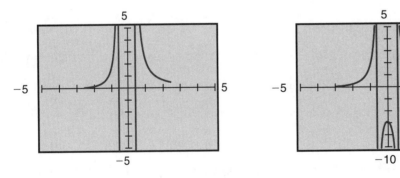

Figure 1.31   Figure 1.32

EXAMPLE 1.5.3   Graphically approximate the solutions to the inequality $x^2 + 3x - 2 < 0$.

SOLUTION: In (a), the equation $y = x^2 + 3x - 2$ is graphed in a viewing rectangle $[-5, 5] \times [-5, 5]$. The inequality is negative, which corresponds to the portion of the graph that lies below the x-axis. Clicking on the x-intercepts of the graph gives values of $x \approx -3.63$ and $x \approx 0.56$. So, $x^2 + 3x - 2 < 0$ for $x > -3.63$ or $x < 0.56$.

In (b), the viewing rectangle is $[0, 1] \times [-1, 1]$. Each tickmark between 0.4 and 0.6 is 0.04 apart, so our answer of 0.56 is accurate to within 2 decimal places. For more accuracy, *zoom* closer in to the x-intercept by shrinking the viewing rectangle.   ◇

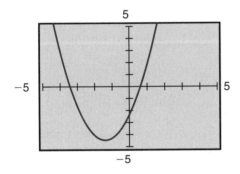

Figure 1.33

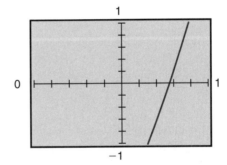

Figure 1.34

EXAMPLE 1.5.4   The number of birds in a sanctuary at time t is given by

$$n = 5000 \left(\frac{4t^2 + 1}{2t^2 + 1} \right).$$

As the time t increases, does the size of the population become stable? If so, what is the stabilizing level?

SOLUTION: The equation along with the horizontal line at 10000 are plotted using a viewing rectangle of $[0, 25] \times [0, 12000]$. The figure indicates that the population appears to level off to a stabilizing value of 10000.   ◇

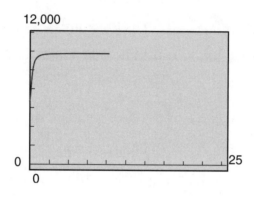

Figure 1.35

Solutions for Exercise Set 1.5

1. The graphs in the viewing rectangles are shown below. Part (d) gives the best representation.

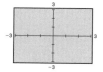 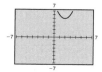 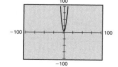 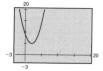

3. The graphs in the viewing rectangles are shown below. Part (c) gives the best representation.

  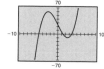 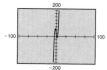

5. $[-10, 10] \times [-10, 10]$

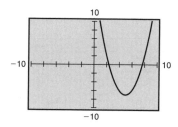

7. $[0, 20] \times [0, 10]$

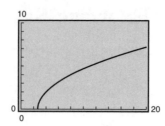

9. $[-5, 10] \times [-10, 10]$

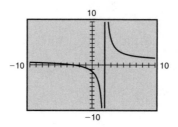

11. The graph crosses the x-axis at $x \approx 0.86$.

13. The graph crosses the x-axis at $x \approx -0.93$ and $x \approx 2.82$.

15. The points of intersection are approximately $(1, 6)$ and $(-1, 0)$.

17. The points of intersection are approximately $(-0.04, -1.02)$ and $(0.48, -0.79)$.

19. In part (a) we approximate where the graph of the expression is above the x-axis, and in part (b) we approximate where the graph of the expression is below the x-axis. This gives

  (a) $x \leq -3.6$ or $x \geq 0.6$ (b) $x < -2.3$ or $1.3 < x < 3$

21. For x very large the graphs are almost identical. That is, as x grows without bound $x^4 - 4x^3 + 3x^2$ is approximately x^4.

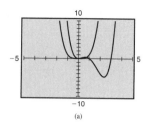

(a)

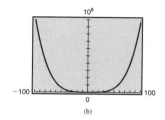

(b)

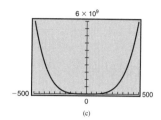

(c)

23. The graph appears to approach the horizontal line $y = \frac{a}{b}$ and the vertical line $x = \frac{1}{b}$.

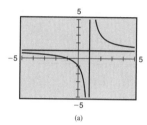

(a)

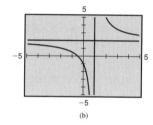

(b)

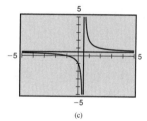

(c)

25. If $a = b = 0$, the graph of $y = (x - a)^2 + b$ is the standard parabola $y = x^2$. The constants a and b shift $y = x^2$ either horizontally or vertically. If $a > 0$, the shift is to the right a units. If $a < 0$, the shift is to the left $|a|$ units. If $b > 0$, the shift is upward b units, and if $b < 0$, the shift is downward $|b|$ units.

1.6 Functions

A *function* is a process that for each admissable input returns another *unique* real number. If a function is called f and x is an input value, then the output is denoted as $f(x)$, read "f of x." Functions will play a central role in all quantitative study, and you should become comfortable with the key ideas and notation that is used.

EVALUATION OF FUNCTIONS

EXAMPLE 1.6.1   Find each value of the function f defined by $f(x) = x^2 - 2x + 1$.

(a) $f(3)$ (b) $f(-2)$ (c) $f\left(\frac{1}{2}\right)$

SOLUTION:

(a) $f(3) = (3)^2 - 2(3) + 1 = 4$

(b) $f(-2) = (-2)^2 - 2(-2) + 1 = 9$

(c)

$$f\left(\frac{1}{2}\right) = \left(\frac{1}{2}\right)^2 - 2\left(\frac{1}{2}\right) + 1 = \left(\frac{1}{4}\right) - 1 + 1 = \frac{1}{4}$$

◇

The input to a function can be any expression that represents a real number. The variable used in the definition of the function is simply a place holder, and evaluating a function at a specific value means replace the variable everywhere it appears with the input value. For example, if

$$f(x) = \frac{x^3 - x + 1}{2x - 1},$$

then the function can be thought of as

$$f(\square) = \frac{\square^3 - \square + 1}{2\square - 1}.$$

EXAMPLE 0.6.2   If $f(t) = |t - 4|$, find each of the following.

(a) $f(5)$ (b) $f(1)$ (c) $f(0)$ (d) $f(t+3)$ (e) $f(4 - t^2)$ (f) $f(-t)$

SOLUTION:

(a) $f(5) = |5 - 4| = 1$

(b) $f(1) = |1 - 4| = |-3| = 3$

(c) $f(0) = |0 - 4| = |-4| = 4$

(d) $f(t + 3) = |(t + 3) - 4| = |t - 1|$

(e) $f(4 - t^2) = |(4 - t^2) - 4| = |-t^2| = t^2$

(f) $f(-t) = |-t - 4| = |-(t + 4)| = |t + 4|$

◇

DOMAIN AND RANGE

The *domain* of a function is the collection of all real numbers that can be input to the function. The *range* is the collection of all outputs from the function. If the domain of a function is not explicitly stated, we assume it is the largest set of real numbers for which the function is defined.

EXAMPLE 1.6.3   Find the domain and range of the function.

(a) $f(x) = \dfrac{3}{x+1}$ (b) $f(x) = \sqrt{x+3}$

SOLUTION:

(a) The only values of x for which the function is not defined are the values that make the denominator $x+1 = 0$. So the domain is all real numbers except $x = -1$. The domain is $(-\infty, -1) \cup (-1, \infty)$.

The only value the function can not assume is 0, so the range is $(-\infty, 0) \cup (0, \infty)$.

(b) For the function to be defined, the expression under the radical must be greater than or equal to 0. That is, the domain is all x with $x + 3 \geq 0$, and $x \geq -3$. In interval notation this is $[-3, \infty)$.

Since $y = \sqrt{x+3} \geq 0$ for all $x \geq -3$, the range is $[0, \infty)$.   ◇

EXAMPLE 1.6.4   Find the domain of each function.

(a) $f(x) = \sqrt{x(x+2)}$ (b) $f(x) = \sqrt{\dfrac{x}{x+2}}$

SOLUTION:

(a) For the function to be defined, the expression under the radical must be greater than or equal to 0. To solve the inequality

$$x(x+2) \geq 0$$

make the line chart

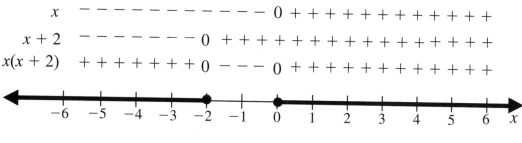

Figure 1.36

The chart indicates that the inequality is positive on $(-\infty, -2) \cup (0, \infty)$. Since $\sqrt{0} = 0$, the values -2 and 0 are also in the domain, so the domain is $(-\infty, -2] \cup [0, \infty)$.

(b) For x to be in the domain of f, we must have the quotient   ◇

$$\frac{x}{x+2} \geq 0.$$

The line chart

$$x \quad - - - - - - - - - - - - - \; 0 + + + + + + + + + + + + +$$
$$x+2 \quad - - - - - - - - \; 0 + + + + + + + + + + + + + + + +$$
$$\frac{x}{(x+2)} \quad + + + + + + + \; \Box \; - - - \; 0 + + + + + + + + + + + +$$

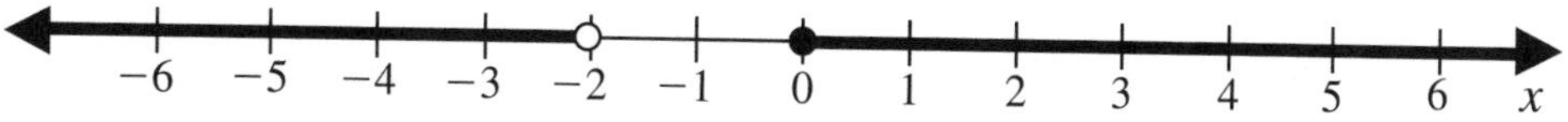

Figure 1.37

shows that $\frac{x}{x+2} > 0$, if $x < -2$ or $x > 0$. In addition, $\frac{x}{x+2} = 0$ when $x = 0$, which is permissible, but $\frac{x}{x+2}$ does not exist, indicated by the $\Box$, when

$x = -2$. Hence the domain of f is $(-\infty, -2) \cup [0, \infty)$.

VERTICAL AND HORIZONTAL LINE TESTS

Suppose that you are given the graph of an equation. If every vertical line crosses the graph at most once, then the equation defines a function. If a vertical line crosses the graph of a function, then the x-intercept is in the domain. If a horizontal line crosses the graph of a function, then the y-intercept is in the range of the function.

EXAMPLE 1.6.5   Determine if the graph defines a function and if so find the domain and the range.   ◇

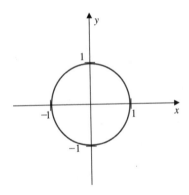

Figure 1.38

SOLUTION:

(a) Any vertical line drawn between $x = -1$ and $x = 1$ crosses the curve in exactly two places, so the graph does not define a function. In this example, the curve is the circle with center the origin and radius 1, which has the equation $x^2 + y^2 = 1$. Solving for y gives

$$y = \pm\sqrt{1 - x^2}$$

so

$$y = \sqrt{1 - x^2} \quad \text{or} \quad y = -\sqrt{1 - x^2}.$$

Each equation defines a function of y in terms of x. Choosing the positive square root gives us a function that describes the upper semi-circle. Choosing the negative square root gives us a function that describes the lower semi-circle.

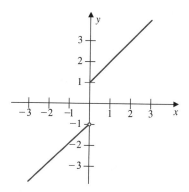

Figure 1.39

(b) Every vertical line crosses the curve in at most only one place, so the graph does define a function. Since every vertical line crosses the curve, the domain is the set of all real numbers, $(-\infty, \infty)$. The horizontal lines that do *not* cross the curve are those between -1 and 1 including the line $y = -1$. The range is consequently $(-\infty, -1) \cup [1, \infty)$.

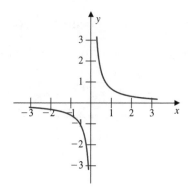

Figure 1.40

(c) The curve defines a function with domain all real numbers except $x = 0$. The range is also all real numbers except 0. So the domain is $(-\infty, 0) \cup (0, \infty)$ and the range is $(-\infty, 0) \cup (0, \infty)$.

DIFFERENCE QUOTIENTS

In calculus we frequently encounter an important quantity called the *difference quotient* of a function. The difference quotient for the function f at a number x in its domain is

$$\frac{f(x+h) - f(x)}{h},$$

where h represents an arbitrary small positive number.

EXAMPLE 1.6.6   Let $f(x) = 3x^2 + 4x + 1$. Find

$$f(x+h) \quad \text{and} \quad \frac{f(x+h) - f(x)}{h},$$

assuming that $h \neq 0$.

SOLUTION: Since

$$f(x+h) = 3(x+h)^2 + 4(x+h) + 1$$
$$= 3(x^2 + 2hx + h^2) + 4x + 4h + 1$$
$$= 3x^2 + 6hx + 3h^2 + 4x + 4h + 1,$$

we have

$$\frac{f(x+h) - f(x)}{h} = \frac{3(x+h)^2 + 4(x+h) + 1 - (3x^2 + 4x + 1)}{h}$$
$$= \frac{3(x^2 + 2hx + h^2) + 4x + 4h + 1 - 3x^2 - 4x - 1}{h}$$
$$= \frac{3x^2 - 3x^2 + 4x - 4x + 1 - 1 + 6hx + 3h^2 + 4h}{h}$$
$$= \frac{6hx + 3h^2 + 4h}{h} = \frac{h(6x + 3h + 4)}{h}$$
$$= 6x + 3h + 4.$$

◇

EXAMPLE 1.6.7   Let $f(x) = \sqrt{2x + 1}$. Find

$$f(x+h) \quad \text{and} \quad \frac{f(x+h) - f(x)}{h},$$

assuming that $h \neq 0$.

SOLUTION:

Since

$$f(x+h) = \sqrt{2(x+h) + 1} = \sqrt{2x + 2h + 1},$$

we have

$$\frac{f(x+h) - f(x)}{h} = \frac{\sqrt{2x + 2h + 1} - \sqrt{2x + 1}}{h}.$$

Multiplying the numerator and denominator by $\sqrt{2x+2h+1}+\sqrt{2x+1}$ permits us to rewrite the difference quotient as

$$
\begin{aligned}
\frac{f(x+h)-f(x)}{h} &= \frac{\sqrt{2x+2h+1}-\sqrt{2x+1}}{h} \cdot \frac{\sqrt{2x+2h+1}+\sqrt{2x+1}}{\sqrt{2x+2h+1}+\sqrt{2x+1}} \\
&= \frac{\left(\sqrt{2x+2h+1}\right)^2 - \left(\sqrt{2x+1}\right)^2}{h\left(\sqrt{2x+2h+1}+\sqrt{2x+1}\right)} \\
&= \frac{2x+2h+1-2x-1}{h\left(\sqrt{2x+2h+1}+\sqrt{2x+1}\right)} \\
&= \frac{2}{\sqrt{2x+2h+1}+\sqrt{2x+1}}.
\end{aligned}
$$

You might argue that this final result is not really simpler than the original form. In a sense this is true, but in calculus you will want to determine the value of the difference quotient as h approaches zero. This final result shows that as h approaches zero, the difference quotient approaches

$$
\frac{2}{\sqrt{2x+0+1}+\sqrt{2x+1}} = \frac{2}{2\sqrt{2x+1}} = \frac{1}{\sqrt{2x+1}}.
$$

Notice if we substitute $h=0$ in the original expression for the difference quotient, the result is a quantity of the form $\frac{0}{0}$, which is an undefined quantity. So we can not determine what happens to the difference quotient as h approaches 0 from the original form.   ◇

ODD AND EVEN FUNCTIONS

A function f is *odd* provided that for all x in its domain $f(-x) = -f(x)$, and a function is *even* provided $f(-x) = f(x)$. Geometrically, this means

> **an odd function is symmetric with respect to the origin**

and

> **an even function is symmetric with respect to the y−axis.**

EXAMPLE 1.6.8   Determine whether each of the following functions is even, odd, or neither even nor odd.

(a) $f(x) = x^2 + 3$ (b) $f(x) = x^3 + 1$ (c) $f(x) = x^3 + 2x$ (d) $f(x) = \sqrt[3]{x}$

SOLUTION:

(a) Since

$$f(-x) = (-x)^2 + 3 = x^2 + 3 = f(x), \ f \text{ is even,}$$

and

$$-f(x) = -x^2 - 3 \neq f(-x), \text{ so } f \text{ is not odd.}$$

(b) Since

$$f(-x) = (-x)^3 + 1 = -x^3 + 1 \neq f(x), \ f \text{ is not even,}$$

and

$$-f(x) = -x^3 - 1 \neq f(-x), \text{ so } f \text{ is not odd.}$$

(c) Since

$$f(-x) = (-x)^3 + 2(-x) = -x^3 - 2x \neq f(x), \ f \text{ is not even,}$$

and

$$-f(x) = -x^3 - 2x = f(-x), \text{ so } f \text{ is odd.}$$

(d) Since

$$f(-x) = \sqrt[3]{-x} = -\sqrt[3]{x} \neq f(x), \ f \text{ is not even,}$$

and

$$-f(x) = -\sqrt[3]{x} = f(-x), \text{ so } f \text{ is odd.}$$

◇

EXAMPLE 1.6.9   The graph of a function f is given for $x \geq 0$. Extend the graph for $x < 0$ if

(a) f is even   (b) f is odd.

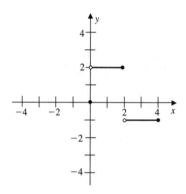

Figure 1.41

SOLUTION:

(a) The graph must be symmetric with respect to the y-axis, which gives the following graph.

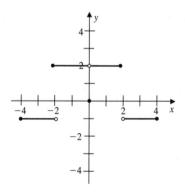

Figure 1.42

(b) The graph must be symmetric with respect to the origin, which gives the following graph.

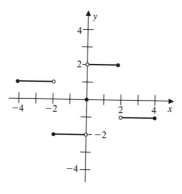

Figure 1.43

◇

APPLICATIONS

EXAMPLE 1.6.10   A rectangular plot of ground containing 100 feet^2 is to be fenced within a large plot.

(a) Express the perimeter of the plot as a function of the width. What is the domain of the function?

(b) Use a graphing device to approximate the dimensions of the plot that requires the least amount of fence.

SOLUTION:

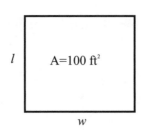

Figure 1.44

(a) The perimeter of the plot is

$$P = 2\ell + 2w.$$

To eliminate the variable ℓ from the equation, we use the information from the area of the plot. That is,

$$A = \ell w = 100, \quad \text{so} \quad \ell = \frac{100}{w}.$$

Substituting into the equation for P gives

$$P(w) = 2w + \frac{200}{w}.$$

The domain of the function is all real numbers $w > 0$.

(b) To find the exact dimensions of the plot that uses the least amount of fencing requires concepts from calculus. However, we can approximate the dimensions using a graphing device to plot the perimeter P with respect to the width w. The figure shows a low point at approximately $w \approx 10$ feet. That is, the perimeter is smallest, and hence the amount of fencing required is minimized, at this point. Then $\ell = w \approx 10$ feet.

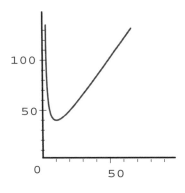

Figure 1.45

◇

EXAMPLE 1.6.11   A manufacturer estimates the profit on producing x units of their product at $P(x) = 200x - x^2$.

(a) What is the average rate of change in the profit as the number of units changes from x to $x + h$?

(b) Use the result in part (a) to find the average rate of change in the profit as the number of units produced changes from 40 to 60?

(c) Sketch the graph of $y = P(x)$ and the line that passes through the points $(40, P(40))$ and $(60, P(60))$.

SOLUTION: (a) The average rate of change is the change in the values of the function as the independent variable varies over some interval. That is, the average rate of change is the difference quotient,

$$\frac{P(x+h) - P(x)}{h} = \frac{200(x+h) - (x+h)^2 - (200x - x^2)}{h}$$

$$= \frac{200x + 200h - (x^2 + 2hx + h^2) - 200x + x^2}{h}$$

$$= \frac{200h - 2hx - h^2}{h} = \frac{h(200 - 2x - h)}{h}$$

$$= 200 - 2x - h.$$

(b) If $x = 40$ and $h = 20$, then $x + h = 60$, so the average rate of change can be computed using the final formula from part (a) as

$$200 - 2(40) - (20) = 100.$$

So, if the number of units produced increases from 40 to 60, the profit increases at an average rate of 100 per unit increase in production.

(c)

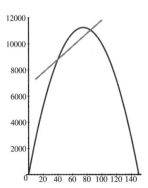

Figure 1.46

◇

Solutions for Exercise Set 1.6

1. For $f(x) = 4x^2 + 1$ we have

(a) $f(2) = 4(2)^2 + 1 = 17$ (b) $f(\sqrt{3}) = 4(\sqrt{3})^2 + 1 = 13$

(c) $f(2 + \sqrt{3}) = 4(2 + \sqrt{3})^2 + 1 = 4(4 + 4\sqrt{3} + 3) + 1 = 29 + 16\sqrt{3}$

(d) $f(2) + f(\sqrt{3}) = 17 + 13 = 30$

(e) $f(2x) = 4(2x)^2 + 1 = 4(4x^2) + 1 = 16x^2 + 1$

(f) $f(1 - x) = 4(1 - x)^2 + 1 = 4(1 - 2x + x^2) + 1 = 4x^2 - 8x + 5$

(g) $f(x + h) = 4(x + h)^2 + 1 = 4(x^2 + 2hx + h^2) + 1 = 4x^2 + 8hx + 4h^2 + 1$

(h)

$$f(x + h) - f(x) = 4x^2 + 8hx + 4h^2 + 1 - (4x^2 + 1)$$

$$= 8hx + 4h^2$$

3. For $f(t) = |t - 2|$ we have

(a) $f(4) = |4 - 2| = |2| = 2$ (b) $f(1) = |1 - 2| = |-1| = 1$

(c) $f(0) = |0 - 2| = |-2| = 2$ (d) $f(t + 2) = |t + 2 - 2| = |t|$

(e) $f(2 - t^2) = |2 - t^2 - 2| = |-t^2| = |t^2| = t^2$

(f) $f(-t) = |-t - 2| = |-(t + 2)| = |t + 2|$

5. Reading from the graph given in the exercise we have

(a) $f(-2) = -3$, $f(0) = \frac{1}{2}$, $f(2) = 0$, and $f(3) = 2$

(b) $D_f = [-2, 3]$ and $R_f = [-3, 2]$

7. The graph satisfies the vertical line test, that is, each vertical line crosses the curve in at most one place, so it is the graph of a function.

9. The graph satisfies the vertical line test, that is, each vertical line crosses the curve in at most one place, so it is the graph of a function.

11. Since the only vertical line that does not intersect the curve is $x = 0$, the domain is $(-\infty, 0) \cup (0, \infty)$. Since the only horizontal line that does not intersect the curve in at least one place is $y = 0$, the range is $(-\infty, 0) \cup (0, \infty)$.

13. (a) Domain: the set of all real numbers, that is, $(-\infty, \infty)$; (b) Range: $(1, \infty) \cup \{-1\}$.

15. For $f(x) = x^2 - 1$ we have:

(a) Since $f(x) = x^2 - 1$ is defined for all real numbers, the domain is $(-\infty, \infty)$. (b) The range is $[-1, \infty)$.

17. For $f(x) = \sqrt{x} + 3$ we have:

(a) The domain is $[0, \infty)$. (b) The range is $[3, \infty)$.

19. For $x^2 - 2x + 1 = (x-1)^2$ we have:

(a) The domain is $(-\infty, \infty)$. (b) The range is $[0, \infty)$.

21. For $f(x) = \begin{cases} 1, & \text{if } x \geq 0 \\ -1, & \text{if } x < 0, \end{cases}$ we have:

(a) The domain is $(-\infty, \infty)$. (b) The only values taken by the function are -1 and 1, so the range is $\{-1, 1\}$.

23. For $f(x) = x^2 - 1$, $b = 0$, and $f(a) = a^2 - 1$ we have

$$f(a) = 0 \Leftrightarrow a^2 - 1 = 0 \Leftrightarrow a^2 = 1 \Rightarrow a = \pm 1.$$

25. For $f(x) = \sqrt{x-1}$, $b = \frac{1}{2}$, and $f(a) = \sqrt{a-1}$ we have

$$f(a) = \frac{1}{2} \Leftrightarrow \sqrt{a-1} = \frac{1}{2} \Leftrightarrow a - 1 = \frac{1}{4} \Leftrightarrow a = \frac{5}{4}.$$

27. (a) $(-\infty, \infty)$

(b) $D_f = \{x : 2 - x \neq 0\} = \{x : x \neq 2\} = (-\infty, 2) \cup (2, \infty)$.

(c) $D_f = \{x : 2 - x \geq 0\} = \{x : x \leq 2\} = (-\infty, 2]$.

(d) $D_f = \{x : 2 - x > 0\} = (-\infty, 2)$.

29. (a) $D_f = \{x : x^2 - 1 \neq 0\} = \{x : x \neq \pm 1\} = (-\infty, -1) \cup (-1, 1) \cup (1, \infty)$.

(b) For $f(x) = \dfrac{x+1}{x^2-1} = \dfrac{x+1}{(x-1)(x+1)} = \dfrac{1}{x+1}$, the domain is still $(-\infty, -1) \cup (-1, 1) \cup (1, \infty)$.

(c)

$$D_f = \left\{ x : \dfrac{x^2}{x^2-1} \geq 0 \right\} = \{0\} \cup \left\{ x : \dfrac{x^2}{x^2-1} > 0 \right\}$$
$$= \{0\} \cup \{x : x^2 - 1 > 0\}$$
$$= \{0\} \cup (-\infty, -1) \bigcup (1, \infty).$$

31. (a) For $f(x) = \sqrt{x(x-2)}$, the domain is $\{x : x(x-2) \geq 0\} = (-\infty, 0] \cup [2, \infty)$.

(b) For $f(x) = \sqrt{x(2-x)}$, the domain is $\{x : x(2-x) \geq 0\} = \{x : x(x-2) \leq 0\} = [0, 2]$.

(c) For $f(x) = \dfrac{x^2}{x^2-2x} = \dfrac{x^2}{x(x-2)}$, the domain is $\{x : x(x-2) \neq 0\} = \{x : x \neq 0, x \neq 2\} = (-\infty, 0) \cup (0, 2) \cup (2, \infty)$.

(d) For $f(x) = \sqrt{\dfrac{x^2}{x^2-2x}} = \sqrt{\dfrac{x^2}{x(x-2)}}$, the domain is $\{x : x(x-2) > 0\} = (-\infty, 0) \cup (2, \infty)$.

33. For $f(x) = x^2 + 2$ we have

$$f(-x) = (-x)^2 + 2 = x^2 + 2; \quad -f(x) = -(x^2 + 2) = -x^2 - 2;$$
$$f\left(\dfrac{1}{x}\right) = \left(\dfrac{1}{x}\right)^2 + 2 = \dfrac{1}{x^2} + 2; \quad \dfrac{1}{f(x)} = \dfrac{1}{x^2+2};$$
$$f(\sqrt{x}) = (\sqrt{x})^2 + 2 = x + 2; \quad \sqrt{f(x)} = \sqrt{x^2+2}.$$

35. For $f(x) = \dfrac{1}{x}$ we have

$$f(-x) = \dfrac{1}{-x} = -\dfrac{1}{x}; \quad -f(x) = -\dfrac{1}{x};$$
$$f\left(\dfrac{1}{x}\right) = \dfrac{1}{\frac{1}{x}} = x; \quad \dfrac{1}{f(x)} = \dfrac{1}{\frac{1}{x}} = x;$$
$$f(\sqrt{x}) = \dfrac{1}{\sqrt{x}} = \dfrac{1}{\sqrt{x}} \cdot \dfrac{\sqrt{x}}{\sqrt{x}} = \dfrac{\sqrt{x}}{x}; \quad \sqrt{f(x)} = \sqrt{\dfrac{1}{x}} = \dfrac{1}{\sqrt{x}} = \dfrac{\sqrt{x}}{x}.$$

37. For $f(x) = 2x - 4$

(a) $f(x + h) = 2(x + h) - 4 = 2x + 2h - 4$

(b)

$$\frac{f(x + h) - f(x)}{h} = \frac{2x + 2h - 4 - (2x - 4)}{h}$$

$$= \frac{2x + 2h - 4 - 2x + 4}{h} = \frac{2h}{h} = 2$$

(c) As $h \to 0$, $\dfrac{f(x + h) - f(x)}{h} \to 2$.

39. For $f(x) = x^2$

(a) $f(x + h) = (x + h)^2 = x^2 + 2hx + h^2$

(b) $\frac{f(x+h)-f(x)}{h} = \frac{x^2+2hx+h^2-x^2}{h} = \frac{2hx+h^2}{h} = \frac{h(2x+h)}{h} = 2x + h$

(c) As $h \to 0$, $\dfrac{f(x + h) - f(x)}{h} \to 2x$.

41. For $f(x) = 2 - x - x^2$

(a) $f(x + h) = 2 - (x + h) - (x + h)^2 = 2 - x - h - x^2 - 2hx - h^2$

(b)

$$\frac{f(x + h) - f(x)}{h} = \frac{2 - x - h - x^2 - 2hx - h^2 - (2 - x - x^2)}{h}$$

$$= \frac{-h - 2hx - h^2}{h} = \frac{h(-1 - 2x - h)}{h} = -1 - 2x - h$$

(c) As $h \to 0$, $\dfrac{f(x + h) - f(x)}{h} \to -1 - 2x$.

43. For $f(x) = \dfrac{1}{x}$

(a) $f(x + h) = \dfrac{1}{x + h}$

(b)

$$\frac{f(x + h) - f(x)}{h} = \frac{\frac{1}{x+h} - \frac{1}{x}}{h} = \frac{\frac{x-(x+h)}{x(x+h)}}{h}$$

$$= \frac{-h}{hx(x + h)} = -\frac{1}{x(x + h)}$$

(c) As $h \to 0$, $\dfrac{f(x + h) - f(x)}{h} \to -\dfrac{1}{x^2}$.

45. For $f(x) = \dfrac{x}{x-3}$

(a) $f(x+h) = \dfrac{x+h}{x+h-3}$

(b)

$$\frac{f(x+h) - f(x)}{h} = \frac{\frac{x+h}{x+h-3} - \frac{x}{x-3}}{h}$$

$$= \frac{\frac{(x+h)(x-3) - x(x+h-3)}{(x+h-3)(x-3)}}{h}$$

$$= \frac{x^2 - 3x + hx - 3h - x^2 - hx + 3x}{h(x+h-3)(x-3)}$$

$$= \frac{-3h}{h(x+h-3)(x-3)} = -\frac{3}{(x+h-3)(x-3)}$$

(c) As $h \to 0$, $\dfrac{f(x+h) - f(x)}{h} \to -\dfrac{3}{(x-3)^2}$.

47. For $f(x) = x^3$

(a) $f(x+h) = (x+h)^3 = x^3 + 3x^2 h + 3xh^2 + h^3$

(b)

$$\frac{f(x+h) - f(x)}{h} = \frac{x^3 + 3x^2 h + 3xh^2 + h^3 - x^3}{h}$$

$$= \frac{3x^2 h + 3xh^2 + h^3}{h} = \frac{h(3x^2 + 3xh + h^2)}{h}$$

$$= 3x^2 + 3xh + h^2$$

(c) As $h \to 0$, $\dfrac{f(x+h) - f(x)}{h} \to 3x^2$.

49. The function is even, since the graph is symmetric with respect to the y-axis.

51. The function is neither even nor odd, since the graph is not symmetric with respect to either the x-axis or the y-axis.

53. $f(x) = x^2 + 1$ is even, since $f(-x) = (-x)^2 + 1 = x^2 + 1 = f(x)$.

55. $f(x) = x^3 + x^5$ is odd, since $f(-x) = (-x)^3 + (-x)^5 = -x^3 - x^5 = -f(x)$.

57. $f(x) = x^3 + x^2$ is neither even nor odd, since

$$f(-x) = (-x)^3 + (-x)^2 = -x^3 + x^2 \neq f(x)$$

and

$$-f(x) = -x^3 - x^2 \neq f(x).$$

59. $f(x) = x^{-1} = \dfrac{1}{x}$ is odd, since $f(-x) = -\dfrac{1}{x} = -f(x)$.

61. (a)

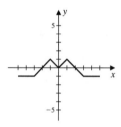

(b)

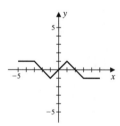

63. (a)

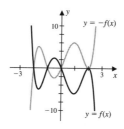

(b)

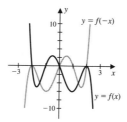

(c)

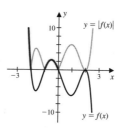

65.   (a) For $f(x) = \sqrt{x}$, we have $f(x+h) = \sqrt{x+h}$ and

$$\frac{f(x+h) - f(x)}{h} = \frac{\sqrt{x+h} - \sqrt{x}}{h}$$

$$= \frac{\sqrt{x+h} - \sqrt{x}}{h} \cdot \frac{\sqrt{x+h} + \sqrt{x}}{\sqrt{x+h} + \sqrt{x}} = \frac{(x+h) - x}{h(\sqrt{x+h} + \sqrt{x})}$$

$$= \frac{h}{h(\sqrt{x+h} + \sqrt{x})} = \frac{1}{\sqrt{x+h} + \sqrt{x}}.$$

(b) As $h \to 0$, the value of the difference quotient $\dfrac{f(x+h) - f(x)}{h}$ approaches

$$\frac{1}{\sqrt{x} + \sqrt{x}} = \frac{1}{2\sqrt{x}}.$$

67. At time t after 3:00 P.M. the first ship has traveled $t+3$ hours and the second ship has traveled t hours. As a consequence, the distances of the ships from port are, respectively, $10(t+3)$ and $15t$ nautical miles. The distance between the ships is

$$d = \sqrt{(10(3+t))^2 + (15t)^2} = \sqrt{325t^2 + 600t + 900}$$

nautical miles.

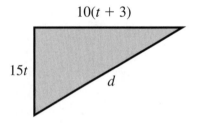

Figure 1.47

69. $P(x) = 300x - 2x^2$

(a) The average rate of change is the difference quotient

$$\frac{P(x+h) - P(x)}{h} = \frac{300(x+h) - 2(x+h)^2 - (300x - 2x^2)}{h}$$

$$= \frac{300x + 300h - 2(x^2 + 2hx + h^2) - 300x + 2x^2}{h}$$

$$= \frac{300h - 4hx - 2h^2}{h} = \frac{h(300 - 4x - 2h)}{h}$$

$$= 300 - 4x - 2h.$$

(b) If $h = 25$ and $x = 25$, then $x + h = 50$, and the average rate of change in the profit as the number of units changes from 25 to 50 is

$$300 - 4(25) - 2(25) = 150.$$

(c) As $h \to 0$, $\frac{P(x+h) - P(x)}{h} \to 300 - 4x$, which is the instantaneous rate of change. And when $x = 25$, the instantaneous rate of change is $300 - 4(25) = 200$.

(d)

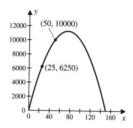

Figure 1.48

1.7 Linear Functions

THE SLOPE OF A LINE

The *slope* of a line describes the inclination of the line and is a number that can be determined from any two points on the line. If two points on the line are (x_1, y_1) and (x_2, y_2), then the slope is

$$m = \frac{y_2 - y_1}{x_2 - x_1}.$$

It does not matter which point is considered first and which is second in the slope formula, provided we are consistent in the order of subtraction in both the numerator and the denominator. This follows from the fact that

$$\frac{y_1 - y_2}{x_1 - x_2} = \frac{-(y_2 - y_1)}{-(x_2 - x_1)} = \frac{y_2 - y_1}{x_2 - x_1}.$$

EXAMPLE 1.7.1   Find the slope of the line that passes through the points, and sketch the graph of the line.

(a) $(5, 2)$ and $(-3, 1)$ (b) $(3, 4)$ and $(6, 1)$

SOLUTION:

(a) $m = \dfrac{1 - 2}{-3 - 5} = \dfrac{-1}{-8} = \dfrac{1}{8}$ ◇

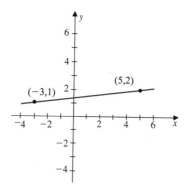

Figure 1.49

(b) $m = \dfrac{4-1}{3-6} = \dfrac{3}{-3} = -1$

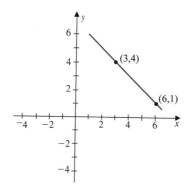

Figure 1.50

EXAMPLE 1.7.2   Find the slope of the line that passes through the points, and sketch the line.

(a) $(-2, -1)$ and $(7, -1)$ (b) $(-1, -4)$ and $(-1, 8)$

SOLUTION:

(a) $m = \dfrac{-1-(-1)}{-2-7} = \dfrac{-1+1}{-9} = 0$

Since the slope of a line is 0, the line is horizontal. This line has y-intercept $(0, -1)$.

◇

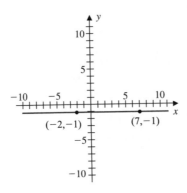

Figure 1.51

(b) $m = \dfrac{8+4}{-1+1} = \dfrac{12}{0}$ which is an undefined quantity.

Since the slope of a line is undefined, the line is vertical. This line has x-intercept $(-1, 0)$.

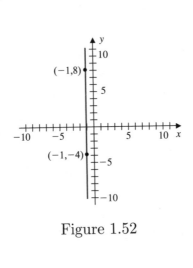

Figure 1.52

POINT-SLOPE EQUATION OF A LINE

If a line passes through the point (x_0, y_0) and has slope m, then a point (x, y) will lie on the line precisely when

$$m = \frac{y - y_0}{x - x_0},$$

that is, when

$$y - y_0 = m(x - x_0).$$

This gives an equation describing all points that are on the line, called the *point-slope equation* of the line.

EXAMPLE 1.7.3   Find an equation of the line that passes through the point $(-1, 1)$ and has slope 2.

SOLUTION: Substituting directly into the point-slope form gives

$$y - 1 = 2(x - (-1)) \quad \text{or} \quad y = 2x + 3.$$

◇

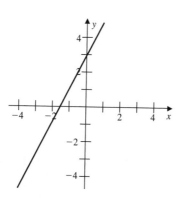

Figure 1.53

EXAMPLE 1.7.4 Find the equation of the line passing through the two points $(-1, 1)$ and $(2, -3)$.

SOLUTION: First use the two points to find the slope. Then either of the given points along with the slope can be substituted into the point-slope form to find the equation. So,

$$m = \frac{-3 - 1}{2 - (-1)} = -\frac{4}{3}$$

$$y - 1 = -\frac{4}{3}(x + 1)$$

and

$$y = -\frac{4}{3}x - \frac{1}{3}.$$

◇

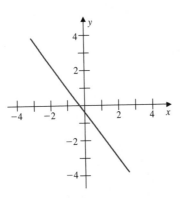

Figure 1.54

SLOPE-INTERCEPT EQUATION OF A LINE

The *slope-intercept* equation of a line is a special case of the point-slope equation of the line where the y-intercept $(0, b)$ is the given point on the line.

The slope-intercept equation then has the form

$$y = mx + b.$$

This was the form given in the final reduction in Examples 1.7.3 and 1.7.4. When the equation is in this form, the slope is m and the line crosses the y-axis at b.

EXAMPLE 1.7.5   Find an equation of the line that has slope -2 and y-intercept 1.

SOLUTION: Substituting directly into the slope-intercept form of a line gives

$$y = -2x + 1.$$

◇

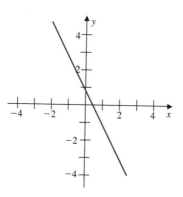

Figure 1.55

PARALLEL AND PERPENDICULAR LINES

Two lines are *parallel* provided they have the same slope, or both have undefined slopes (so are both vertical lines). Two lines are *perpendicular* if their slopes are negative reciprocals. That is, if the slopes are m_1 and m_2, then

$$m_1 m_2 = -1 \quad \text{or} \quad m_2 = -\frac{1}{m_1}.$$

EXAMPLE 1.7.6   Find the line that passes through $(1,1)$ and is parallel to the line $2x + 3y = -2$. Sketch the two lines.

SOLUTION: First find the slope of the given line by writing the equation in the form $y = mx + b$. So,

$$2x + 3y = -2$$

$$3y = -2x - 2$$

and

$$y = -\frac{2}{3}x - \frac{2}{3}$$

and the slope of the line is $-\frac{2}{3}$. Since the desired line is parallel to the given line, it also has slope $-\frac{2}{3}$. Since it also passes through the point $(1,1)$ the equation is

$$y - 1 = -\frac{2}{3}(x - 1)$$

and

$$y = -\frac{2}{3}x + \frac{5}{3}.$$

◇

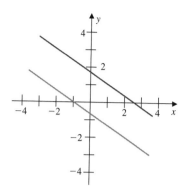

Figure 1.56

EXAMPLE 1.7.7   Find the line that passes through $(-1, -1)$ and is perpendicular to the line $-3x + 5y = 3$. Sketch the two lines.

SOLUTION: Since

$$-3x + 5y = 3$$

$$5y = 3x + 3$$

and

$$y = \frac{3}{5}x + \frac{3}{5}.$$

The two lines are perpendicular, so the slope of the desired line is the negative reciprocal of $\frac{3}{5}$, that is, $-\frac{5}{3}$. The line we want passes through $(-1, -1)$ so its equation is

$$y + 1 = -\frac{5}{3}(x + 1)$$

and

$$y = -\frac{5}{3}x - \frac{8}{3}.$$

◇

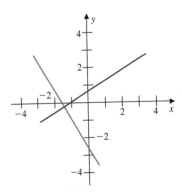

Figure 1.57

EXAMPLE 1.7.8   Find an equation of the line that is tangent to the circle $x^2 + y^2 = 9$ at the point $(2\sqrt{2}, 1)$. At what other point on the circle will the tangent line be parallel to this line?

SOLUTION: The slope of the radius is $m = \frac{1}{2\sqrt{2}} = \frac{\sqrt{2}}{4}$, so the tangent line, which is perpendicular to the radius, has slope $-\frac{4}{\sqrt{2}} = -2\sqrt{2}$. The equation of the tangent line is then

$$y - 1 = -2\sqrt{2}(x - 2\sqrt{2}),$$

so

$$y = -2\sqrt{2}x + 8 + 1 = -2\sqrt{2}x + 9.$$

The other point on the circle where the tangent line is parallel is at the point diametrically opposite $(2\sqrt{2}, 1)$, which is the point $(-2\sqrt{2}, -1)$.   ◇

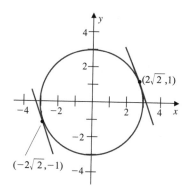

Figure 1.58

APPLICATION

EXAMPLE 1.7.9   A new Corvette costs $45,000.00$. At the end of 5 years, it will be worth an estimated $25,000.00$. Assume that the value of the car depreciates linearly.

(a) Find a linear equation that expresses the value V of the car as a function of time t, where $0 \le t \le 5$.

(b) How much will the car be worth after 3 years?

(c) What is the average rate of change in the value of the car from 2 to 4 years?

SOLUTION:

(a) At time 0 the car is worth 45000 dollars, and at time 5 it is worth 25000 dollars. If a linear model is used to represent value against time, the two points $(0, 45000)$ and $(5, 25000)$ lie on the line. The slope is then

$$m = \frac{45000 - 25000}{0 - 5} = -\frac{20000}{5} = -4000,$$

and the equation of the line is

$$y = -4000t + 45000.$$

So the value at time t is

$$V(t) = -4000t + 45000.$$

(b) Substitute $t = 3$ in the equation obtained in part (a) to get the value of the equipment after 3 years to be

$$V(3) = -4000(3) + 45000 = 33000 \text{ dollars.}$$

(c) The average rate of change is the difference quotient

$$\frac{V(4) - V(2)}{4 - 2} = \frac{29000 - 37000}{2} = -4000.$$

◇

Solutions for Exercise Set 1.7

1. Slope: $m = \dfrac{2-0}{1-0} = 2$; equation: $y = 2x$

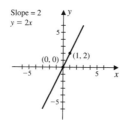

3. Slope: $m = \dfrac{-4-1}{2-(-2)} = -\dfrac{5}{4}$; equation: $y - 1 = -\dfrac{5}{4}(x+2) \Leftrightarrow y = -\dfrac{5}{4}x - \dfrac{3}{2}$

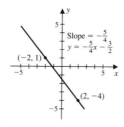

5. Slope: $m = \dfrac{3-3}{-1-2} = 0$; equation: $y = 3$

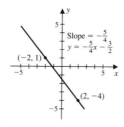

7. Slope: $m = \dfrac{6-(-3)}{-1-(-1)}$ which is undefined; equation: $x = -1$

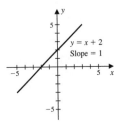

9. For $y = x + 2$, the slope is $m = 1$.

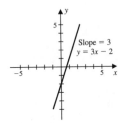

11. For $3x - y = 2 \Leftrightarrow y = 3x - 2$, the slope is $m = 3$.

13. For $-2x - 3y = 1 \Leftrightarrow -3y = 2x + 1 \Leftrightarrow y = -\frac{2}{3}x - \frac{1}{3}$, the slope is $m = -\frac{2}{3}$.

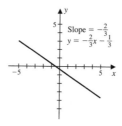

15. For $y = 2$, the slope is 0.

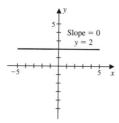

17. (a) $y - 2 = x - 3 \Leftrightarrow y = x - 1$

(b) $y - 2 = -(x - 3) \Leftrightarrow y = -x + 5$

(c) $y - 2 = 2(x - 3) \Leftrightarrow y = 2x - 4$

(d) $y - 2 = \frac{1}{2}(x - 3) \Leftrightarrow y = \frac{1}{2}x + \frac{1}{2}$

19. The slopes of the lines are:

(a) 0   (b) 1   (c) −1   (d) 1   (e) 2

(f) 0   (g) −1   (h) 2   (i) 1   (j) 1

Lines that are parallel have the same slope.

Parallel Lines: (a) and (f); (c) and (g); (b), (d), (i), and (j); (e) and (h).

21. For $y = 2x + 1$ the slope is $m = 2$.

(a) The line parallel through $(0, 0)$ has same slope $m = 2$, so the equation is $y = 2x$.

(b) The line perpendicular through $(0,0)$ has slope negative reciprocal $m = -\frac{1}{2}$, so the equation is $y = -\frac{1}{2}x$.

23. For $y = -2x + 3$ the slope is $m = -2$.

(a) The line parallel through $(-1, 2)$ has same slope $m = -2$, so the equation is $y - 2 = -2(x + 1) \Leftrightarrow y = -2x$.

(b) The line perpendicular through $(-1, 2)$ has slope negative reciprocal $m = \frac{1}{2}$, so the equation is $y - 2 = \frac{1}{2}(x + 1) \Leftrightarrow y = \frac{1}{2}x + \frac{5}{2}$.

25. Since the y-intercept is 2, the line passes through the point $(0, 2)$, so the equation is $y - 2 = -x$ or $y = -x + 2$.

27. $y - (-2) = 3(x - 1) \Leftrightarrow y = 3x - 5$

29. The line passes through $(2, 0)$ and $(0, 4)$, so the slope is $m = \frac{4-0}{0-2} = -2$, and the equation is $y - 0 = -2(x - 2)$ or $y = -2x + 4$.

31. Since the line is parallel to the x-axis its slope is 0. The line is horizontal with equation $y = -1$.

33. We have $2x - 3y = 2 \Leftrightarrow -3y = -2x + 2 \Leftrightarrow y = \frac{2}{3}x - \frac{2}{3}$ and the line has slope $m = \frac{2}{3}$. The parallel line also has slope $\frac{2}{3}$ and equation $y - 3 = \frac{2}{3}(x - 4)$ or $y = \frac{2}{3}x + \frac{1}{3}$.

35. A line perpendicular to $y = 2 - x$ has slope 1, and if it passes through $(1, 1)$, has equation $y - 1 = x - 1$ or $y = x$.

37. (a) The line tangent to the circle $x^2 + y^2 = 3$ at the point $(1, \sqrt{2})$ is perpendicular to the radius line passing through $(0, 0)$ and $(1, \sqrt{2})$. The slope of the radius line is $m = \frac{\sqrt{2}-0}{1-0} = \sqrt{2}$. A line perpendicular to the circle has slope $-\frac{1}{\sqrt{2}} = -\frac{\sqrt{2}}{2}$, and if it passes through $(1, \sqrt{2})$ has equation

$$y - \sqrt{2} = -\frac{\sqrt{2}}{2}(x - 1) \text{ or } y = -\frac{\sqrt{2}}{2}x + \frac{3\sqrt{2}}{2}.$$

(b) The other point on the circle where the tangent line is parallel to this line is at the point diametrically opposite the point $(1, \sqrt{2})$, which is the point $(-1, -\sqrt{2})$. The tangent line at this point has equation

$$y + \sqrt{2} = -\frac{\sqrt{2}}{2}(x+1) \quad \text{or} \quad y = -\frac{\sqrt{2}}{2}x - \frac{3\sqrt{2}}{2}.$$

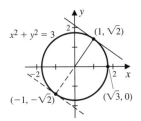

39. Since $v(t) = -32t \Rightarrow v(7) = -224$, and the rock is traveling downward at a speed of 224 ft/sec.

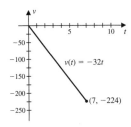

41. (a) For $W(n) = 500 - 0.5n$ we have the following graph.

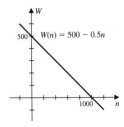

(b) The total fish weight is

$$T(n) = n \cdot W(n) = n(500 - 0.5n) = 500n - 0.5n^2.$$

(c) At $n = 1000$ the total weight is 0, so the fish population has disappeared. This is the predicted limiting value of the number of fish in the pond. For $n > 1000$, the total weight given by the equation is negative, which is physically unreasonable.

1.8 Quadratic Functions

A *quadratic function* is a function of the form $f(x) = ax^2 + bx + c$, where a, b, and c are real numbers and $a \neq 0$. Every quadratic function can be rewritten in the *standard form* $f(x) = a(x - h)^2 + k$. The graph of a quadratic function is called a *parabola*.

COMPLETING THE SQUARE

The process of writing a quadratic function in standard form is called *completing the square*.

EXAMPLE 1.8.1   Write the quadratic function in standard form.

(a) $f(x) = x^2 - 4x + 5$ (b) $f(x) = 2x^2 - 4x - 1$

SOLUTION:

(a) To complete the square, take half the coefficient of the x term, square it and both add and subtract the value, so the net result is to add 0. So,

$$
\begin{aligned}
f(x) &= x^2 - 4x + 5 \\
&= x^2 - 4x + \left(\frac{4}{2}\right)^2 - \left(\frac{4}{2}\right)^2 + 5 \\
&= x^2 - 4x + 4 + (-4 + 5) \\
&= (x - 2)^2 + 1.
\end{aligned}
$$

(b) If the coefficient of the x^2 term is not 1, then first factor the coefficient from both the x^2 term and the x term and proceed as in part (a). So,

$$f(x) = 2x^2 - 4x - 1$$
$$= 2(x^2 - 2x) - 1$$
$$= 2\left(x^2 - 2x + \left(\frac{2}{2}\right)^2 - \left(\frac{2}{2}\right)^2\right) - 1$$
$$= 2\left(x^2 - 2x + 1 - 1\right) - 1$$
$$= 2\left(x - 1\right)^2 - 2 - 1$$
$$= 2\left(x - 1\right)^2 - 3.$$

◇

HORIZONTAL AND VERTICAL SHIFTS

If h, $k > 0$, then

(i) The graph of $y = f(x - h)$ is the graph of $y = f(x)$ shifted to the right h units;

(ii) The graph of $y = f(x + h)$ is the graph of $y = f(x)$ shifted to the left h units;

(iii) The graph of $y = f(x) + k$ is the graph of $y = f(x)$ shifted upward k units;

(iv) The graph of $y = f(x) - k$ is the graph of $y = f(x)$ shifted downward k units.

EXAMPLE 1.8.2   Use the graph of $y = f(x) = x^2$ given in the figure to sketch the graphs of

(a) $y = f(x - 1)$ (b) $y = f(x + 2)$ (c) $y = f(x) + 1$

(d) $y = f(x) - 2$ (e) $y = f(x - 2) + 3$.

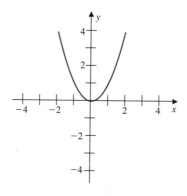

Figure 1.59

SOLUTION: We have   ◇

   (a) $y = f(x - 1)$ (b) $y = f(x + 2)$ (c) $y = f(x) + 1$

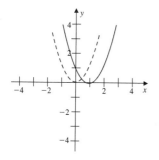
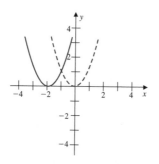
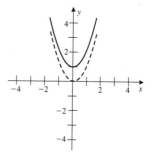

Figure 1.60   Figure 1.61   Figure 1.62

(d) $y = f(x) - 2$ (e) $y = f(x - 2) + 3$.

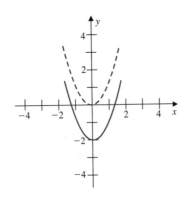

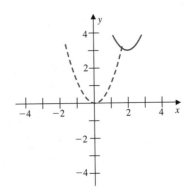

Figure 1.63   Figure 1.64

EXAMPLE 1.8.3   Sketch the graph of the quadratic function.

(a) $f(x) = x^2 - 2x - 1$ (b) $f(x) = x^2 + 4x + 5$

SOLUTION: To use the basic graph $y = x^2$ and the shifting properties, the quadratic function is first put in standard form by completing the square.

(a)

$$f(x) = x^2 - 2x - 1$$
$$= (x^2 - 2x + 1 - 1) - 1$$
$$= (x - 1)^2 - 2$$

To obtain the graph, first shift $y = x^2$, to the right 1 unit and then downward 2 units. The vertex of the parabola is at $(1, -2)$.

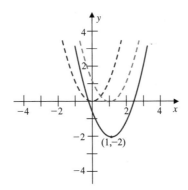

Figure 1.65

(b)

$$f(x) = x^2 + 4x + 5$$

$$= (x^2 + 4x + 4 - 4) + 5$$

$$= (x + 2)^2 + 1$$

To obtain the graph first shift $y = x^2$, to the left 2 units and then upward 1 unit. The vertex of the parabola is at $(-2, 1)$.   ◇

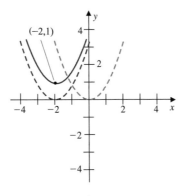

Figure 1.66

HORIZONTAL AND VERTICAL SCALING AND REFLECTION

If $a > 1$, the graph of

    i. $y = af(x)$, is a vertical stretching, by a factor of a, of the graph of $y = f(x)$;

    ii. $y = \dfrac{1}{a}f(x)$, is a vertical shrinking, by a factor of a, of the graph of $y = f(x)$.

    iii. The graph of $y = -f(x)$ is the reflection through the x-axis of the graph of $y = f(x)$.

For example, if $y = x^2$, then the point $(1,1)$ on the curve lies

$$\text{directly below the point } (1,2) \text{ on } y = 2x^2,$$
$$\text{directly above the point } \left(1, \frac{1}{2}\right) \text{ on } y = \frac{1}{2}x^2,$$

and when $x = 1$, the point $(1,-1)$ on $y = -x^2$ is one unit below the x-axis.

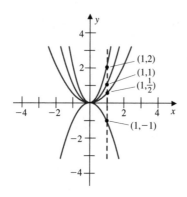

Figure 1.67

EXAMPLE 1.8.4 Sketch the graph of the function and find the range.

(a) $f(x) = 2x^2 - 8x + 10$ (b) $f(x) = -4x^2 - 4x + 3$

SOLUTION:

(a)First we write

$$f(x) = 2x^2 - 8x + 10$$

$$= 2(x^2 - 4x) + 10$$

$$= 2(x^2 - 4x + 4 - 4) + 10$$

$$= 2(x - 2)^2 + 2.$$

To obtain the graph, start with $y = x^2$, stretch it by a factor of 2, and shift the result to the right 2 units and upward 2 units. The range of the function is then $[2, \infty)$, with a minimum point at $(2, 2)$.

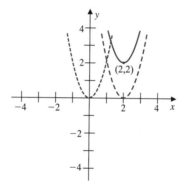

Figure 1.68

(b) Similarly,

$$f(x) = -4x^2 - 4x + 3$$

$$= -4(x^2 + x) + 3$$

$$= -4\left(x^2 + x + \frac{1}{4} - \frac{1}{4}\right) + 3$$

$$= -4\left(x + \frac{1}{2}\right)^2 + 4.$$

To obtain the graph, start with $y = x^2$, stretch it by a factor of 4, reflect the result about the x-axis, and shift the new result to the left $\frac{1}{2}$ unit and upward 4 units. The range of the function is then $(-\infty, 4]$, with a maximum point at $\left(-\frac{1}{2}, 4\right)$.

◇

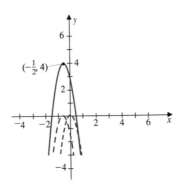

Figure 1.69

EXAMPLE 1.8.5   Given the quadratic function $f(x) = x^2 - 4x + 5$.

(a) Express the quadratic in standard form.

(b) Find any x- and y-intercepts.

(c) Find the maximum or minimum value of the function.

SOLUTION:

(a) Completing the square

$$f(x) = x^2 - 4x + 5$$
$$= x^2 - 4x + 4 - 4 + 5$$
$$= (x - 2)^2 + 1.$$

(b) <u>x-intercepts:</u> Solve $f(x) = 0$.

$$(x - 2)^2 + 1 = 0$$

$$(x - 2)^2 = -1$$

Since the left hand side is always positive, there are no x-intercepts. (See the figure.)

<u>y-intercepts:</u> Set $x = 0$, then $y = 5$ and the y-intercept is $(0, 5)$.

(c) Since the leading coefficient of the quadratic is 1, the parabola opens upward, and the vertex $(2, 1)$ is the minimum point on the curve. The minimum value of the function is $f(2) = 1$.   ◇

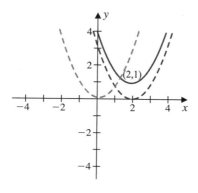

Figure 1.70

EXAMPLE 1.8.6   Use the quadratic formula to find any x-intercepts of the parabola $y = 2x^2 - 4x + 1$.

SOLUTION:

The x-intercepts are those values of x that make $y = 0$, so solve $2x^2 - 4x + 1 = 0$. Notice the quadratic can not be factored directly, so we will use the quadratic formula with $a = 2, b = -4$, and $c = 1$.

Since

$$2x^2 - 4x + 1 = 0,$$

$$x = \frac{-(-4) \pm \sqrt{(-4)^2 - 4(2)(1)}}{2(2)}$$

$$= \frac{4 \pm \sqrt{8}}{4} = \frac{4 \pm 2\sqrt{2}}{4}$$

$$= \frac{2 \pm \sqrt{2}}{2} = 1 \pm \frac{\sqrt{2}}{2}.$$

The x-intercepts can also be found by writing the quadratic in standard form,

$$y = 2x^2 - 4x + 1$$

$$= 2(x^2 - 2x) + 1$$

$$= 2(x^2 - 2x + 1 - 1) + 1$$

$$= 2(x - 1)^2 - 1.$$

Then

$$2(x - 1)^2 - 1 = 0$$

$$2(x - 1)^2 = 1$$

$$(x - 1)^2 = \frac{1}{2}$$

$$x - 1 = \pm\sqrt{\frac{1}{2}} = \pm\frac{\sqrt{2}}{2},$$

and

$$x = 1 \pm \frac{\sqrt{2}}{2}.$$

◇

EXAMPLE 1.8.7 Match the curve with the equation in the figure.

(a) $y = x^2 + 1$ (b) $y = x^2 - 2x + 1$ (c) $y = (x+1)^2$

(d) $y = (x+1)^2 - 1$ (e) $y = x^2 - 2x - 1$ (f) $y = -x^2 + 1$

(i)

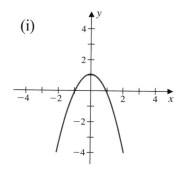

Figure 1.71

(ii)

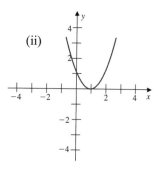

Figure 1.72

(iii)

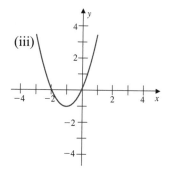

Figure 1.73

(iv)

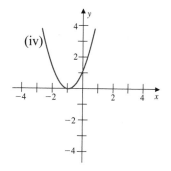

Figure 1.74

(v)

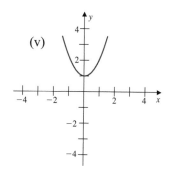

Figure 1.75

(vi)

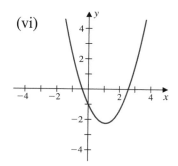

Figure 1.76

SOLUTION: The curves for equations (a), (c), (d), and (e) can be recognized as transformations of $y = x^2$.

(a) An upward shift by 1 unit of $y = x^2$, so it is (v).

(c) A shift left by 1 unit of $y = x^2$, so it is (iv).

(d) A shift left by 1 unit of $y = x^2$, followed by a shift downward of 1 unit, so it is (iii).

(f) A reflection of $y = x^2$ about the x-axis followed by a shift upward by 1 unit, so it is (i). Note also (f) is the only equation with a negative leading coefficient, so it is the only equation with a curve that opens downward.

(b) Since

$$y = x^2 - 2x + 1 = (x - 1)^2,$$

the curve is a shift right by 1 unit of $y = x^2$, so it is (ii).

(e) By process of elimination, the equation has graph (vi). Note also

$$y = x^2 - 2x - 1$$
$$= \left(x^2 - 2x + 1 - 1\right) - 1$$
$$= (x - 1)^2 - 2.$$

◇

EXAMPLE 1.8.8   What can you say about a, b, and c in $f(x) = ax^2 + bx + c$ if

(a) $(1, 0)$ is on the graph?

(b) the y-intercept is 3?

(c) $(1, 0)$ is the vertex?

(d) conditions (a), (b), and (c) are all satisfied?

SOLUTION:

(a) If $(1,0)$ is on the graph, then

$$0 = f(1,$$

$$0 = a(1)^2 + b(1) + c,$$

and

$$a + b + c = 0.$$

(b) If $(0,3)$ is on the graph, then

$$3 = f(0)$$

$$3 = a(0)^2 + b(0) + c$$

and

$$c = 3.$$

(c) The vertex of a parabola is given by the point

$$\left(-\frac{b}{2a}, \frac{4ac - b^2}{4a} \right)$$

so

$$-\frac{b}{2a} = 1, \quad \text{and} \quad \frac{4ac - b^2}{4a} = 0.$$

(d) Using (a), (b), and (c),

$$a + b + c = 0$$

$$c = 3$$

$$b = -2a$$

$$a - 2a + 3 = 0$$

$$-a = -3$$

and

$$a = 3.$$

So,

$$a = 3$$

$$b = -2(3) = -6$$

$$c = 3$$

$$f(x) = 3x^2 - 6x + 3.$$

◇

Solutions for Exercise Set 1.8

1. $y = x^2 + 1$

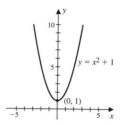

3. $y = -x^2 + 1$

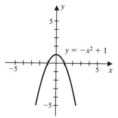

5. $y = (x - 2)^2$

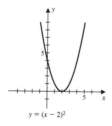

7. $y = (x + 1)^2 - 1$

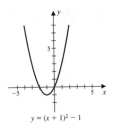

$y = (x + 1)^2 - 1$

9. Completing the square gives $y = x^2 - 4x + 3 = x^2 - 4x + 4 - 4 + 3 = (x - 2)^2 - 1$.

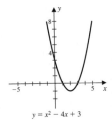

$y = x^2 - 4x + 3$

11. Completing the square gives $y = -x^2 - 2x = -(x^2 + 2x) = -(x^2 + 2x + 1 - 1) = -(x + 1)^2 + 1$.

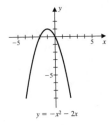

$y = -x^2 - 2x$

13. Completing the square gives $y = 3x^2 + 6x = 3(x^2 + 2x) = 3(x^2 + 2x + 1 - 1) = 3(x + 1)^2 - 3$

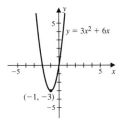

15. $y = \frac{1}{2}x^2 - 1$

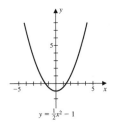

17. Completing the square gives $y = \frac{1}{2}x^2 - 2x + 2 = \frac{1}{2}\left(x^2 - 4x + 4\right) = \frac{1}{2}(x - 2)^2.$

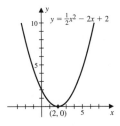

19. (a) $f(x) = x^2 - 6x + 7 = (x^2 - 6x + 9 - 9) + 7 = (x - 3)^2 - 2$

(b) To find the x-intercepts solve

$$(x-3)^2 - 2 = 0$$

$$(x-3)^2 = 2$$

$$x = \pm\sqrt{2} + 3$$

so the intercepts are $(3+\sqrt{2}, 0)$ and $(3-\sqrt{2}, 0)$. To find the y-intercept set $x = 0 \Rightarrow y = 7$ and the y-intercept is $(0, 7)$.

(c) Since the parabola opens upward the vertex $(3, -2)$ is a minimum point so the minimum value of the function is $f(3) = -2$.

21. (a) $f(x) = -x^2 + 6x - 8 = -(x^2 - 6x + 9 - 9) - 8 = -(x-3)^2 + 1$

(b) To find the x-intercepts solve

$$-(x-3)^2 + 1 = 0$$

$$(x-3)^2 = 1$$

$$x = \pm 1 + 3$$

$$x = 4, \ x = 2$$

so the intercepts are $(4, 0)$ and $(2, 0)$. To find the y-intercept set $x = 0 \Rightarrow y = -8$ and the y-intercept is $(0, -8)$.

(c) Since the parabola opens downward the vertex $(3, 1)$ is a maximum point so the maximum value of the function is $f(3) = 1$.

23.

$$6x^2 - 5x + 1 = 0$$

$$x = \frac{-(-5) \pm \sqrt{(-5)^2 - 4(6)(1)}}{2(6)} = \frac{5 \pm \sqrt{25 - 24}}{12}$$

$$= \frac{5 \pm 1}{12}$$

$$x = \frac{1}{2}, \ x = \frac{1}{3}$$

25.

$$2x^2 - 4x + 1 = 0$$

$$x = \frac{-(-4) \pm \sqrt{(-4)^2 - 4(2)(1)}}{2(2)} = \frac{4 \pm \sqrt{16 - 8}}{4}$$

$$= \frac{4 \pm \sqrt{8}}{4} = \frac{4 \pm 2\sqrt{2}}{4}$$

$$x = 1 + \frac{\sqrt{2}}{2}, x = 1 - \frac{\sqrt{2}}{2}$$

27.

$$2x^2 + 4x + 3 = 0$$

$$x = \frac{-4 \pm \sqrt{(4)^2 - 4(2)(3)}}{2(2)} = \frac{-4 \pm \sqrt{16 - 24}}{4}$$

$$= \frac{-4 \pm \sqrt{-8}}{4}$$

Since the square root of a negative number is not a real number the quadratic has no solutions and there are no x-intercepts.

29. (a) An upward shift by 2 units of $y = x^2$, so is (v).

(b) Since $y = x^2 - 4x + 4 = (x - 2)^2$, the curve is a right shift by 2 units of $y = x^2$, so is (ii).

(c) A left shift by 2 units of $y = x^2$, so is (iv).

(d) A left shift by 1 unit of $y = x^2$, followed by a downward shift of 2 units so is (iii).

(e) Since $y = x^2 - 2x + 3 = (x - 1)^2 + 2$, the curve is a left shift by 1 unit of $y = x^2$, followed by an upward shift of 2 units so is (vi).

(f) A reflection of $y = x^2$ about the x-axis followed by a downward shift of 1 unit so is (i).

31. The function will have the form $f(x) = a(x - 1)^2 + 2$ and if the graph passes through the point $(-2, 5)$ we have

$$5 = a(-2 - 1)^2 + 3$$

$$5 = 9a + 3$$

$$a = \frac{2}{9}.$$

So $f(x) = \frac{2}{9}(x - 1)^2 + 3$.

33. (a) The domain of $f(x) = \sqrt{x^2 - 3}$ is all x with

$$x^2 - 3 \geq 0 \Leftrightarrow \left(x - \sqrt{3}\right)\left(x + \sqrt{3}\right) \geq 0 \Leftrightarrow x \leq -\sqrt{3} \quad \text{or} \quad x \geq \sqrt{3},$$

which can be written $\left(-\infty, -\sqrt{3}\right] \cup \left[\sqrt{3}, \infty\right)$.

(b) The domain of $f(x) = \sqrt{x^2 - \frac{1}{2}x} = \sqrt{x\left(x - \frac{1}{2}\right)}$ is all x with $x\left(x - \frac{1}{2}\right) \geq 0 \Leftrightarrow x \leq 0$ or $x \geq \frac{1}{2}$, which can be written $(-\infty, 0] \cup \left[\frac{1}{2}, \infty\right)$.

35. Let $f(x) = ax^2 + bx + c$.

(a) $(1, 1)$ is on the graph $\Leftrightarrow 1 = f(1) = a + b + c$. So $a + b + c = 1$.

(b) The y-intercept is $6 \Leftrightarrow 6 = f(0) = c$. So $c = 6$.

(c) The vertex of a parabola is $\left(-\frac{b}{2a}, \frac{4ac - b^2}{4a}\right)$. The vertex is $(1, 1) \Leftrightarrow -\frac{b}{2a} = 1$ and $\frac{4ac - b^2}{4a} = 1 \Leftrightarrow$. This is true precisely when

$$b = -2a$$

and

$$\frac{4ac - b^2}{4a} = 1 \Leftrightarrow \frac{4ac - (-2a)^2}{4a} = 1$$

$$\frac{4ac - 4a^2}{4a} = 1 \Leftrightarrow \frac{4a(c - a)}{4a} = 1$$

$$c - a = 1 \Leftrightarrow c = a + 1.$$

So $b = -2a$ and $c = a + 1$.

(d) $a + b + c = 1$ and $c = 6 \Rightarrow a + b = -5$. Then $c = a + 1 \Rightarrow a = 5$ and $b = -10$. So $a = 5, b = -10, c = 6$ and $f(x) = 5x^2 - 10x + 6$.

37. (a) Let $v(t) = 144 - 32t$.

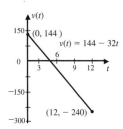

(b) From Exercise 36(b), the rock strikes the ground at $t = 12$, so the velocity of the rock is $v(12) = 144 - 32(12) = -240$ ft/sec. The minus sign indicates the rock is falling to the earth.

(c) Domain: $[0, 12]$; range: $[-240, 144]$

39. Let $P(x) = -0.1x^2 + 160x - 20000$. The parabola opens downward, so the vertex is the maximum point on the curve. To find the number of terminals to produce in order to maximize the profit, complete the square and find the vertex.

$$-0.1x^2 + 160x - 20000 = -0.1(x^2 - 1600x) - 20000$$
$$= -0.1(x^2 - 1600x + 640000 - 640000) - 20000$$
$$= -0.1(x - 800)^2 + 64000 - 20000$$
$$= -0.1(x - 800)^2 + 44000$$

The vertex is at $(800, 44000)$, so the company should (a) produce 800 terminals to maximize the profit and (b) the maximum profit is $\$44,000.00$.

41. (a) Let $P(x) = 200x - x^2$. Completing the square gives

$$P(x) = -x^2 + 200x = -(x^2 - 200x)$$
$$= -(x^2 - 200x + 10000 - 10000)$$
$$= -(x - 100)^2 + 10000.$$

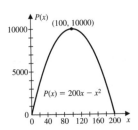

(b) The vertex of the parabola will give the maximum profit, so from part (a) the vertex is at $(100, 10000)$ and hence 100 units should be produced giving a maximum profit is $\$10,000.00$.

(c) We have

$$\frac{P(x+h) - P(x)}{h} = \frac{200(x+h) - (x+h)^2 - (200x - x^2)}{h}$$

$$= \frac{200x + 200h - (x^2 + 2hx + h^2) - 200x + x^2}{h}$$

$$= \frac{200h - 2hx - h^2}{h} = \frac{h(200 - 2x - h)}{h}$$

$$= 200 - 2x - h.$$

As h approaches 0, the difference quotient approaches $200 - 2x$.

1.9 Other Common Functions

THE ABSOLUTE VALUE FUNCTION

The absolute value function is defined as

$$f(x) = |x| = \begin{cases} x, & \text{if } x \geq 0 \\ -x, & \text{if } x < 0. \end{cases}$$

For $x \geq 0$, the graph of the absolute value function is the same as the straight line $y = x$. For $x < 0$, the graph is the same as $y = -x$, as shown in the figure.

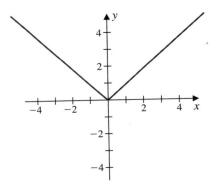

Figure 1.77

EXAMPLE 1.9.1   Use the graph of $y = |x|$ to sketch the graph of the function.

(a) $f(x) = |x - 2| + 2$ (b) $f(x) = |2x - 5|$

SOLUTION:

(a) The shifting properties from the previous section applied to quadratic functions work equally well for any function. So the graph is just a horizontal shift of 2 units to the right of $y = |x|$, followed by a vertical shift 2 units upward.

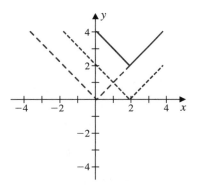

Figure 1.78

(b) First note that

$$f(x) = |2x - 5| = 2\left|x - \frac{5}{2}\right|.$$

The graph is a vertical stretching, by a factor of 2, of the graph of $y = |x|$, followed by a shift to the right of $\frac{5}{2}$ units.   ◇

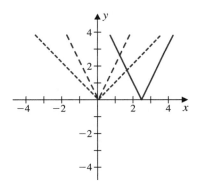

Figure 1.79

EXAMPLE 1.9.2   Sketch the graph of $f(x) = |x^2 - 2x - 1|$.

SOLUTION: First sketch the graph of $y = x^2 - 2x - 1$. Completing the square gives

$$y = x^2 - 2x - 1$$
$$= x^2 - 2x + 1 - 1 - 1$$
$$= (x - 1)^2 - 2.$$

Taking the absolute value of this term will leave any portions of the graph that lie above the x-axis unchanged and will reflect about the x-axis any portions of the graph that lie below the x-axis. For example, if $y = x^2 - 2x - 1$ and $x = 1$, then $y = -2$ and if $y = |x^2 - 2x - 1|$ and $x = 1$, then $y = 2$. So, $(1, -2)$ is on the graph of $y = x^2 - 2x - 1$, and $(1, 2)$ is on the graph of $y = |x^2 - 2x - 1|$.   ◇

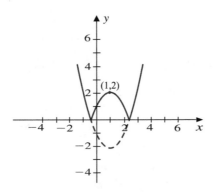

Figure 1.80

THE SQUARE ROOT FUNCTION

EXAMPLE 1.9.3   Use the graph of $y = \sqrt{x}$ to sketch the graph of the function $f(x) = 1 - \sqrt{x + 1}$. Find the domain and range of the function.

SOLUTION: The graph of $y = -\sqrt{x}$ is the reflection of the graph of $y = \sqrt{x}$ about the x-axis. Then shift the graph of $y = -\sqrt{x}$ to the left 1 unit and upward 1 unit. The domain of the function is all real numbers such that

$$x + 1 \geq 0,$$

that is,

$$x \geq -1.$$

In interval notation, the domain is $[-1, \infty)$. From the graph, we see that the range is the interval $(-\infty, 1]$.   ◊

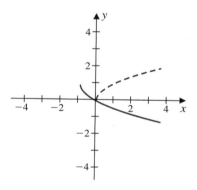

Figure 1.81

The Greatest Integer Function

The greatest integer function is defined by

$$\lfloor x \rfloor = \text{ the largest integer less than or equal to } x.$$

EXAMPLE 1.9.4   Evaluate each of the following.

(a) $\lfloor 2 \rfloor$ (b) $\lfloor 1.6 \rfloor$ (c) $\lfloor -4 \rfloor$ (d) $\lfloor -3.5 \rfloor$

SOLUTION:

(a) The greatest integer less than or equal to an integer is the integer itself, so $\lfloor 2 \rfloor = 2$.

(b) Since $1 \leq 1.6 < 2$, we have $\lfloor 1.6 \rfloor = 1$.

(c) -4 is an integer, so $\lfloor -4 \rfloor = -4$.

(d) Since $-4 \leq -3.5 < -3$, and since the smaller integer is -4, we have $\lfloor -3.5 \rfloor = -4$.

◇

EXAMPLE 1.9.5 Sketch the graph of $f(x) = x + \lfloor x \rfloor$.

SOLUTION: First write the definition for $f(x)$ on a selection of intervals. The meaning of $f(x)$ depends on the values of $\lfloor x \rfloor$. Then we have

$$\lfloor x \rfloor = \begin{cases} 0, & \text{if } 0 \leq x < 1, \\ 1, & \text{if } 1 \leq x < 2, \\ 2, & \text{if } 2 \leq x < 3, \\ 3, & \text{if } 3 \leq x < 4, \\ -1, & \text{if } -1 \leq x < 0, \\ -2, & \text{if } -2 \leq x < -1, \end{cases}$$

$$f(x) = x + \lfloor x \rfloor = \begin{cases} x, & \text{if } 0 \leq x < 1, \\ x+1, & \text{if } 1 \leq x < 2, \\ x+2, & \text{if } 2 \leq x < 3, \\ x+3, & \text{if } 3 \leq x < 4, \\ x-1, & \text{if } -1 \leq x < 0, \\ x-2, & \text{if } -2 \leq x < -1. \end{cases}$$

If a represents an integer, then the function is equivalent to

$$f(x) = x + a, \quad \text{for} \quad a \leq x < a+1.$$

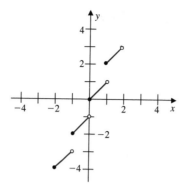

Figure 1.82

Solutions for Exercise Set 1.9

1. $f(x) = |x - 3|$

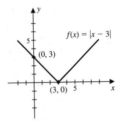

3. $f(x) = |x + 2| - 2$

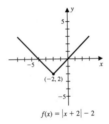

5. $f(x) = -2|x|$

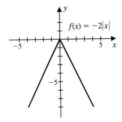

7. $f(x) = -3|x - 1| + 1$

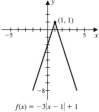

$f(x) = -3|x - 1| + 1$

9. (a)

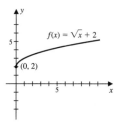

$f(x) = \sqrt{x} + 2$

(0, 2)

(b) Domain: $g(x) = \sqrt{x} + 2$ is defined when $x \geq 0$, so the domain is $[0, \infty)$.

Range: Since $\sqrt{x} + 2 \geq 2$, the range is $[2, \infty)$.

11. (a)

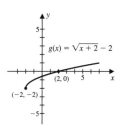

$g(x) = \sqrt{x + 2} - 2$

(2, 0)

(−2, −2)

(b) Domain: $g(x) = \sqrt{x + 2} - 2$ is defined when $x + 2 \geq 0$, so $x \geq -2$ and the domain is $[-2, \infty)$.

Range: Since $\sqrt{x + 2} - 2 \geq -2$, the range is $[-2, \infty)$.

13. (a)

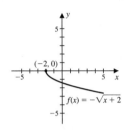

(b) Domain: $g(x) = -\sqrt{x+2}$ is defined when $x + 2 \geq 0$, so $x \geq -2$ and the domain is $[-2, \infty)$.

Range: Since $-\sqrt{x+2} \leq 0$, the range is $(-\infty, 0]$.

15. (a) $g(x) = \sqrt{2-x} - 1 = \sqrt{-(x-2)} - 1$

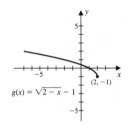

(b) Domain: $g(x) = \sqrt{2-x} - 1$ is defined when $2 - x \geq 0 \Leftrightarrow x - 2 \leq 0$, so $x \leq 2$ and the domain is $(-\infty, 2]$.

Range: Since $\sqrt{2-x} - 1 \geq -1$, the range is $[-1, \infty)$.

17. (a) $g(x) = x^3 + 1$ (b) $g(x) = x^3 - 1$ (c) $g(x) = (x+1)^3$

     (d) $g(x) = (x-1)^3$ (e) $g(x) = 2x^3$ (f) $g(x) = -2x^3$

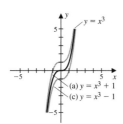
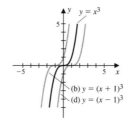

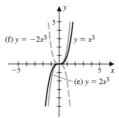

19. $f(x) = 2x - 3$

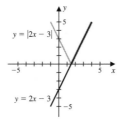

21. $f(x) = -x^2 - 2$

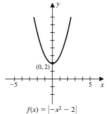

23. $f(x) = -(x-1)^2 + 1$

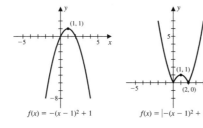

25. $f(x) = x^2 - 4x + 2 = x^2 - 4x + 4 - 4 + 2 = (x-2)^2 - 2$

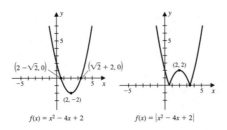

27. $f(x) = \lfloor x - 2 \rfloor$

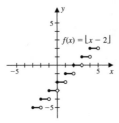

29. $f(x) = \lfloor x + 1 \rfloor - 2$

31. $f(x) = 2 - \lfloor x \rfloor$

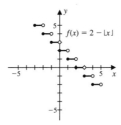

$f(x) = 2 - \lfloor x \rfloor$

33. (a) $y = f(x) + 1$ (b) $y = f(x - 2)$ (c) $y = 2f(x)$ (d) $y = f(2x)$

(e) $y = -f(x)$ (f) $y = f(-x)$ (g) $y = |f(x)|$ (h) $y = f(|x|)$

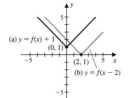

(a) $y = f(x) + 1$
(0, 1)
(2, 1)
(b) $y = f(x - 2)$

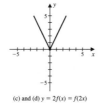

(c) and (d) $y = 2f(x) = f(2x)$

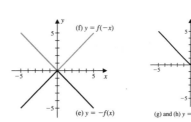

(f) $y = f(-x)$
(e) $y = -f(x)$

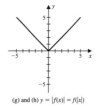

(g) and (h) $y = |f(x)| = f(|x|)$

35. (a)

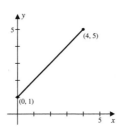

(4, 5)

(0, 1)

(b) The area under the curve is the area of a trapezoid, which equals

$$A(t) = \frac{1}{2}(1 + (t + 1)) \cdot t = t + \frac{1}{2}t^2.$$

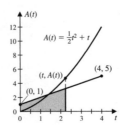

(c)

$$d(x) = d((0,0),(x,f(x)) = d((0,0),(x,x+1))$$
$$= \sqrt{(x-0)^2 + (x+1-0)^2} = \sqrt{x^2 + x^2 + 2x + 1}$$
$$= \sqrt{2x^2 + 2x + 1}$$

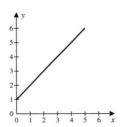

37. (a) Domain: $[0, 13]$;   Range: $\{.33, .55, .77, .., 2.97\} = \{.33 + 0.22n, n = 0, 1, ..., 12\}$

(b) $P(w) = \begin{cases} 0.33 + 0.22\lfloor w \rfloor, & \text{if } w \text{ is not a positive integer} \\ 0.33 + 0.22(w-1) = 0.11 + 0.22w, & \text{if } w \text{ is a positive integer.} \end{cases}$

Also,

$$0.11-0.22\lfloor -w\rfloor = \begin{cases} 0.11 - 0.22(-\lfloor w\rfloor - 1), & \text{if } w \text{ is not a positive integer} \\ 0.11 - 0.22(-\lfloor w\rfloor), & \text{if } w \text{ is a positive integer.} \end{cases}$$

$$= \begin{cases} 0.33 + 0.22\lfloor w\rfloor, & \text{if } w \text{ is not a positive integer} \\ 0.11 + 0.22\lfloor w\rfloor, & \text{if } w \text{ is a positive integer,} \end{cases}$$

so $P(w) = 0.11 + 0.22\lfloor -w\rfloor$.

39. (a) For approximating $\sqrt{2}$, start with $a = 1.4$. Then we have

a	$b = \frac{1}{2}\left(a + \frac{2}{a}\right)$
1.4	1.414285714
1.414285714	1.414213564
1.414213564	1.414213562
1.414213562	1.414213562

So $\sqrt{2} \approx 1.414313562 \approx 1.414$.

(b) For approximating $\sqrt{13}$ start with $a = 3.5$. Then we have

a	$b = \frac{1}{13}\left(a + \frac{13}{a}\right)$
3.5	1.607142857
3.607142857	3.605551627
3.605551627	3.605551276
3.605551276	3.605551276

So $\sqrt{3} \approx 3.605551276 \approx 3.606$.

Notice that the number of decimal places of accuracy approximately doubles with each iteration of the technique. This provides an extremely efficient technique for determining square roots.

1.10 Arithmetic Combinations of Functions

If f and g are functions, and x is an admissible input for each of the functions, then $f(x)$ and $g(x)$ are both real numbers. The quantities $f(x) + g(x)$, $f(x) - g(x)$, $f(x) \cdot g(x)$, and if $g(x) \neq 0$, $\frac{f(x)}{g(x)}$, are all real numbers. We can define the *arithmetic combinations* of two functions by the formulas

$$\text{Addition}: \ (f + g)(x) = f(x) + g(x)$$

$$\text{Subtraction}: \ (f - g)(x) = f(x) - g(x)$$

$$\text{Multiplication}: \ (fg)(x) = f(x) \cdot g(x)$$

$$\text{Division}: \ \left(\frac{f}{g}\right)(x) = \frac{f(x)}{g(x)}$$

As you might expect, the symbols $f + g, f - g, fg$, and $\frac{f}{g}$ are the names used for the new functions, whereas the expressions on the right of the equation indicate how to evaluate these functions at specific inputs.

For x to be an admissable input value for $f + g, f - g$ or fg, the value x must be a valid input for both f and g. So, if the domain of the function f is A and the domain of g is B, the domain of $f + g, f - g$ and fg consists of all real numbers common to both domains which is the intersection $A \cap B$. The domain of $\frac{f}{g}$ consists of those values in $A \cap B$, excluding any values of x for which $g(x) = 0$.

EXAMPLE 1.10.1   Let $f(x) = x^2$ and $g(x) = x + 1$. Find $f + g, f - g, fg$, and f/g and give the domains of each new function.

SOLUTION:

$$(f+g)(x) = f(x) + g(x) = x^2 + x + 1$$

$$(f-g)(x) = f(x) - g(x) = x^2 - (x+1) = x^2 - x - 1$$

$$(fg)(x) = f(x) \cdot g(x) = x^2(x+1) = x^3 + x^2$$

and

$$\left(\frac{f}{g}\right)(x) = \frac{f(x)}{g(x)} = \frac{x^2}{x+1}.$$

Since the domain of f is the set of all real numbers, and the domain of g is the set of all real numbers, the domain of $f+g, f-g$ and fg is the set of all real numbers. The domain of f/g is the set of all real numbers satisfying $x + 1 \neq 0$, that is, the domain is $(-\infty, -1) \cup (-1, \infty)$.   ◇

EXAMPLE 1.10.2   Let $f(x) = \dfrac{1}{x}$ and $g(x) = \sqrt{x+2}$. Find $f+g, f-g, fg$, and f/g and give the domains of each new function.

SOLUTION:

$$(f+g)(x) = f(x) + g(x) = \frac{1}{x} + \sqrt{x+2}$$

$$(f-g)(x) = f(x) - g(x) = \frac{1}{x} - \sqrt{x+2}$$

$$(fg)(x) = f(x) \cdot g(x) = \frac{1}{x} \cdot \sqrt{x+2} = \frac{\sqrt{x+2}}{x}$$

and

$$\left(\frac{f}{g}\right)(x) = \frac{f(x)}{g(x)} = \frac{\frac{1}{x}}{\sqrt{x+2}} = \frac{1}{x\sqrt{x+2}}.$$

<u>Domain of f</u> : All real numbers except $x = 0$, that is, $(-\infty, 0) \cup (0, \infty)$.

<u>Domain of g</u> : All x for which $\sqrt{x+2}$ is defined, so

$$x + 2 \geq 0.$$

That is,

$$x \geq 2.$$

Domain of $f+g$, $f-g$, fg: The intersection of the domains of f and g. Since the interval $[2, \infty)$ is contained in the interval $(0, \infty)$, the domain is $[2, \infty)$.

Domain of f/g: All x in $[2, \infty)$, except those values that make the denominator of f/g equal to 0. Since

$$x\sqrt{x+2} = 0 \Rightarrow x = 0 \quad \text{or} \quad x = -2,$$

and 0 and -2 are not in the interval $[2, \infty)$, the domain of f/g is also $[2, \infty)$.   ◇

EXAMPLE 1.10.3   Let $f(x) = x^2+x-2$ and $g(x) = x-1$. Find $f+g$, $f-g$, fg, and f/g and give the domains of each new function.

SOLUTION:

$$(f+g)(x) = f(x) + g(x) = x^2 + x - 2 + x - 1 = x^2 + 2x - 3$$

$$(f-g)(x) = f(x) - g(x) = x^2 + x - 2 - (x-1)$$

$$= x^2 + x - 2 - x + 1 = x^2 - 1$$

$$(fg)(x) = f(x) \cdot g(x) = (x^2 + x - 2)(x - 1)$$

$$= x^3 - x^2 + x^2 - x - 2x + 2 = x^3 - 3x + 2$$

and

$$\left(\frac{f}{g}\right)(x) = \frac{f(x)}{g(x)} = \frac{x^2 + x - 2}{x - 1} = \frac{(x-1)(x+2)}{x-1} = x + 2.$$

Domains of f and g: All real numbers.

Domain of $f+g$, $f-g$, fg: All real numbers.

Domain of f/g : It appears from the simplification for f/g, that the domain is all real numbers. This is *not* the case since for x to be in the domain of f/g it must be in both the domains of f and g and $g(x)$ *must be nonzero*. So, $x = 1$, must be removed from the domain of f/g, since $g(1) = 0$. The domain of f/g is $(-\infty, 1) \cup (1, \infty)$.

◇

EXAMPLE 1.10.4   Let

$$f(x) = \begin{cases} x^2 + 1, & \text{if } x \geq 0 \\ x, & \text{if } x < 0 \end{cases} \quad \text{and} \quad g(x) = \begin{cases} x + 1, & \text{if } x \geq 0 \\ -1, & \text{if } x < 0. \end{cases}$$

Find $f + g, f - g, fg$, and f/g and give the domains of each new function.

SOLUTION: The functions are plotted in the figure.

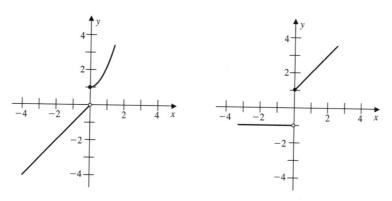

Figure 1.83   Figure 1.84

To combine these two functions, consider the definitions in two separate parts of the domains of f and g, for $x \geq 0$ and for $x < 0$. So,

$$(f + g)(x) = \begin{cases} x^2 + x + 2, & \text{if } x \geq 0 \\ x - 1, & \text{if } x < 0, \end{cases}$$

$$(f - g)(x) = \begin{cases} x^2 - x, & \text{if } x \geq 0 \\ x + 1, & \text{if } x < 0, \end{cases}$$

$$(fg)(x) = \begin{cases} x^3 + x^2 + x + 1, & \text{if } x \geq 0 \\ -x, & \text{if } x < 0, \end{cases}$$

and

$$\left(\frac{f}{g}\right)(x) = \begin{cases} \frac{x^2+1}{x+1}, & \text{if } x \geq 0 \\ -x, & \text{if } x < 0. \end{cases}$$

<u>Domains of f and g</u> : All real numbers, since all real numbers are specified in the two part definitions.

<u>Domain of $f + g, f - g, fg$</u> : All real numbers.

<u>Domain of f/g</u> : The quotient in the first part of the definition is undefined when $x = -1$. However, this part of the definition is only valid for $x \geq 0$, so this is not a problem. Notice this can also be seen from the fact the $g(x)$ is never zero. The domain is the set of all real numbers. ◇

EXAMPLE 1.10.5   Let $f(x) = \sqrt{x^2 - 1}$ and $g(x) = \sqrt{9 - x^2}$. Find $f + g, f - g, fg$, and f/g and give the domains of each new function.

SOLUTION: First we have

$$(f + g)(x) = f(x) + g(x) = \sqrt{x^2 - 1} + \sqrt{9 - x^2}$$
$$(f - g)(x) = f(x) - g(x) = \sqrt{x^2 - 1} - \sqrt{9 - x^2}$$
$$(fg)(x) = f(x) \cdot g(x) = \sqrt{x^2 - 1} \cdot \sqrt{9 - x^2}$$

and

$$\left(\frac{f}{g}\right)(x) = \frac{f(x)}{g(x)} = \frac{\sqrt{x^2 - 1}}{\sqrt{9 - x^2}}.$$

<u>Domain of f</u> : The expression under the radical must be greater than or equal to 0. To solve the inequality, factor and set up a chart. So we must have

$$x^2 - 1 = (x - 1)(x + 1) \geq 0.$$

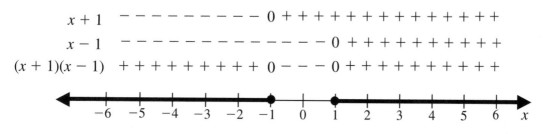

Figure 1.85

Since $x^2 - 1 = 0$, for $x = \pm 1$, the domain of f is $(-\infty, -1] \cup [1, \infty)$.

<u>Domain of g</u> : The expression under the radical must be greater than or equal to 0. So, solve

$$9 - x^2 = (3 - x)(3 + x) \geq 0.$$

$$
\begin{array}{ll}
(3 - x) & +\,+\,+\,+\,+\,+\,+\,+\,+\,+\,+\,+\,+\,+\,+\,0\,-\,-\,-\,-\,-\,- \\
(3 + x) & -\,-\,-\,0\,+ \\
(3 - x)(3 + x) & -\,-\,-\,0\,+\,+\,+\,+\,+\,+\,+\,+\,+\,+\,+\,0\,-\,-\,-\,-\,-\,-
\end{array}
$$

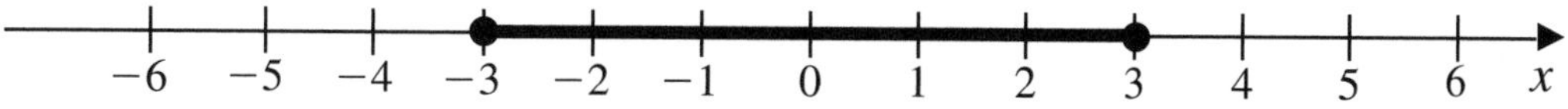

Figure 1.86

Since $9 - x^2 = 0$, for $x = \pm 3$, the domain of g is $[-3, 3]$.

<u>Domain of $f + g$, $f - g$, fg</u> : $((-\infty, -1] \cup [1, \infty)) \cap [-3, 3] = [-3, -1] \cup [1, 3]$.

<u>Domain of f/g</u> : Remove the values which make the denominator 0. So,

$$\sqrt{9 - x^2} = 0$$

$$9 - x^2 = 0$$

and

$$x = \pm 3.$$

The domain of f/g is $(-3, -1] \cup [1, 3)$.   ◇

EXAMPLE 1.10.6   Functions f and g are defined by $f(x) = x^2 - 9$ and $g(x) = x + 3$. Sketch the graph of $y = f(x)/g(x)$. Sketch the graph of $y = x - 3$. How do the graphs differ?

SOLUTION: We have

$$\frac{f(x)}{g(x)} = \frac{x^2 - 9}{x + 3} = \frac{(x + 3)(x - 3)}{x + 3} = x - 3, \text{ for } x \neq -3.$$

The important observation here is that the quotient function $f(x)$ is not defined at $x = -3$ since the denominator is 0 when $x = -3$. The fact that the fraction simplifies does not change the fact that $\frac{x^2 - 9}{x + 3}$ is undefined for $x = -3$. The only difference in the graphs of $y = x - 3$ and $y = \frac{f(x)}{g(x)}$, is that the point $(-3, -6)$ is removed from the graph of $y = \frac{f(x)}{g(x)}$.   ◇

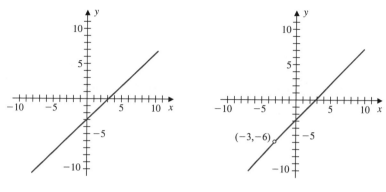

Figure 1.87   Figure 1.88

THE RECIPROCAL GRAPHING TECHNIQUE

Properties of the function $g(x) = \frac{1}{f(x)}$, called the *reciprocal of f*, can be obtained from the graph of the function f. The most important values of x are those that make $f(x) = 0$.

EXAMPLE 1.10.7   Let $g(x) = x^2 + 2x - 3$. Use the results about the graph of the reciprocal of a function to sketch the graph of $h(x) = 1/g(x)$.

SOLUTION: The points where $g(x) = 0$, which is where the graph of $y = g(x)$ crosses the x–axis, are the points where $h(x)$ is undefined and are the essential points in determining the graph. So,

$$x^2 + 2x - 3 = 0$$

$$(x + 3)(x - 1) = 0$$

and

$$x = -3 \quad \text{or} \quad x = 1.$$

The graph of $g(x) = x^2 + 2x - 3 = (x + 1)^2 - 4$ in the figure shows

$$g(x) > 0, \text{ for } x > 1 \text{ or } x < -3,$$

and

$$g(x) < 0, \text{ for } -3 < x < 1.$$

As x gets close to -3 or 1, $g(x)$ gets close to 0, so the magnitude of $h(x) = \frac{1}{g(x)}$ becomes arbitrarily large. This implies that

$$x \to -3^-, h(x) \to \infty$$

$$x \to -3^+, h(x) \to -\infty$$

$$x \to 1^-, h(x) \to -\infty$$

and

$$x \to 1^+, h(x) \to \infty.$$

As $g(x)$ gets large in magnitude, $h(x) = \dfrac{1}{g(x)}$ gets very small. So

$$\text{as} \quad x \to \infty, \quad g(x) \to \infty \text{ and } h(x) \to 0,$$

and

$$\text{as} \quad x \to -\infty, \quad g(x) \to \infty \text{ and } h(x) \to 0.$$

The vertex of the parabola is $(-1, -4) = (-1, g(-1))$, so the point $\left(-1, \frac{1}{g(-1)}\right) = (-1, -\frac{1}{4})$ is on the graph of $y = h(x)$. This is generally a good additional point to plot.

◇

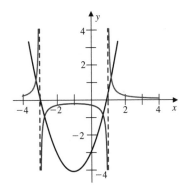

Figure 1.89

EXAMPLE 1.10.8   Sketch the graphs of the given functions in the order given, and observe the difference in the graph that each successive complication introduces.

(a) $f_1(x) = x + 1$ (b) $f_2(x) = (x + 1)^2$ (c) $f_3(x) = x^2 + 2x$

(d) $f_4(x) = |x^2 + 2x|$ (e) $f_5(x) = \frac{1}{|x^2+2x|}$ (f) $f_6(x) = \frac{-1}{|x^2+2x|}$

SOLUTION:

(a) The graph of $y = f_1(x)$ is a straight line passing through $(0, 2)$ and $(-1, 0)$.

(b) The graph of $y = f_2(x)$ is a parabola with vertex at $(-1, 0)$ and opening upward.

(c) Since

$$f_3(x) = x^2 + 2x$$
$$= x^2 + 2x + 1 - 1$$
$$= (x + 1)^2 - 1,$$

the graph of $y = f_3(x)$ is obtained by shifting the graph of $y = f_2(x)$ downward 1 unit.

(d) The graph of $y = f_4(x) = |x^2 + 2x|$ is obtained by reflecting any portions of $y = f_3(x) = x^2 + 2x$ that lie below the $x-$axis above it.

(e) To obtain the graph of

$$y = f_5(x) = \frac{1}{|x^2 + 2x|},$$

use the reciprocal graphing technique.

(f) The graph of

$$y = f_6(x) = \frac{-1}{|x^2 + 2x|}$$

is a reflection about the x-axis of the graph of

$$y = f_5(x) = \frac{1}{|x^2 + 2x|}.$$

◇

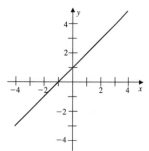

Figure 1.90

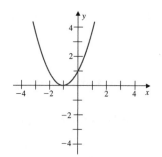

Figure 1.91

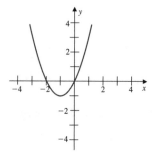

Figure 1.92

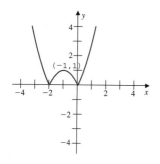

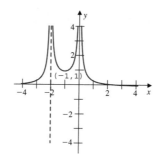

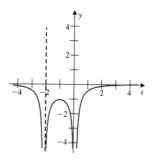

Figure 1.93   Figure 1.94   Figure 1.95

Solutions for Exercise Set 1.10

In this problem set, the notation D_f stands for "the domain of the function f."

1. For $f(x) = x$ and $g(x) = 3x - 2$ we have

$(f + g)(x) = f(x) + g(x) = x + (3x - 2) = 4x - 2;$

$(f - g)(x) = f(x) - g(x) = x - (3x - 2) = x - 3x + 2 = -2x + 2;$

$(f \cdot g)(x) = f(x)g(x) = x(3x - 2) = 3x^2 - 2x;$

$(f/g)(x) = \frac{f(x)}{g(x)} = \frac{x}{3x-2}.$

The domain of $f + g$, $f - g$, and $f \cdot g$ is $D_f \cap D_g = (-\infty, \infty) \cap (-\infty, \infty) = (-\infty, \infty)$.

The domain of f/g is $D_f \cap D_g$ excluding any values that make the denominator 0. The denominator is 0 when $3x - 2 = 0 \Leftrightarrow x = \frac{2}{3}$, so

$D_{f/g} = \left(-\infty, \frac{2}{3}\right) \cup \left(\frac{2}{3}, \infty\right).$

3. For $f(x) = \frac{1}{x}$ and $g(x) = \frac{x}{x-2}$ we have

$(f + g)(x) = f(x) + g(x) = \frac{1}{x} + \frac{x}{x-2} = \frac{(x-2)+x^2}{x(x-2)} = \frac{x^2+x-2}{x^2-2x};$

$(f - g)(x) = f(x) - g(x) = \frac{1}{x} - \frac{x}{x-2} = \frac{(x-2)-x^2}{x(x-2)} = \frac{-x^2+x-2}{x^2-x};$

$(f \cdot g)(x) = f(x)g(x) = \frac{x}{x(x-2)} = \frac{1}{x-2};$

$(f/g)(x) = \frac{f(x)}{g(x)} = \frac{\frac{1}{x}}{\frac{x}{x-2}} = \frac{1}{x} \cdot \frac{x-2}{x} = \frac{x-2}{x^2}.$

Domain of $f : x \neq 0$; Domain of $g : x - 2 \neq 0 \Leftrightarrow x \neq 2$

So $D_f \cap D_g = \{x : x \neq 0 \text{ and } x \neq 2\} = (-\infty, 0) \cup (0, 2) \cup (2, \infty).$

The domain of $f + g$, $f - g$, and $f \cdot g$ is $D_f \cap D_g = (-\infty, 0) \cup (0, 2) \cup (2, \infty).$

The domain of f/g is $D_f \cap D_g$ excluding any values that make the denominator 0, which is $x = 0$, which has already been excluded. So

$D_{f/g} = (-\infty, 0) \cup (0, 2) \cup (2, \infty).$

5. For $f(x) = \sqrt{x+2}$ and $g(x) = \sqrt{2-x}$ we have

$(f + g)(x) = f(x) + g(x) = \sqrt{x+2} + \sqrt{2-x};$

$(f - g)(x) = f(x) - g(x) = \sqrt{x + 2} - \sqrt{2 - x};$

$(f \cdot g)(x) = f(x)g(x) = \sqrt{x + 2}\sqrt{2 - x} = \sqrt{(x + 2)(2 - x)} = \sqrt{4 - x^2};$

$(f/g)(x) = \frac{f(x)}{g(x)} = \frac{\sqrt{x+2}}{\sqrt{2-x}}.$

Domain of $f : x + 2 \geq 0 \Leftrightarrow x \geq -2$; Domain of $g : 2 - x \geq 0 \Leftrightarrow x \leq 2$

So $D_f \cap D_g = [-2, 2]$.

The domain of $f + g, f - g$, and $f \cdot g$ is $D_f \cap D_g = [-2, 2]$.

The domain of f/g is $D_f \cap D_g$, excluding any values that make the

denominator 0, which is $x = 2$, so $D_{f/g} = [-2, 2)$.

7. For $f(x) = \begin{cases} -1, & \text{if } x < 0 \\ 1, & \text{if } x \geq 0 \end{cases}$ and $g(x) = \begin{cases} 1, & \text{if } x < 0 \\ 0, & \text{if } x \geq 0 \end{cases}$ we perform the

operations separately for $x < 0$ and for $x \geq 0$.

$$(f + g)(x) = f(x) + g(x) = \begin{cases} 0, & \text{if } x < 0 \\ 1, & \text{if } x \geq 0; \end{cases}$$

$$(f - g)(x) = f(x) - g(x) = \begin{cases} -2, & \text{if } x < 0 \\ 1, & \text{if } x \geq 0; \end{cases}$$

$$(f \cdot g)(x) = f(x)g(x) = \begin{cases} -1, & \text{if } x < 0 \\ 0, & \text{if } x \geq 0; \end{cases}$$

The quotient is defined only for $x < 0$, since $g(x) = 0$ for $x \geq 0$. So

$(f/g)(x) = -1$, for $x < 0$.

The domain of $f + g, f - g, f \cdot g$ is $(-\infty, \infty)$ and the domain of f/g is

$(-\infty, 0)$.

9. $g(x) = x - 1; h(x) = \frac{1}{x-1}$

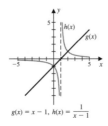

$g(x) = x - 1, h(x) = \frac{1}{x-1}$

11. $g(x) = |x|; h(x) = \frac{1}{|x|}$

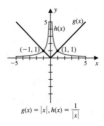

$$g(x) = |x|, h(x) = \frac{1}{|x|}$$

13. $g(x) = x^2 - 1; h(x) = \frac{1}{x^2 - 1}$

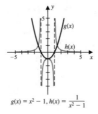

$$g(x) = x^2 - 1, h(x) = \frac{1}{x^2 - 1}$$

15. $g(x) = x^2 - 4x + 3 = x^2 - 4x + 4 - 4 + 3 = (x-2)^2 - 1; h(x) = \frac{1}{(x-2)^2 - 1}$

$$g(x) = x^2 - 4x + 3$$
$$h(x) = \frac{1}{x^2 - 4x + 3}$$

17. For the graph shown we have

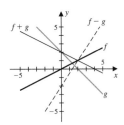

19. For the graph shown we have

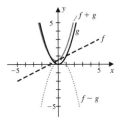

21. (a) $y = x - 1$ (b) $y = (x - 1)^2$ (c) $y = x^2 - 2x = (x - 1)^2 - 1$

(d) $y = |x^2 - 2x|$ (e) $y = \frac{1}{|x^2 - 2x|}$ (f) $y = -\frac{1}{|x^2 - 2x|}$

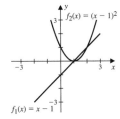   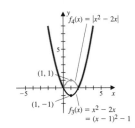   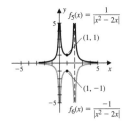

23. We have $f(x) = \frac{x^2 - 4}{x - 2} = \frac{(x-2)(x+2)}{x-2} = x + 2$, for $x \neq 2$. The denominator of the original expression is 0 at $x = 2$, making the quotient undefined. To graph $y = f(x) = \frac{x^2 - 4}{x - 2}$, simply graph $y = x + 2$ and remove the point with x-coordinate 2. That is, remove the point $(2, 4)$. So the graphs of

$f(x) = \frac{x^2-4}{x-2}$ and $g(x) = x + 2$ are identical except for the point $(2, 4)$ removed from the graph of $y = f(x)$.

25.

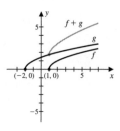

27.

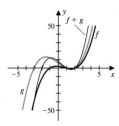

1.11 Composition of Functions

The *composition* of two functions is a special way of combining the two processes. For example, the function given by $f(x) = (x^2 + 1)^3$ can be interpreted as

$$x \rightarrow \boxed{x}^{\,2} + 1 \rightarrow x^2 + 1 \rightarrow \boxed{x^2 + 1}^{\,3} \rightarrow (x^2 + 1)^3.$$

So x is input to the process that squares the number and adds one, and then this new value is used as input to the process that cubes a number. The final result is the process that does both operations in the prescribed order.

The composition of the two functions f and g is written as

$$(f \circ g)(x) = f(g(x)),$$

with the domain being all x in the domain of g so that $g(x)$ is in the domain of f.

EXAMPLE 1.11.1   Let $f(x) = x^2 + 1$ and $g(x) = \sqrt{x}$.

(a) Find $(f \circ g)(2)$ and $(g \circ f)(-3)$.

(b) Write expressions for $f(g(x))$ and $g(f(x))$.

(c) Can $f \circ g$ be evaluated at -2?

SOLUTION:

(a) $\underline{(f \circ g)(2)}$: First evaluate g at 2, so $g(2) = \sqrt{2}$. Then substitute this value into f, so

$$(f \circ g)(2) = f(g(2)) = f(\sqrt{2}) = (\sqrt{2})^2 + 1 = 3.$$

$(g \circ f)(-3)$:

$$(g \circ f)(-3) = g(f(-3)) = g(10) = \sqrt{10}$$

(b) We have

$$f(g(x)) = f(\sqrt{x}) = (\sqrt{x})^2 + 1 = x + 1,$$

and

$$g(f(x)) = g(x^2 + 1) = \sqrt{x^2 + 1}.$$

(c) No, $f \circ g$ can not be evaluated at -2, since -2 is not in the domain of g.   ◇

EXAMPLE 1.11.2   Let $f(x) = \sqrt{x - 6}$ and $g(x) = x^2 + 5x$. Find $f(g(x))$ and specify the domain of $f \circ g$.

SOLUTION:

$$f(g(x)) = f(x^2 + 5x) = \sqrt{x^2 + 5x - 6}$$

The domain of $f \circ g$ is the set of all x in the domain of g, so that $g(x)$ is also in the domain of f. The domain of g is all real numbers and the domain of f is the interval $[6, \infty)$. So x is in the domain of $f \circ g$ provided that

$$x^2 + 5x \geq 6$$
$$x^2 + 5x - 6 \geq 0$$

and

$$(x - 1)(x + 6) \geq 0.$$

To solve the inequality we have,

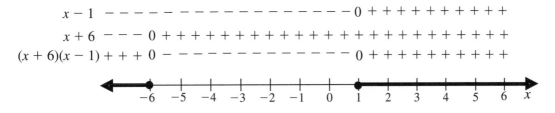

Figure 1.96

The domain of $f \circ g$ is $(-\infty, -6] \cup [1, \infty)$.   ◇

EXAMPLE 1.11.3   (a) Use the graphs of f and g in the figure to evaluate each expression.

(i) $(f \circ g)(0)$ (ii) $(g \circ f)(0)$ (iii) $(g \circ f)(2)$ (iv) $(f \circ g)(2)$

(b) Determine all values of x for which

(i) $(f \circ g)(x) = 0$ (ii) $(g \circ f)(x) = 0$

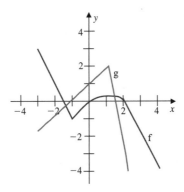

Figure 1.97

SOLUTION:

(a) (i) Reading from the graph, we have $g(0) = 1$. So

$$(f \circ g)(0) = f(g(0)) = f(1) = \frac{1}{4}.$$

(ii) Similarly,

$$(g \circ f)(0) = g(f(0)) = g(0) = 1.$$

(iii)

$$(g \circ f)(2) = g(f(2)) = g(0) = 1.$$

(iv)

$$(f \circ g)(2) = f(g(2)) = f(-3) = 3.$$

(b) (i) To solve $(f \circ g)(x) = f(g(x)) = 0$, first observe the values which make f zero, and then find those x that make $g(x)$ one of these values. Then

$$f(a) = 0 \text{ when } a = 0 \quad \text{or} \quad a = 2, \quad \text{or} \quad a = -\frac{3}{2}$$

and

$$g(x) = 0 \text{ when } x = -1, \quad \text{and} \quad g(x) = 2 \text{ when } x = 1,$$

and

$$g(x) = -\frac{3}{2} \text{ when } x = -3.$$

(ii) Since $g(-1) = 0$ and $g(\frac{3}{2}) = 0$, $(g \circ f)(x) = g(f(x)) = 0$ when $f(x) = -1$ or $f(x) = \frac{3}{2}$. This is when $x = -1$, $x = \frac{5}{2}$, or $x = -\frac{9}{4}$.   ◇

EXAMPLE 1.11.4   Write each function h as a composition $f \circ g$.

(a) $h(x) = (x^5 + 3x^3 - 2x + 1)^8$ (b) $h(x) = \sqrt{x^2 - 2x + 1}$

SOLUTION:

(a) A natural separation of the process h is to first perform the *inside* operation of $x^5 + 3x^3 - 2x + 1$ and then perform the *outside* operation of raising a number to the 8th power. Since we want g to be the inside operation and f to be the outside operation, we can let

$$f(x) = x^8 \quad \text{and} \quad g(x) = x^5 + 3x^3 - 2x + 1.$$

Then

$$f(g(x)) = (x^5 + 3x^3 - 2x + 1)^8 = h(x).$$

(b) Let $f(x) = \sqrt{x}$ and $g(x) = x^2 - 2x + 1$. Then

$$f(g(x)) = f(x^2 - 2x + 1)$$
$$= \sqrt{x^2 - 2x + 1}$$
$$= h(x).$$

$\diamond$

EXAMPLE 1.11.5 (a) Show that $f(x) = \frac{1}{|x^2 - 2x|}$ can be written as $f(x) = g_4(g_3(g_2(g_1(x))))$, where $g_1(x) = (x-1)^2$, $g_2(x) = x - 1$, $g_3(x) = |x|$, and $g_4(x) = \frac{1}{x}$.

(b) Sketch the graphs of (i) $y = g_1(x)$, (ii) $y = g_2(g_1(x))$, (iii) $y = g_3(g_2(g_1(x)))$, (iv) $y = f(x)$.

SOLUTION: A strategy that can be used to graph $y = f(x)$ is to first recognize that the quadratic in the denominator can be graphed easily by writing it in standard form. Then use the reciprocal graphing technique to graph the quotient.

(a) First write the quadratic in the denominator in standard form, giving us

$$x^2 - 2x = x^2 - 2x + 1 - 1 = (x-1)^2 - 1.$$

So,

$$f(x) = \frac{1}{|(x-1)^2 - 1|}.$$

Now carefully take the successive compositions, starting on the inside, and re-member that the parameter x in any intermediate step is a place holder that can accept an arbitrary expression. Then

$$g_1(x) = (x-1)^2,$$

$$g_2(g_1(x)) = g_2((x-1)^2) = (x-1)^2 - 1,$$

$$g_3(g_2(g_1(x))) = g_3((x-1)^2 - 1) = |(x-1)^2 - 1|,$$

and

$$g_4(g_3(g_2(g_1(x)))) = \frac{1}{|(x-1)^2 - 1|} = \frac{1}{|x^2 - 2x|}.$$

(b)

(i) $y = g_1(x) = (x-1)^2$ is obtained by shifting $y = x^2$ to the right 1 unit.

(ii) $y = g_2(g_1(x)) = (x-1)^2 - 1$ is obtained by shifting $y = (x-1)^2$ downward 1 unit.

(iii) $y = g_3(g_2(g_1(x))) = |(x-1)^2 - 1|$ is obtained by reflecting the portions of the graph of $y = g_2(g_1(x))$ that lie below the x-axis above the x-axis.

(iv) Use the reciprocal graphing technique to plot

$$\frac{1}{|(x-1)^2 - 1|}.$$

The important points are those where $y = |(x-1)^2 - 1|$ crosses the x-axis. Solving the quadratic equal to 0 gives

$$(x-1)^2 - 1 = 0$$

$$(x-1)^2 = 1$$

$$x - 1 = \pm 1$$

and

$$x = 0 \quad \text{or} \quad x = 2.$$

So

$$\text{As} \quad x \to 0, \quad \text{we have} \quad \frac{1}{|x^2 - 2x|} \to \infty,$$

$$\text{and as} \quad x \to 2, \quad \text{we have} \quad \frac{1}{|x^2 - 2x|} \to \infty.$$

One other useful point is where x is equal to 1. This is the vertex of the parabola $y = (x-1)^2 - 1$. If $x = 1$, then $|(x-1)^2 - 1| = |-1| = 1$, so $\frac{1}{|(x^2-2x)|} = 1$, and the point $(1,1)$ is on the final curve. $\diamond$

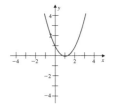

Figure 1.98

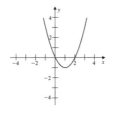

Figure 1.99

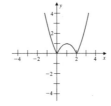

Figure 1.100

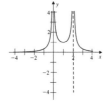

Figure 1.101

EXAMPLE 1.11.6   A cube with sides of length 2 feet is being filled with water, so that the height of the water at the end of t seconds is $h(t) = \sqrt{t}$ feet, $0 \le t \le 4$. Express the volume of the water in the cube as a function of time.

SOLUTION:

The volume of the water in the cube as a function of the height of the water is

$$V(h) = 2 \cdot 2 \cdot h = 4h.$$

Since $h(t) = \sqrt{t}$, the volume as a function of time is the composition

$$V(t) = V(h(t)) = 4h(t) = 4\sqrt{t} \text{ feet}^3.$$

◇

Solutions for Exercise Set 1.11

1. For $f(x) = 2x - 3$ and $g(x) = x^2 + 2$ we have

$$(f \circ g)(2) = f(g(2)) = f(2^2 + 2) = f(6) = 2(6) - 3 = 9.$$

3. For $f(x) = 2x - 3$ and $g(x) = x^2 + 2$ we have

$$(f \circ g)(-3) = f(g(-3)) = f((-3)^2 + 2) = f(11) = 2(11) - 3 = 19.$$

5. For $f(x) = 2x - 3$ and $g(x) = x^2 + 2$ we have

$$(f \circ f)(-2) = f(f(-2)) = f(2(-2) - 3) = f(-7) = 2(-7) - 3 = -17.$$

7. For $f(x) = 2x + 1$ and $g(x) = 3x - 1$ we have

$$(f \circ g)(x) = f(g(x)) = f(3x - 1) = 2(3x - 1) + 1 = 6x - 1;$$
$$(g \circ f)(x) = g(f(x)) = g(2x + 1) = 3(2x + 1) - 1 = 6x + 2.$$

Since the domain of f and g is the set of all real numbers, the domain of each of the compositions is also the set of all real numbers.

9. For $f(x) = \frac{1}{x}$ and $g(x) = x^2 + 2x$ we have

$(f \circ g)(x) = f(g(x)) = f(x^2 + 2x) = \frac{1}{x^2 + 2x};$

$(g \circ f)(x) = g(f(x)) = g\left(\frac{1}{x}\right) = \left(\frac{1}{x}\right)^2 + 2\left(\frac{1}{x}\right) = \frac{1}{x^2} + \frac{2}{x} = \frac{1 + 2x}{x^2}.$

Domain $f : \{x : x \neq 0\} = (-\infty, 0) \cup (0, \infty)$; Domain $g : (-\infty, \infty)$;

Domain $f \circ g : \{x : x^2 + 2x = x(x + 2) \neq 0\} = \{x : x \neq 0, x \neq -2\} = (-\infty, -2) \cup (-2, 0) \cup (0, \infty)$;

Domain $g \circ f : \{x \neq 0 : 1/x \text{ is defined}\} = \{x : x \neq 0\} = (-\infty, 0) \cup (0, \infty).$

11. For $f(x) = \sqrt{x - 1}$ and $g(x) = x^2 - 3$ we have

$$(f \circ g)(x) = f(g(x)) = f(x^2 - 3) = \sqrt{x^2 - 3 - 1} = \sqrt{x^2 - 4};$$
$$(g \circ f)(x) = g(f(x)) = f(\sqrt{x - 1}) = (\sqrt{x - 1})^2 - 3 = x - 4.$$

Domain $f : \{x : x - 1 \geq 0\} = [1, \infty)$; Domain $g : (-\infty, \infty)$;

Domain $f \circ g : \{x : x^2 - 3 \geq 1\} = \{x : (x - 2)(x + 2) \geq 0\} = (-\infty, -2] \cup [2, \infty)$;

Domain $g \circ f : \{x : x \geq 1 \text{ and } f(x) \text{ is defined}\} = [1, \infty)$.

13. For $f(x) = \frac{1}{x}$ and $g(x) = \frac{1}{x+1}$ we have

$$(f \circ g)(x) = f(g(x)) = f\left(\frac{1}{x+1}\right) = \frac{1}{\frac{1}{x+1}} = x + 1;$$

$$(g \circ f)(x) = g(f(x)) = g\left(\frac{1}{x}\right) = \frac{1}{\frac{1}{x} + 1} = \frac{1}{\frac{1+x}{x}} = \frac{x}{x+1}.$$

Domain $f : \{x : x \neq 0\} = (-\infty, 0) \cup (0, \infty)$;

Domain $g : \{x : x + 1 \neq 0\} = (-\infty, -1) \cup (-1, \infty)$;

Domain $f \circ g : \{x \neq -1 : \frac{1}{x+1} \neq 0\} = (-\infty, -1) \cup (-1, \infty)$;

Domain $g \circ f : \{x \neq 0 : \frac{1}{x} \neq -1\} = (-\infty, -1) \cup (-1, 0) \cup (0, \infty)$.

15. The inside operation of $h(x) = (3x^2 - 2)^4$ is $3x^2 - 2$ and the outside operation is raising to the fourth power, x^4. So define $g(x) = 3x^2 - 2$ and $f(x) = x^4$. This gives

$$(f \circ g)(x) = f(g(x)) = f(3x^2 - 2) = (3x^2 - 2)^4 = h(x).$$

17. The inside operation of $h(x) = \sqrt{x - 2}$ is $x - 2$ and the outside operation is the square root of a number, $\sqrt{x}$. So define $g(x) = x - 2$ and $f(x) = \sqrt{x}$. This gives

$$(f \circ g)(x) = f(g(x)) = f(x - 2) = \sqrt{x - 2} = h(x).$$

19. Define the first operation to be $x + 2$ and then take the reciprocal, so let $g(x) = x + 2$ and $f(x) = \frac{1}{x}$. Then

$$(f \circ g)(x) = f(g(x)) = f(x + 2) = \frac{1}{x + 2} = h(x).$$

21. (a) (i) $(f \circ g)(-2) = f(g(-2)) = f(1) = 2$

(ii) $(g \circ f)(-2) = g(f(-2)) = g(0) = 0$

(iii) $(g \circ f)(2) = g(f(2)) = g(0) = 0$

(iv) $(f \circ g)(2) = f(g(2)) = f(-1) = -2$

(b) (i) $(f \circ g)(x) = f(g(x)) = 0 \Leftrightarrow g(x) = -2, 0, 2 \Leftrightarrow x = -4, 0, 4$

(ii) $(g \circ f)(x) = g(f(x)) = 0 \Leftrightarrow f(x) = 0 \Leftrightarrow x = -2, 0, 2$

23.

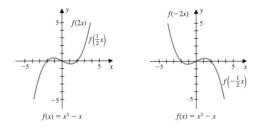

25. (a) For $f(x) = \frac{1}{|x^2+2x-1|}$, completing the square on the quadratic in the denominator gives

$$f(x) = \frac{1}{|x^2 + 2x + 1 - 1 - 1|} = \frac{1}{|(x+1)^2 - 2|}.$$

Then letting $g_1(x) = (x+1)^2, g_2(x) = x - 2, g_3(x) = |x|$ and $g_4(x) = \frac{1}{x}$ we have

$$g_4(g_3(g_2(g_1(x)))) = g_4(g_3(g_2((x+1)^2)))$$
$$= g_4(g_3((x+1)^2 - 2))$$
$$= g_4\left(|(x+1)^2 - 2|\right)$$
$$= \frac{1}{|(x+1)^2 - 2|}$$
$$= f(x).$$

(b)

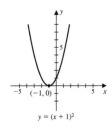

$y = (x + 1)^2$

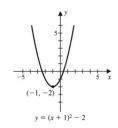

$y = (x + 1)^2 - 2$

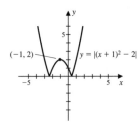

$y = |(x + 1)^2 - 2|$

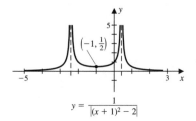

$$y = \frac{1}{|(x + 1)^2 - 2|}$$

27. Let $f(x) = x$ and $g(x) = x + 1$. Then $(f \circ g)(x) = f(g(x)) = f(x+1) = x + 1$ and $(g \circ f)(x) = g(f(x)) = g(x) = x + 1$. There are many other examples.

29. If f is an odd function and g is an even function, then $f(-x) = -f(x)$ and $g(-x) = g(x)$. So

$$(f \circ g)(-x) = f(g(-x)) = f(g(x)) = (f \circ g)(x)$$

and

$$(g \circ f)(-x) = g(f(-x)) = g(-f(x)) = g(f(x)) = (g \circ f)(x).$$

31. Let $f(x) = ax + b$ and $g(x) = cx + d$. Then

$$(f \circ g)(x) = a(cx + d) + b = acx + ad + b$$

$$(g \circ f)(x) = c(ax + b) + d = acx + bc + d.$$

(a) $(f \circ g)(x) = (g \circ f)(x) \Leftrightarrow ad + b = bc + d$

(b) $(f \circ g)(x) = f(x) \Leftrightarrow ac = a$ and $ad + b = b \Leftrightarrow c = 1$ and $d = 0$

(c) $(f \circ g)(x) = g(x) \Leftrightarrow ac = c$ and $ad + b = d \Leftrightarrow a = 1, b = 0$

33. $g(x) = 3f(x - 1) + 2$

35. The volume of a sphere of radius r is $V(r) = \frac{4}{3}\pi r^3$. So $V(t) = \frac{4}{3}\pi(3 + 0.01t)^3$.

37. (a) The volume of a sphere of radius r is $V(r) = \frac{4}{3}\pi r^3$ so

$$V(t) = V(r(t)) = \frac{4}{3}\pi \left(3\sqrt{t} + 5\right)^3 \text{ centimeters}^3.$$

(b) The surface area is $S(r) = 4\pi r^2$ so

$$S(t) = S(r(t)) = 4\pi \left(3\sqrt{t} + 5\right)^2 \text{ centimeters}^2.$$

1.12 Inverse Functions

The *inverse* function, when it exists (not all functions have an inverse function), is the process that undoes the original operation of the function. For example, the function $f(x) = 2x - 1$ sends the real number $x = 2$ to the real number $f(2) = 3$. The inverse process should send the value 3 back to the originating value 2.

ONE-TO-ONE FUNCTIONS

Functions that have inverses are those functions that are *one-to-one*. A function is one-to-one provided no two different x values are sent to the same y value. This can be written as

$$f(x_1) \neq f(x_2), \text{ if } x_1 \neq x_2,$$

or equivalently

$$\text{if } f(x_1) = f(x_2), \text{ then } x_1 = x_2.$$

Geometrically, a function will be one-to-one provided any horizontal line that crosses the graph of the function crosses it in only one place. For example, $f(x) = 2x - 1$ is one-to-one, but $g(x) = x^2$ is not one-to-one.

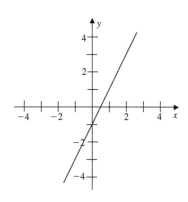

Figure 1.102   Figure 1.103

EXAMPLE 1.12.1   Determine if the function is one-to-one.

(a) $f(x) = x^3 - 1$ (b) $f(x) = 2 - |x|$

SOLUTION:

(a) Use the second equivalent statement to verify the function is one-to-one. If

$$f(x_1) = f(x_2),$$

then

$$x_1^3 - 1 = x_2^3 - 1$$

so

$$x_1^3 = x_2^3$$

and

$$x_1 = x_2.$$

Therefore, the function is one-to-one.

(b) We can determine this function is not one-to-one by inspection. From its graph, we see for example, that

$$f(1) = 2 - |1| = 1 = 2 - |-1| = f(-1),$$

so $f(1) = f(-1)$ and the function is not one-to-one.   ◇

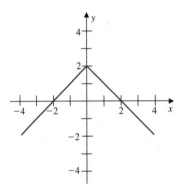

Figure 1.104

EXAMPLE 1.12.2   Assume the function is one-to-one, and find the indicated value.

(a) If $f(x) = x + 3$, find $f^{-1}(1)$.   (b) If $f(x) = \dfrac{1}{x - 1}$, find $f^{-1}(1)$.

SOLUTION:

(a) For any function f and its inverse f^{-1},

$$f^{-1}(f(x)) = x.$$

So to find $f^{-1}(1)$, we need the x with $f(x) = 1$. If

$$f(x) = x + 3 = 1, \quad \text{then} \quad x = -2,$$

and

$$f^{-1}(1) = -2.$$

(b) If

$$\frac{1}{x-1} = 1, \quad \text{then} \quad 1 = x - 1 \quad \text{and} \quad x = 2.$$

So

$$f^{-1}(1) = 2.$$

◇

PROCESS FOR FINDING AN INVERSE FUNCTION

When a function is one-to-one, the following process can be used to find the inverse. The inverse of the function f is denoted f^{-1}.

<u>Procedure for finding the Inverse of f</u>

1. Set $y = f(x)$.

2. Solve for x in terms of y.

3. Interchange the variables x and y.

An important relationship between the graph of a function and its inverse is they are reflections of each other across the line $y = x$.

EXAMPLE 1.12.3   Find the inverse of the one-to-one function. Sketch $y = f(x)$ and $y = f^{-1}(x)$.

(a) $f(x) = \frac{2x-1}{3}$ (b) $f(x) = \sqrt{x-1}$

SOLUTION:

(a) <u>Step 1</u>:

$$y = \frac{2x - 1}{3}.$$

Step 2:

$$y = \frac{2x - 1}{3}$$

$$3y = 2x - 1$$

$$3y + 1 = 2x$$

and

$$x = \frac{3y + 1}{2}.$$

Step 3:

$$f^{-1}(x) = \frac{3x + 1}{2}.$$

Note that if, for example, $x = -2$, then

$$f(-2) = -\frac{5}{3},$$

and

$$f^{-1}(f(-2)) = f^{-1}\left(-\frac{5}{3}\right) = \frac{3\left(-\frac{5}{3}\right) + 1}{2} = \frac{-4}{2} = -2.$$

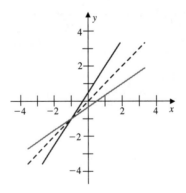

Figure 1.105

(b) <u>Step 1</u>:

$$y = \sqrt{x - 1}.$$

<u>Step 2</u>:

$$y^2 = x - 1, \quad \text{so} \quad x = y^2 + 1.$$

<u>Step 3</u>:

$$f^{-1}(x) = x^2 + 1.$$

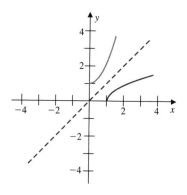

Figure 1.106

In the figure, the inverse is shown only for $x \geq 0$. This is due to the domain restriction. Another important relationship between a function and its inverse is:

The domain of f is the range of f^{-1}.

The range of f is the domain of f^{-1}.

In the example, we have the information in the table.   ◇

Function	Domain	Range
$f(x) = \sqrt{x-1}$	$[1, \infty)$	$[0, \infty)$
$f^{-1}(x) = x^2 + 1$	$[0, \infty)$	$[1, \infty)$

EXAMPLE 1.12.4   Let $f(x) = x^2 - 4x$. (a) Show that the function is not one-to-one. (b) Determine a subset of the domain of the function on which it is one-to-one, and find its inverse on this restricted domain.

SOLUTION: The function can be rewritten as

$$f(x) = x^2 - 4x = x^2 - 4x + 4 - 4 = (x-2)^2 - 4.$$

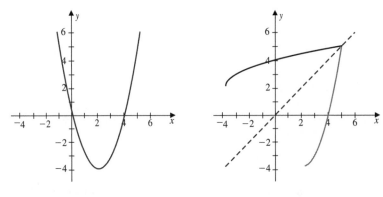

Figure 1.107   Figure 1.108

(a) The figure clearly shows that the function does not satisfy the horizontal line test since many horizontal lines will cross the graph two times. For example, $f(0) = 0 = f(4)$. If the domain is restricted to $[2, \infty)$, then the new function will be one-to-one. The inverse is then found by:

Step 1:

$$y = x^2 - 4x = (x-2)^2 - 4.$$

Step 2:

$$y + 4 = (x - 2)^2, \quad \text{so} \quad x - 2 = \sqrt{y + 4}$$

and

$$x = \sqrt{y + 4} + 2.$$

Step 3:

$$f^{-1}(x) = \sqrt{x + 4} + 2.$$

The domain of f^{-1}, which is the same as the range of f, is the interval $[-4, \infty)$.

◇

Solutions for Exercise Set 1.12

1. The function is $1-1$ since every horizontal line will cross the curve in only one point.

3. The function is not $1-1$ since many horizontal lines cross the curve in more than one point.

5. The function is not $1-1$ since every horizontal line that crosses the curve intersects the curve in two points.

7. The function defined by $f(x) = 3x - 4$ is $1-1$, since

$$f(x_1) = f(x_2) \Leftrightarrow 3x_1 - 4 = 3x_2 - 4 \Leftrightarrow 3x_1 = 3x_2 \Leftrightarrow x_1 = x_2.$$

9. The function defined by $f(x) = |x - 1| + 1$ is not $1-1$. For example, if $x_1 = 0$ and $x_2 = 2$, then

$$f(x_1) = f(0) = |0 - 1| + 1 = 2$$

and

$$f(x_2) = f(2) = |2 - 1| + 1 = 2.$$

So we have found x_1 and x_2, with $x_1 \neq x_2$ and $f(x_1) = f(x_2)$.

11. The function defined by $f(x) = x^4 + 1$ is not $1-1$, since, for example, $f(-1) = 2 = f(1)$.

13. The function defined by $f(x) = \begin{cases} x^2, & \text{if } x \geq 0 \\ x - 1 & \text{if } x < 0 \end{cases}$ is $1-1$, since the graph satisfies the horizontal line test. That is, every horizontal line that crosses the graph does so in only one point.

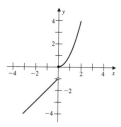

15. The graph of the inverse function is

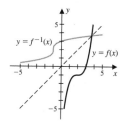

17. The graph of the inverse function is

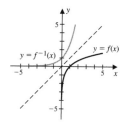

19. Since $f(x) = 2 \Leftrightarrow f^{-1}(2) = x$, solve $x + 1 = 2 \Leftrightarrow x = 1$. So $f^{-1}(2) = 1$.

21. Since $f(x) = 1 \Leftrightarrow f^{-1}(1) = x$, solve $3x - 2 = 1 \Leftrightarrow x = 1$. So

$$f^{-1}(1) = 1.$$

23. Since $f(x) = 2 \Leftrightarrow f^{-1}(2) = x$, solve $\frac{1}{x} = 2 \Leftrightarrow x = \frac{1}{2}$. So

$$f^{-1}(2) = \frac{1}{2}.$$

25. Let $f(x) = 2x - 1$. Then f is $1 - 1$ since

$$f(x_1) = f(x_2) \Leftrightarrow 2x_1 - 1 = 2x_2 - 1 \Leftrightarrow x_1 = x_2.$$

To find f^{-1} : $y = 2x - 1 \Leftrightarrow 2x = y + 1 \Leftrightarrow x = \frac{y+1}{2}$, so

$$f^{-1}(x) = \frac{x+1}{2} = \frac{1}{2}x + \frac{1}{2}.$$

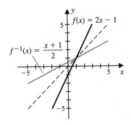

27. Let $f(x) = \sqrt{x - 3}$. Then f is $1 - 1$ since

$$f(x_1) = f(x_2) \Leftrightarrow \sqrt{x_1 - 3} = \sqrt{x_2 - 3} \Leftrightarrow x_1 - 3 = x_2 - 3 \Leftrightarrow x_1 = x_2.$$

To find f^{-1} : $y = \sqrt{x - 3} \Leftrightarrow x - 3 = y^2 \Leftrightarrow x = y^2 + 3$, so

$$f^{-1}(x) = x^2 + 3.$$

Note $D_f = [3, \infty) = R_{f^{-1}}$ and $R_f = [0, \infty) = D_{f^{-1}}$.

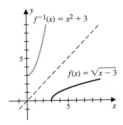

29. $f(x) = \frac{1}{x}$. Then f is $1 - 1$ since

$$f(x_1) = f(x_2) \Leftrightarrow \frac{1}{x_1} = \frac{1}{x_2} \Leftrightarrow x_1 = x_2.$$

To find f^{-1} : $y = \frac{1}{x} \Leftrightarrow xy = 1 \Leftrightarrow x = \frac{1}{y}$, so

$$f^{-1}(x) = \frac{1}{x}.$$

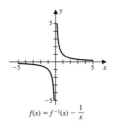

$f(x) = f^{-1}(x) - \frac{1}{x}$

31. Let $f(x) = \frac{1}{\sqrt{x}}$. Then f is $1 - 1$ since

$$f(x_1) = f(x_2) \Leftrightarrow \frac{1}{\sqrt{x_1}} = \frac{1}{\sqrt{x_2}} \Leftrightarrow \sqrt{x_2} = \sqrt{x_1} \Leftrightarrow x_1 = x_2.$$

To find f^{-1} : $y = \frac{1}{\sqrt{x}} \Leftrightarrow \sqrt{x} = \frac{1}{y} \Leftrightarrow x = \frac{1}{y^2}$, so

$$f^{-1}(x) = \frac{1}{x^2}.$$

Note $D_f = (0, \infty) = R_{f^{-1}}$ and $R_f = (0, \infty) = D_{f^{-1}}$.

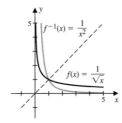

$f^{-1}(x) = \frac{1}{x^2}$

$f(x) = \frac{1}{\sqrt{x}}$

33. Let $f(x) = 1 + x^3$. Then f is $1 - 1$ since

$$f(x_1) = f(x_2) \Leftrightarrow 1 + x_1^3 = 1 + x_2^3 \Leftrightarrow x_1^3 = x_2^3 \Leftrightarrow x_1 = x_2.$$

To find f^{-1}: $y = 1 + x^3 \Leftrightarrow x^3 = y - 1 \Leftrightarrow x = (y-1)^{\frac{1}{3}}$, so

$$f^{-1}(x) = (x-1)^{\frac{1}{3}}.$$

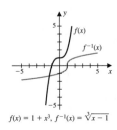

$f(x) = 1 + x^3$, $f^{-1}(x) = \sqrt[3]{x-1}$

35. Let $f(x) = x^2 + 1, x \geq 0$. Then f is $1 - 1$ since

$$f(x_1) = f(x_2) \Leftrightarrow x_1^2 + 1 = x_2^2 + 1 \Leftrightarrow x_1^2 = x_2^2 \Leftrightarrow x_1 = x_2,$$

since $x \geq 0$ (otherwise we could only conclude that $|x_1| = |x_2|$).
To find f^{-1}: $y = x^2 + 1 \Leftrightarrow y - 1 = x^2 \Leftrightarrow x = \sqrt{y-1}$, so

$$f^{-1}(x) = \sqrt{x-1}.$$

Note $D_f = [0, \infty) = R_{f^{-1}}$ and $R_f = [1, \infty) = D_{f^{-1}}$.

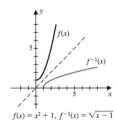

$f(x) = x^2 + 1$, $f^{-1}(x) = \sqrt{x-1}$

37. (a) The graph of the function $f(x) = |2 - x|$ is V-shaped, opens upward and the point is at $(2, 0)$. The line $y = 1$, for example, must cross the curve two times, and $|2 - x| = 1 \Leftrightarrow 2 - x = 1$ or $2 - x = -1 \Leftrightarrow x = 1$ or $x = 3$. So the function is not $1 - 1$.

(b) The function is $1 - 1$ on the interval $[2, \infty)$.

f^{-1} for f restricted on $[2, \infty)$:

$$y = -(2 - x) \Leftrightarrow y = x - 2 \Leftrightarrow x = y + 2, \text{ so } f^{-1}(x) = x + 2.$$

The domain of f^{-1} is the range of f which equals $[0, \infty)$.

39. (a) We have $f(x) = x^2 - 2x = x(x - 2)$, and $f(x) = 0$ for both $x = 0$ and $x = 2$, so f is not $1 - 1$. The graph is a parabola that opens upward and has vertex $(1, -1)$.

(b) The function is $1 - 1$ on the interval $[1, \infty)$.

f^{-1} for f restricted on $[1, \infty)$:

$$y = x^2 - 2x = x^2 - 2x + 1 - 1 = (x - 1)^2 - 1 \Leftrightarrow y + 1 = (x - 1)^2$$

if and only if

$$x - 1 = \sqrt{y + 1} \Leftrightarrow x = \sqrt{y + 1} + 1, \text{ so } f^{-1}(x) = \sqrt{x + 1} + 1.$$

The domain of f^{-1} is $[-1, \infty)$ which is also the range of f.

41. For $f(x) = mx + b$,

$$f(x_1) = f(x_2) \Leftrightarrow mx_1 + b = mx_2 + b \Leftrightarrow mx_1 = mx_2$$

if and only if

$$x_1 = x_2 \text{ provided } m \neq 0.$$

So for all $m \neq 0$ the function $f(x) = mx + b$ is $1 - 1$. Solving for x gives

$$y = mx + b \Leftrightarrow y - b = mx \Leftrightarrow x = \frac{y - b}{m}, \text{ so } f^{-1}(x) = \frac{x - b}{m}.$$

Solutions for Exercise Set 1 Review

1. (a) $-1 \le x \le 7$

   (b)

3. (a) $x < 7$

   (b)

   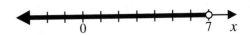

5. (a) $(-\infty, -5)$

   (b)

   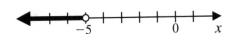

7. (a) $[2, 10)$

   (b)

9. $2x + 3 \ge 4 \Leftrightarrow 2x \ge 1 \Leftrightarrow x \ge \frac{1}{2}$

11. $x^2 + 2x + 1 \ge 1 \Leftrightarrow x^2 + 2x \ge 0 \Leftrightarrow x(x + 2) \ge 0$

```
x          - - - - - - - - - - 0 + + + + + + + + + +
x + 2      - - - - - - 0 + + + + + ++ + + + + + + +
x (x + 2)  + + + + + + 0 - - - 0 + + + + + + + + + +
```

```
        -5 -4 -3 -2 -1  0  1  2  3  4  5   x
```

The solution set $(-\infty, -2] \cup [0, \infty)$.

13. $(x - 2)(x + 2)(x + 3) \geq 0$

```
x + 3             - - - - 0 + + + + + + + + + + + + + + + +
x - 2             - - - - - - - - - - - - - 0 + + + + + +
x + 2             - - - - - - 0 + + + + + + + + + + + + + +
(x + 3)(x - 2)(x + 2)  - - - - 0 + 0 - - - - - - - 0 + + + + + +
```

```
        -5 -4 -3 -2 -1  0  1  2  3  4  5   x
```

The solution set is $[-3, -2] \cup [2, \infty)$.

15. $x^2 + 3x > 0 \Leftrightarrow x(x + 3) \geq 0$

```
x          - - - - - - - - - - - 0 + + + + + + + + + +
x + 3      - - - - 0 + + + + + + + ++ + + + + + + + +
x (x + 3)  + + + + 0 - - - - - - 0 + + + + + + + + + +
```

```
        -5 -4 -3 -2 -1  0  1  2  3  4  5   x
```

The solution set is $(-\infty, -3] \cup [0, \infty)$.

17. $\frac{2x-1}{x+1} \leq -2 \Leftrightarrow \frac{2x-1}{x+1} + 2 \leq 0 \Leftrightarrow \frac{2x-1+2(x+1)}{x+1} \leq 0 \Leftrightarrow \frac{4x+1}{x+1} \leq 0$

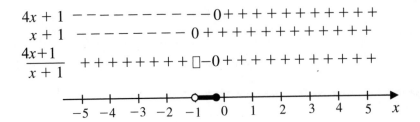

The solution set is $\left(-1, -\frac{1}{4}\right]$.

19. (a) $|2x - 3| < 5 \Leftrightarrow -5 < 2x - 3 < 5 \Leftrightarrow -2 < 2x < 8 \Leftrightarrow -1 < x < 4$

(b)

21. (a) $|2 - x| \geq 2 \Leftrightarrow 2 - x \geq 2$ or $2 - x \leq -2 \Leftrightarrow -x \geq 0$ or $-x \leq -4 \Leftrightarrow$ $x \leq 0$ or $x \geq 4$. That is, $(-\infty, 0] \bigcup [4, \infty)$.

(b)

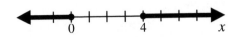

23. $2 < y \leq 3$

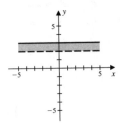

25. $|x - 1| < 2 \Leftrightarrow -2 < x - 1 < 2 \Leftrightarrow -1 < x < 3$

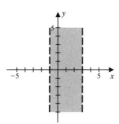

27. $|x| + |y| = 1$

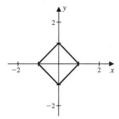

29. $|x| > |y|$

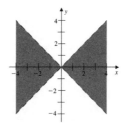

31. (a) For $f(x) = x^2 - 3$ the domain is the set of all real numbers. (b) Since the graph is obtained from the basic graph of $y = x^2$, shifted 3 units downward, the vertex is $(0, -3)$, so the range is $[-3, \infty)$.

33. (a) For $f(x) = \sqrt{x - 2} + 2$ the domain is the set of all real numbers for which the expression under the radical is nonnegative, so the domain is $\{x : x - 2 \geq 0\} = [2, \infty)$.

(b) The range is $[2, \infty)$.

35. (a) For $f(x) = \frac{1}{x^2-6x+8}$ the domain is

$$\{x \ : \ x^2-6x+8 \neq 0\} = \{x : (x-4)(x-2) \neq 0\} = (-\infty, 2) \bigcup (2, 4) \bigcup (4, \infty).$$

(b) For $f(x) = \frac{x-2}{x^2-6x+8} = \frac{x-2}{(x-4)(x-2)} = \frac{1}{x-4}$ the domain is still
$(-\infty, 2) \bigcup (2, 4) \bigcup (4, \infty),$ since $x = 2$ can not be substituted into the original equation.

(c) For $f(x) = \sqrt{\frac{x^2}{x^2-6x+8}},$ the domain is

$$\{x \ : \ (x-4)(x-2) > 0 \text{ or } x = 0\} = (-\infty, 2) \bigcup (4, \infty).$$

37. For $f(x) = 7x + 4$ we have (a) $f(x+h) = 7(x+h) + 4 = 7x + 7h + 4$ and

(b)

$$\frac{f(x+h) - f(x)}{h} = \frac{7x + 7h + 4 - (7x+4)}{h} = \frac{7h}{h} = 7.$$

39. For $f(x) = x^2 - 1$ we have (a) $f(x+h) = (x+h)^2 - 1 = x^2 + 2hx + h^2 - 1$ and

(b)

$$\frac{f(x+h) - f(x)}{h} = \frac{x^2 + 2hx + h^2 - 1 - (x^2 - 1)}{h}$$
$$= \frac{2hx + h^2}{h} = \frac{h(2x+h)}{h} = 2x + h.$$

41. For $f(x) = \frac{1}{x-1}$ we have (a) $f(x+h) = \frac{1}{x+h-1}$ and (b)

$$\frac{f(x+h) - f(x)}{h} = \frac{\frac{1}{x+h-1} - \frac{1}{x-1}}{h} = \frac{\frac{x-1-(x+h-1)}{(x+h-1)(x-1)}}{h}$$
$$= \frac{-h}{h(x+h-1)(x-1)} = \frac{-1}{(x+h-1)(x-1)}.$$

43. (a) (ii)   (b) (i)   (c) (iii)   (d) (iv)   (e) (v)   (f) (vi)

45. (a), (d)

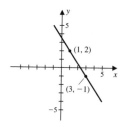

(b) $d = \sqrt{(3-1)^2 + (-1-2)^2} = \sqrt{13}$

(c) $\left(\frac{3+1}{2}, \frac{-1+2}{2}\right) = \left(2, \frac{1}{2}\right)$

(d) $m = \frac{2-(-1)}{1-3} = -\frac{3}{2}; \; y - (-1) = -\frac{3}{2}(x-3) \Leftrightarrow y = -\frac{3}{2}x + \frac{7}{2}$

47. (a), (d)

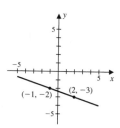

(b) $d = \sqrt{(-1-2)^2 + (-2-(-3))^2} = \sqrt{10}$

(c) $\left(\frac{-1+2}{2}, \frac{-2-3}{2}\right) = \left(\frac{1}{2}, -\frac{5}{2}\right)$

(d) $m = \frac{-3-(-2)}{2-(-1)} = -\frac{1}{3}; \; y - (-2) = -\frac{1}{3}(x-(-1)) \Leftrightarrow y = -\frac{1}{3}x - \frac{7}{3}$

49. (a) $m = 4; y - 0 = 4(x-0) \Leftrightarrow y = 4x$ (b) $m = -\frac{1}{4}; y = -\frac{1}{4}x$

51. $-7x - 5y = -1 \Leftrightarrow y = -\frac{7}{5}x + \frac{1}{5}$

(a) $m = -\frac{7}{5}; y - (-3) = -\frac{7}{5}(x-(-1)) \Leftrightarrow y = -\frac{7}{5}x - \frac{22}{5}$

(b) $m = \frac{5}{7}; y + 3 = \frac{5}{7}(x+1) \Leftrightarrow y = \frac{5}{7}x - \frac{16}{7}$

53. (a) $y = x^2 - 4x = x^2 - 4x + 4 - 4 = (x-2)^2 - 4$

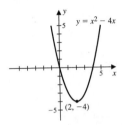

(b) The range is $[-4, \infty)$.

(c) x-intercepts: $(0, 0), (4, 0)$

$$x^2 - 4x = 0 \Leftrightarrow x(x - 4) = 0 \Leftrightarrow x = 0, x = 4$$

y-intercepts: $(0, 0)$; Set $x = 0 \Rightarrow y = 0$.

(d) A minimum occurs at the point $(2, -4)$.

55. (a)

$$y = 2x^2 - 12x + 18 = 2(x^2 - 6x) + 18 = 2(x^2 - 6x + 9 - 9) + 18 = 2(x - 3)^2$$

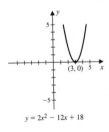

(b) The range is $[0, \infty)$.

(c) x-intercepts: $(3, 0)$; y-intercepts: $(0, 18)$; Set $x = 0 \Rightarrow y = 18$.

(d) A minimum occurs at the point $(3, 0)$.

57. (a)

$$y = -\frac{1}{2}x^2 + 3x - 3 = -\frac{1}{2}(x^2 - 6x) - 3$$

$$= -\frac{1}{2}\left(x^2 - 6x + 9 - 9\right) - 3 = -\frac{1}{2}(x-3)^2 + \frac{3}{2}$$

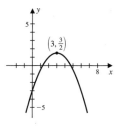

(b) The range is $\left(-\infty, \frac{3}{2}\right]$.

(c) x-intercepts: $\left(3 + \sqrt{3}, 0\right), \left(3 - \sqrt{3}, 0\right)$

$$-\frac{1}{2}(x-3)^2 = -\frac{3}{2} \Leftrightarrow (x-3)^2 = 3 \Leftrightarrow x = 3 \pm \sqrt{3}$$

y-intercepts: $(0, -3)$; Set $x = 0 \Rightarrow y = -3$.

(d) A maximum occurs at the point $\left(3, \frac{3}{2}\right)$.

59. $f(x) = |x + 1|$

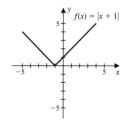

61. $f(x) = |x + 2| - 2$

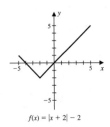

$f(x) = |x + 2| - 2$

63. $g(x) = \sqrt{x + 2}$

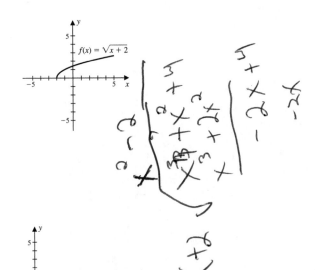

$f(x) = \sqrt{x + 2}$

65. $g(x) = -\sqrt{x} - 2$

$f(x) = -\sqrt{x} - 2$

67. $f(x) = \lfloor x + 2 \rfloor$

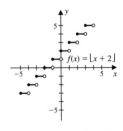

$f(x) = \lfloor x + 2 \rfloor$

69. $f(x) = -\lfloor x \rfloor$

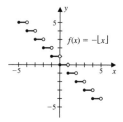

71. center: $(0,0)$; radius: 4

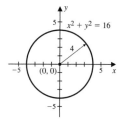

73. center: $(-2, 1)$; radius 3

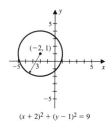

75. Completing the square on the x- and y- terms gives

$$4 = x^2 + 4x + 4 - 4 + y^2 + 2y + 1 - 1 - 6 \Rightarrow (x+2)^2 + (y+1)^2 = 9$$

center: $(-2, -1)$; radius: 3

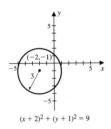

$(x + 2)^2 + (y + 1)^2 = 9$

77. $f(x) = 3x - 2$

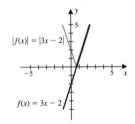

79. $f(x) = -x^2 + 3$

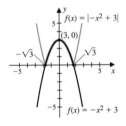

81.   (a) $y = f(x+1)$ (b) $y = f(x+1)+1$ (c) $y = f(x-1)$

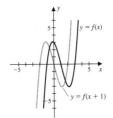
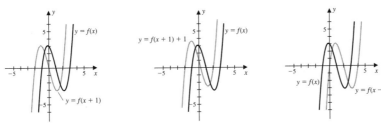
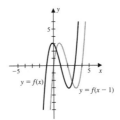

(d) $y = f(x-1)+2$ (e) $y = f(x+1)-1$ (f) $y = f(x-1)-2$

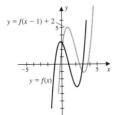
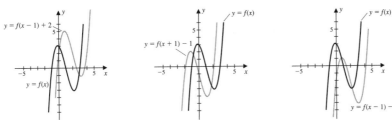
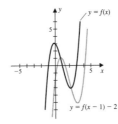

83.   (a) For $f(x) = 2x^2 - 3$ we have $f(-x) = 2(-x)^2 - 3 = 2x^2 - 3$;

$-f(x) = -(2x^2 - 3) = -2x^2 + 3$; $f\left(\frac{1}{x}\right) = 2\left(\frac{1}{x}\right)^2 - 3 = \frac{2}{x^2} - 3 = \frac{2-3x^2}{x^2}$;

$\frac{1}{f(x)} = \frac{1}{2x^2-3}$; $f\left(\sqrt{x}\right) = 2\left(\sqrt{x}\right)^2 - 3 = 2x - 3$; $\sqrt{f(x)} = \sqrt{2x^2 - 3}$.

(b) For $f(x) = \frac{1}{x^2}$ we have $f(-x) = \frac{1}{(-x)^2} = \frac{1}{x^2}$; $-f(x) = -\frac{1}{x^2}$;

$f\left(\frac{1}{x}\right) = \frac{1}{\left(\frac{1}{x}\right)^2} = x^2$; $\frac{1}{f(x)} = \frac{1}{\frac{1}{x^2}} = x^2$;

$f\left(\sqrt{x}\right) = \frac{1}{\left(\sqrt{x}\right)^2} = \frac{1}{x}$; $\sqrt{f(x)} = \sqrt{\frac{1}{x^2}} = \frac{1}{x}$.

85.   The line passes through $(3,0)$ and $(0,2)$, so the slope is $m = \frac{2-0}{0-3} = -\frac{2}{3}$

and the equation is $y = -\frac{2}{3}x + 2$.

87.   A radius for the circle is the line segment between $(-3, -1)$ and $(-5, -3)$, so the radius is

$$r = \sqrt{(-3 - (-5))^2 + (-1 - (-3))^2} = \sqrt{4 + 4} = \sqrt{8}.$$

The equation of the circle is $(x + 3)^2 + (y + 1)^2 = 8$.

89.

$$x^2 - 4x + y^2 - 2y - 1 = 0$$

$$x^2 - 4x + 4 - 4 + y^2 - 2y + 1 - 1 - 1 = 0$$

$$(x - 2)^2 + (y - 1)^2 = 6$$

The circle has center $(2, 1)$ so the line passing through $(2, 1)$ and $(3, 4)$ has slope $m = \frac{4-1}{3-2} = 3$ and equation $y - 1 = 3(x - 2) \Leftrightarrow y = 3x - 5$.

91.   The graph of $y = g(x)$ is shown.

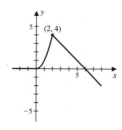

93.   (a) $f(x) = g(h_5(x)) = g(x^5) = \frac{1}{x^5}$

     (b) $f(x) = w(v(h_4(x))) = w(v(x^4)) = w(x^4 - 2) = \sqrt{x^4 - 2}$

     (c) $f(x) = h_6(v(h_2))) = h_6(v(x^2)) = h_6(x^2 - 2) = (x^2 - 2)^6$

     (d) $f(x) = g(w(v(h_3(x)))) = g(w(v(x^3))) = g(w(x^3 - 2)) = g(\sqrt{x^3 - 2}) = $

     $\frac{1}{\sqrt{x^3 - 2}}$

95. $g(x) = f(3(x - 3)) + 1$

97. (a)

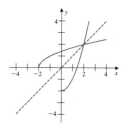

(b) $f(x) = x^2 - 2$

(c) $y = x^2 - 2 \Leftrightarrow x^2 = y + 2 \Rightarrow x = \sqrt{y + 2}$, so $f^{-1}(x) = \sqrt{x + 2}$.

99. (a) $x = 1, 2, 3, 4$ (b) $x \le 1$ or $2 \le x \le 4$ (c) $x = 1, 3$

(d) $0 \le x \le 1.5$ or $3 \le x \le 5$

(e) $(f + g)(3) = f(3) + g(3) = 0 + 3 = 3$;

$(f + g)(4) = f(4) + g(4) = 1.5 + 1.5 = 3$;

$(f - g)(3) = f(3) - g(3) = 0 - 3 = -3$;

$(f - g)(4) = f(4) - g(4) = 1.5 - 1.5 = 0$

(f) $(f \circ g)(4) = f(g(4)) = f(1.5) = 2.25$;

$(g \circ f)(5) = g(f(5)) = g(3) = 3$

(g) $R_f = [0, 3] = R_g$ (h) $f(5) = 3$

101. Let x denote the price over $\$36.00$ for a complete dinner and let $P(x)$ denote the amount taken in by the restaurant, so that

$$P(x) = (36 + x)(200 - 4x) = 7200 + 56x - 4x^2$$

From the graph of $P(x) = 7200 + 56x - 4x^2$ we see that the maximum return occurs at approximately $x = 7$. So the owner should charge $36 +$

7 = $43.00 per dinner. By completing the square on the quadratic we have

$$-4x^2 + 56x + 7200 = -4(x^2 - 14x) + 7200$$

$$= -4(x^2 - 14x + 49 - 49) + 7200$$

$$= -4(x - 7)^2 + 7396$$

so the exact maximum return is when $x = 7$.

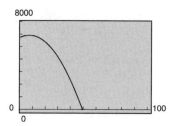

103. (a) The amount of fence required is the perimeter of the plot plus the length of fence down the middle. If ℓ and w are the length and width of the plot, then the amount of fence, F, is $F = 3\ell + 2w$, where the assumption is the length of the middle section of fence is l. Since the area of the plot is 432, we have $432 = \ell w$. This implies that $w = \frac{432}{\ell}$ and that

$$F(\ell) = 3\ell + 2 \cdot \frac{432}{\ell} = 3\ell + \frac{864}{\ell}.$$

(b) The domain of the function is $(0, \infty)$.

105. (a) If the point where the new construction to point B begins is x miles east of point A, we can apply the Pythagorean Theorem to the right triangle to determine that the cost of construction is

$$C(x) = 200000x + 400000\sqrt{(40 - x)^2 + 400}.$$

(b) The graph plots the total cost with respect to the distance x where the new construction begins from point A. The minimum cost corresponds to the x-coordinate of the low point on the graph. So approximately 28.5 miles of the old road should be restored in order to minimize the cost of construction.

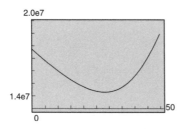

107. (a) If the relationship between cost and number of units produced per day is linear, then the points $(100, 3200)$ and $(500, 9600)$ are on the line. The slope of the line is

$$m = \frac{9600 - 3200}{500 - 100} = \frac{6400}{400} = 16$$

and the equation of the line is

$$y - 3200 = 16(x - 100) \quad \text{or} \quad y = 16x + 1600.$$

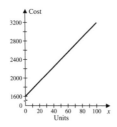

(b) The slope of line indicates that for each one unit increase in the number of units produced, results in an increase of \$16.00 in the cost.

(c) The y-intercept is 1600 and represents the fixed costs of production.

109. We have

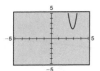 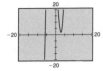 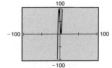 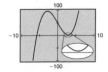

The figure on the right gives the best representation.

111. (a) Between A and B and to the right of E. (b) Between C and D. (c) Between B and C. (d) Between D and E.

Solutions for Exercise Set 1 Calculus

1. For each increase in x of 4 units, the y-coordinate $y = g(x)$ always increases by 12 units, so g is linear and f is quadratic. The slope of the line $y = g(x)$ is $m = \frac{12}{4} = 3$ and the line passes through $(-2, 2)$, so

$$y - 2 = 3(x + 2) \Leftrightarrow y = g(x) = 3x + 8.$$

The quadratic $y = f(x) = ax^2$ and passes through $(-2, 2)$, so $2 = f(-2) = 4a \Rightarrow a = \frac{1}{2}$ and $f(x) = \frac{1}{2}x^2$.

3. (a) The table shows that the solution is $14 \le n \le 16$.

n	12	13	14	15	16	17
$\frac{n(n+1)}{2}$	78	91	105	120	136	153

(b) The table shows that the solution is $8 \le n \le 9$.

n	5	6	7	8	9	10
$\frac{n(n+1)(2n+1)}{6}$	55	91	140	204	285	385

5. The graph of the temperature might be as follows.

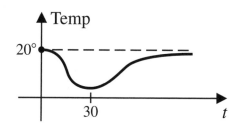

7. Possible graphs for the situations are shown below.

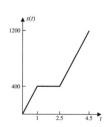 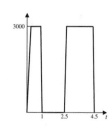 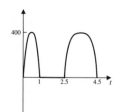 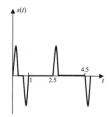

9.   (a) If x pairs of shoes are produced, the average cost for a pair of shoes is

$$\frac{C(x)}{x} = \frac{295 + 3.28x + 0.003x^2}{x} = \frac{295}{x} + 3.28 + 0.003x.$$

The graph of $y = \frac{295}{x} + 3.28 + 0.003x$ has a minimum when $x \approx 310$, so to minimize the average cost about 310 pairs of shoes should be produced.

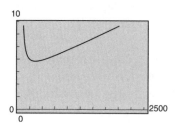

(b) The revenue to the company if x pairs of shoes are sold is given by

$$R(x) = xp(x) = x \left(7.47 + \frac{321}{x} \right) = 7.47x + 321.$$

The profit is then given by

$$P(x) = R(x) - C(x) = 7.47x + 321 - (295 + 3.28x + 0.003x^2)$$

$$= 26 + 4.19x - 0.003x^2.$$

The graph of $y = 26 + 4.19x - 0.003x^2$ has a maximum when $x \approx 700$, so to maximize the profit about 700 pairs of shoes should be produced.

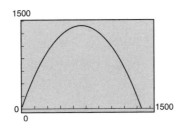

11. a) Since $PV = 8000$ we have $P = \frac{8000}{V}$. The average rate of change of P as V increases from 200 to 250 is

$$\frac{P(250) - P(200)}{250 - 200} = \frac{\frac{8000}{250} - \frac{8000}{200}}{50} = \frac{32 - 40}{50} = -\frac{8}{50} = -\frac{4}{25}\,\text{lb/in}^2.$$

(b) Since

$$\frac{P(200 + h) - P(200)}{h} = \frac{\frac{8000}{200+h} - \frac{8000}{200}}{h}$$

$$= \frac{8000(200) - 8000(200 + h)}{h(200)(200 + h)}$$

$$= \frac{-8000h}{h(200)(200 + h)}$$

$$= -\frac{40}{200 + h}$$

as $h \to 0$, $-\dfrac{40}{200 + h} \to -\dfrac{4}{200} = -\dfrac{1}{25}$, which is the instantaneous rate of change.

CHAPTER 2
ALGEBRAIC FUNCTIONS

2.1 Introduction

A *polynomial function of degree n* is a function of the form

$$P(x) = a_n x^n + a_{n-1} x^{n-1} + \cdots + a_1 x + a_0,$$

where $a_0, a_1, \ldots a_n$ are real numbers and $a_n \neq 0$. The term a_n is called the *leading coefficient*, and a_0 is called the *constant term*. The *algebraic functions* are obtained from the polynomials by any finite combination of the operations of addition, subtraction, multiplication, division, and extracting integral roots.

2.2 Polynomial Functions

EXAMPLE 2.2.1   Specify the degree, leading coefficient, and constant term of the polynomial.

(a) $P(x) = 2x^4 - x^3 + 2x^2 - x + 1$ (b) $P(x) = x^5 - 8$

(c) $P(x) = 3x - 2$ (d) $P(x) = x^{12}$

SOLUTION:

Polynomial	Degree	Leading Coefficient	Constant Term
$2x^4 - x^3 + 2x^2 - x + 1$	4	2	1
$x^5 - 8$	5	1	-8
$3x - 2$	1	3	-2
x^{12}	12	1	0

◇

GRAPHING POLYNOMIAL FUNCTIONS

EXAMPLE 2.2.2   Use the graph of $f(x) = x^3$ shown in the figure to sketch the graph of the function.

(a) $g(x) = (x - 1)^3 + 2$ (b) $g(x) = \frac{1}{2}(x + 2)^3 - 1$

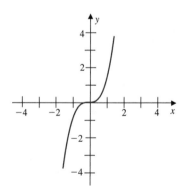

Figure 2.1

SOLUTION:

(a) Knowing the general shape of the graph of $f(x) = x^3$, the graph of $y = g(x)$ can be obtained quickly using the shifting, scaling and reflecting techniques. To sketch $y = g(x) = (x-1)^3 + 2$, first sketch $y = (x-1)^3$ by shifting the graph of $y = x^3$ to the right 1 unit, and then shift the resulting curve upward 2 units.

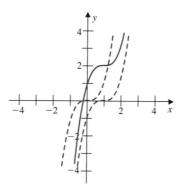

Figure 2.2

(b) First scale $y = x^3$ vertically by a factor of $\frac{1}{2}$. The new curve is below $y = x^3$ for $x > 0$, and above for $x < 0$. For example, the point $(1, 1)$ is on the graph of $y = x^3$, but the point $\left(1, \frac{1}{2}\right)$ is on the graph of $y = \frac{1}{2}x^3$. Similarly, the point $(-1, -1)$ is on the graph of $y = x^3$, but $\left(-1, -\frac{1}{2}\right)$ is on the graph of $y = \frac{1}{2}x^3$. Now shift $y = \frac{1}{2}x^3$ to the left 2 units and downward 1 unit to obtain the final graph.   ◇

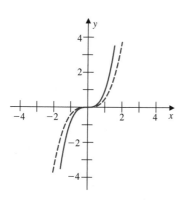

Figure 2.3

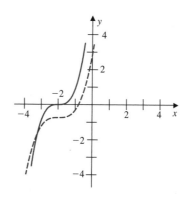

Figure 2.4

EXAMPLE 2.2.3 Use the graph of $f(x) = x^4$ shown in the figure to sketch the graph of the curve.

(a) $y = ax^4$, for $a = 2, 3, \frac{1}{2}, \frac{1}{3}$.

(b) $y = ax^4$, for $a = -1, -2, -3, -\frac{1}{2}, -\frac{1}{3}$.

(c) $y = x^4 - a$, for $a = 1, 2, -1, -2$.

(d) $y = (x - a)^4$, for $a = 1, 2, -1, -2$.

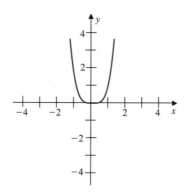

Figure 2.5

SOLUTION:

(a) If the vertical line $x = 1$ is drawn to intersect the curves $y = ax^4$, for $a = 2, 3, \frac{1}{2}, \frac{1}{3}$, then it crosses the curves at the points shown below.

Curve	$y = x^4$	$y = 2x^4$	$y = 3x^4$	$y = \frac{1}{2}x^4$	$y = \frac{1}{3}x^4$
Intersects $x = 1$	$(1, 1)$	$(1, 2)$	$(1, 3)$	$(1, \frac{1}{2})$	$(1, \frac{1}{3})$

Since the curves are symmetric about the y-axis, the same relationship holds for negative x. Using the one point as a model, we can draw the curves in relation to $y = x^4$, as shown in the figure.

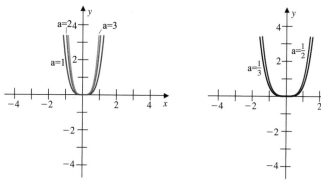

Figure 2.6   Figure 2.7

(b) The curves are the reflection of the curves in part (a) about the x-axis.

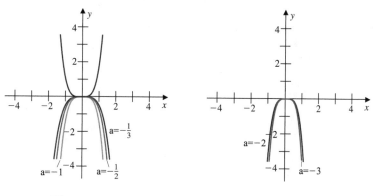

Figure 2.8   Figure 2.9

(c) If $a > 0$, then $y = x^4 - a$ is obtained from a vertical shift of $y = x^4$ downward a units. If $a < 0$, then shift upward $|a|$ units.

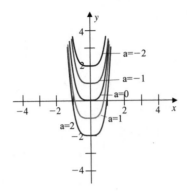

Figure 2.10

(d) If $a > 0$, then $y = (x - a)^4$ is obtained from a horizontal shift of $y = x^4$ to the right a units. If $a < 0$, then shift to the left $|a|$ units.

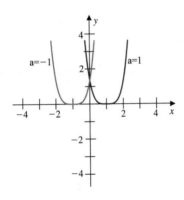

Figure 2.11

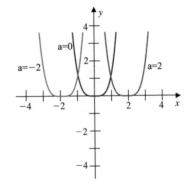

Figure 2.12

◇

ZEROS AND END BEHAVIOR IN SKETCHING POLYNOMIALS

A number c is called a *zero* of a function f if $f(c) = 0$. If a polynomial $P(x)$ has a factor of the form $(x - c)^k$, then $x = c$ is a *repeated* zero of P of *multiplicity* k.

EXAMPLE 2.2.4   Find the zeros of the polynomial and specify the multiplicity of any repeated zeros.

(a) $P(x) = x^3 - x$ (b) $P(x) = x^3 - 2x^2 + x$

SOLUTION:

(a) To find the zeros of a polynomial, factor the polynomial. So

$$P(x) = x^3 - x = x(x^2 - 1) = x(x-1)(x+1).$$

Then if we have $P(x) = 0$,

$$x(x-1)(x+1) = 0$$

and

$$x = 0, \ x = 1, \quad \text{or} \quad x = -1.$$

(b) If $P(x) = 0$, then

$$x^3 - 2x^2 + x = 0$$
$$x(x^2 - 2x + 1) = 0$$
$$x(x-1)(x-1) = 0$$
$$x(x-1)^2 = 0$$

and

$$x = 0 \quad \text{or} \quad x = 1.$$

The zero $x = 1$ is of multiplicity 2.   ◇

The zeros of a polynomial give the x-intercepts of the graph of the function. If the multiplicity of a zero is an even number, then the graph flattens and just touches the x-axis without crossing the axis. If the multiplicity of the zero is an odd number greater than one, the graph flattens and then crosses the x-axis at the zero. This, together with the end behavior of the polynomial, gives enough information to sketch the graph.

The *end behavior* is what happens to the graph of the polynomial as $|x|$ becomes large, that is, at very large values of x on the positive or negative x-axis. The end behavior of a polynomial of degree n depends only on the x^n term. This is written as

$$a_n x^n + a_{n-1} x^{n-1} + \cdots + a_0 \approx a_n x^n.$$

EXAMPLE 2.2.5   Test the end behavior of the polynomial.

(a) $P(x) = x^3 + x^2 + 1$ (b) $P(x) = 5x^4 + 10x^3 + 100x^2 - x + 2$

(c) $P(x) = -3x^5 + x - 1$ (d) $P(x) = -10x^8 + x^7 + 500x^5 + x - 2$

SOLUTION:

(a)

$$P(x) = x^3 + x^2 + 1 \approx x^3$$

As $|x|$ gets large, the graph behaves like the graph $y = x^3$. So, as x goes to ∞, $y = P(x) = x^3 + x^2 + 1$ also goes to ∞, and as x goes to $-\infty$, $y = P(x) = x^3 + x^2 + 1$ goes to $-\infty$. This is written as

$$P(x) \to \infty \quad \text{as} \quad x \to \infty$$

$$P(x) \to -\infty \quad \text{as} \quad x \to -\infty.$$

(b)

$$P(x) = 5x^4 + 10x^3 + 100x^2 - x + 2 \approx 5x^4,$$

so

$$P(x) \to \infty \quad \text{as} \quad x \to \infty$$

$$P(x) \to \infty \quad \text{as} \quad x \to -\infty.$$

(c)

$$P(x) = -3x^5 + x - 1 \approx -3x^5,$$

so

$$P(x) \to -\infty \quad \text{as} \quad x \to \infty$$

$$P(x) \to \infty \quad \text{as} \quad x \to -\infty.$$

(d)

$$P(x) = -10x^8 + x^7 + 500x^5 + x - 2 \approx -10x^8,$$

so

$$P(x) \to -\infty \quad \text{as} \quad x \to \infty$$

$$P(x) \to -\infty \quad \text{as} \quad x \to -\infty.$$

◇

EXAMPLE 2.2.6   Use the zeros of the polynomial and the end behavior to sketch the graph.

(a) $P(x) = x^3 - x^2 - 2x$ (b) $P(x) = x^3 - 2x^2 + x$

(c) $P(x) = (x-1)(x-2)(x+1)^3$ (d) $P(x) = (x+1)^5(x-1)$

SOLUTION:

(a)

Zeros: Set $P(x) = 0$ and factor,

$$x^3 - x^2 - 2x = 0$$

$$x(x^2 - x - 2) = 0$$

$$x(x - 2)(x + 1) = 0,$$

and

$$x = 0, \ x = 2, \quad \text{or} \quad x = -1.$$

End Behavior:

$$P(x) = x^3 - x^2 - 2x \approx x^3,$$

so

$$P(x) \to \infty \quad \text{as} \quad x \to \infty$$

$$P(x) \to -\infty \quad \text{as} \quad x \to -\infty.$$

There is only one way that the curve can cross the x-axis at $-1, 0$, and 2 and satisfy the end behavior. The curve has to come *up* through $x = -1$, rather than down through the point. The curve turns somewhere between $x = -2$ and $x = 0$, passing through $x = 0$ from above, turns again between $x = 0$ and $x = 2$, and goes up through $x = 2$. The curve is shown in the figure. Without the zeros *and* the end behavior, we would not be certain how to sketch the curve.

We still can *not* determine exactly where the turning points are. This is left to calculus. We can approximate the highs and lows, called *local maximums* and *local minimums*, using a graphing device. For example, if after plotting the curve using a graphing device, we click on the local maximum between $x = -1$ and $x = 0$, the point is approximately $(-0.6, 0.6)$.

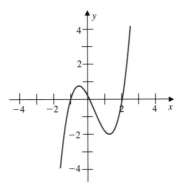

Figure 2.13

(b)

<u>Zeros:</u> Set $P(x) = 0$ and factor,

$$x^3 - 2x^2 + x = 0$$

$$x(x^2 - 2x + 1) = 0$$

$$x(x - 1)(x - 1) = 0$$

$$x(x - 1)^2 = 0,$$

and

$$x = 0 \quad \text{or} \quad x = 1.$$

The zero $x = 1$ has multiplicity 2.

<u>End Behavior:</u>

$$P(x) = x^3 - 2x^2 + x \approx x^3,$$

so

$$P(x) \to \infty \quad \text{as} \quad x \to \infty$$

$$P(x) \to -\infty \quad \text{as} \quad x \to -\infty.$$

The polynomial has a zero of multiplicity 2 at $x = 1$, so the curve flattens and just touches, without crossing, the x-axis at $x = 1$. Applying the end behavior produces the curve as shown in the figure.

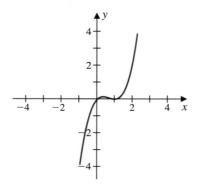

Figure 2.14

(c)

<u>Zeros:</u> The zeros can be read from the function and are $x = 1$, $x = 2$, and $x = -1$. The third is a zero of multiplicity 3.

<u>End Behavior:</u> The factors of $P(x)$ are of degree 1, 1 and 3, so if the factors are expanded, and the terms collected, the degree of the polynomial is 5, and the leading coefficient is 1. Checking, we have

$$P(x) = (x-1)(x-2)(x+1)^3 = x^5 - 4x^3 - 2x^2 + 3x + 2 \approx x^5,$$

so

$$P(x) \to \infty \quad \text{as} \quad x \to \infty$$

$$P(x) \to -\infty \quad \text{as} \quad x \to -\infty.$$

The curve at the zero $x = -1$ flattens and also crosses the x-axis . Putting this information together, the curve is as shown in the figure.

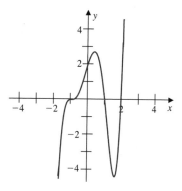

Figure 2.15

(d)

<u>Zeros:</u> The zeros can be read from the function and are $x = 1$, and $x = -1$. The second is a zero of multiplicity 5.

<u>End Behavior:</u> The degree of $P(x)$ is 6, and the leading coefficient is 1.

$$P(x) = (x + 1)^5 (x - 1) = x^6 + 4x^5 + 5x^4 - 5x^2 - 4x - 1 \approx x^6,$$

so

$$P(x) \to \infty \quad \text{as} \quad x \to \infty$$

$$P(x) \to \infty \quad \text{as} \quad x \to -\infty.$$

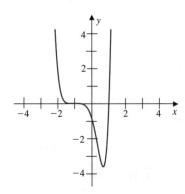

Figure 2.16

◇

EXAMPLE 2.2.7   Match the equation with the correct curve.

(a) $y = (x-1)(x+1)(x-2)$ (b) $y = x(x-1)^2$

(c) $y = (x-1)(x+2)^3$ (d) $y = x(x-1)(x+2)$

(i)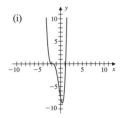

(ii)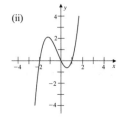

(iii)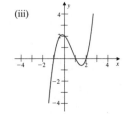

(iv)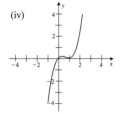

Figure 2.17   Figure 2.18   Figure 2.19   Figure 2.20

SOLUTION:

(a) The only curve with zeros at $x = 1, x = -1$ and $x = 2$ is (iii).

(b) The factor $(x-1)^2$ implies the function has a zero of multiplicity 2 at $x = 1$. Therefore, the curve just touches at $x = 1$ without crossing through it. The only curve that satisfies this condition at $x = 1$ is (iv).

(c) The function has a zero of multiplicity 3 at $x = -2$, so the curve flattens and crosses the x-axis at -2. It also crosses the x-axis at $x = 1$. The graph of the equation matches (i).

(d) By process of elimination, the matching curve is (ii). This agrees with the fact that it is the only function with zeros at $x = 0, x = 1$, and $x = -2$.   ◇

EXAMPLE 2.2.8 The cubic polynomial $P(x)$ has a zero of multiplicity one at $x = 1$, a zero of multiplicity two at $x = 2$, and $P(3) = 6$. Determine $P(x)$, and sketch the graph.

SOLUTION: A polynomial with the specified zeros is

$$Q(x) = (x - 1)(x - 2)^2 = x^3 - 5x^2 + 8x - 4,$$

but

$$Q(3) = (3)^3 - 5(3)^2 + 8(3) - 4 = 2.$$

The problem asks for a polynomial which when evaluated at 3 has a value of 6. So

$$P(x) = 3(x - 1)(x - 2)^2 = 3(x^3 - 5x^2 + 8x - 4) = 3x^3 - 15x^2 + 24x - 12.$$

The end behavior is

$$P(x) \to \infty \quad \text{as} \quad x \to \infty$$

$$P(x) \to -\infty \quad \text{as} \quad x \to -\infty.$$

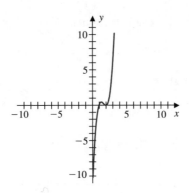

Figure 2.21

Solutions for Exercise Set 2.2

1. (a) (iii) Curve (iii) is the only curve that does not exhibit the behavior of zeros of multiplicity greater than 1, and has zeros at $x = 0, x = 1$, and $x = -1$.

(b) (iv) The factor x^2 implies the function has a zero of multiplicity 2 at the origin, so the curve just touches at the origin without crossing through the origin as in (iv).

(c) (i) The factor $(x + 1)^2$ implies the function has a zero of multiplicity 2 at $x = -1$, so the curve just touches at $x = -1$ without crossing through $x = -1$ as in (i).

(d) (ii) The factor $(x - 1)^2$ implies the function has a zero of multiplicity 2 at $x = 1$, so the curve just touches at $x = 1$ without crossing through $x = -1$ as in (ii).

3. The graph shows a zero of at least multiplicity 2 at the origin and another positive zero, so the lowest possible degree for the polynomial is 3.

5. The graph can be shifted downward so that the shifted graph has 4 distinct zeros, so the lowest possible degree for the polynomial is 4.

7. Vertically stretch the graph of $y = x^3$ by a factor of 3 and then shift the resulting graph downward 2 units.

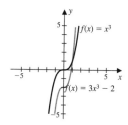

9. Shift the graph of $y = x^4$ to the right 2 units and then shift the result upward 1 unit.

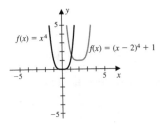

11. Vertically shrink the graph of $y = x^3$ by a factor of 2, flip about the x-axis, shift to the right 3 units and then shift downward 3 units.

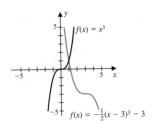

13.

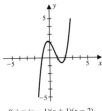

$f(x) = (x - 1)(x + 1)(x - 2)$

15.

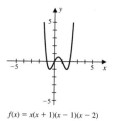

$f(x) = x(x + 1)(x - 1)(x - 2)$

17.

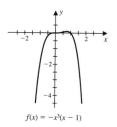

$f(x) = -x^3(x - 1)$

19.

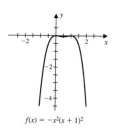

$f(x) = -x^2(x + 1)^2$

21. $f(x) = x^3 - x^2 - 6x = x(x^2 - x - 6) = x(x - 3)(x + 2)$

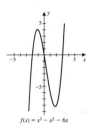

$f(x) = x^3 - x^2 - 6x$

23. $y = f(x - 2)$

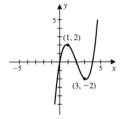

(1, 2)

(3, −2)

25. $y = f(x - 1) + 2$

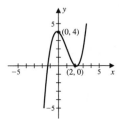

(0, 4)

(2, 0)

27. $y = f(-x)$

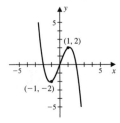

(1, 2)

(−1, −2)

29. $y = |f(x)|$

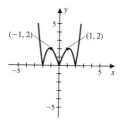

31. (a)

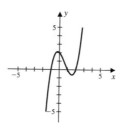

(b) The least possible degree of the polynomial is 3.

33. The polynomial has the form

$$P(x) = a(x-1)^2(x-2)$$

and

$$4 = P(-1) = a(-2)^2(-3) = -12a \Leftrightarrow a = -\frac{1}{3}.$$

So

$$P(x) = -\frac{1}{3}(x-1)^2(x-2)$$
$$= -\frac{1}{3}x^3 + \frac{4}{3}x^2 - \frac{5}{3}x + \frac{2}{3}.$$

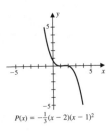

$P(x) = -\frac{1}{3}(x-2)(x-1)^2$

35. First, we must have for some constant a,

$$P(x) = a(x+2)x^2(x-2) = a(x^4 - 4x^2),$$

and since $(1, 2)$ is on the graph,

$$2 = P(1) = a(1-4) = -3a \Leftrightarrow a = -\frac{2}{3}.$$

The polynomial is

$$P(x) = -\frac{2}{3}x^4 + \frac{8}{3}x^2.$$

37. For $f_n(x) = x^n$, if $0 < x < 1$, $f_n(x) = x^n > x^{n+1} = f_{n+1}(x)$ and if $x > 1$, $f_n(x) = x^n < x^{n+1} = f_{n+1}(x)$.

39. (a) $f(x) = x^4 + x^3 - 2x^2$ is increasing on $(-1.5, 0) \cup (0.7, \infty)$ and is decreasing on $(-\infty, -1.5) \cup (0, 0.7)$.

(b) Local minima: $(-1.5, -2.9), (0.7, -0.4)$; local maximum: $(0, 0)$

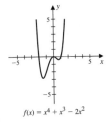

$f(x) = x^4 + x^3 - 2x^2$

41. (a) $f(x) = \frac{1}{5}x^5 - \frac{5}{4}x^4 + \frac{5}{3}x^3 + \frac{5}{2}x^2 - 6x + 1$ is increasing on $(-\infty, -1) \cup (1, 2) \cup (3, \infty)$ and is decreasing on $(-1, 1) \cup (2, 3)$.

(b) Local minima: $(1, -1.9), (3, -2.2)$; local maxima: $(-1, 6.4), (2, -1.2)$

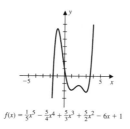

$$f(x) = \tfrac{1}{5}x^5 - \tfrac{5}{4}x^4 + \tfrac{5}{3}x^3 + \tfrac{5}{2}x^2 - 6x + 1$$

43.   For $f(x) = x^3 - 2x^2 + x + 2$ with $a = -3$ and $b = 2$.

(a) $y = f(x)$ (b) $y = -f(x)$ (c) $y = f(-x)$
(d) $y = |f(x)|$ (e) $y = f(x - 3) + 2$

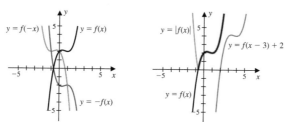

45.   For $f(x) = x^4 - 400x^2$ with $a = -10$ and $b = 30000$.

(a) $y = f(x)$ (b) $y = -f(x)$ (c) $y = f(-x)$
(d) $y = |f(x)|$ (e) $y = f(x - 10) + 30000$

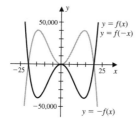
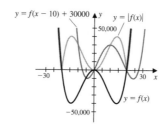

47.   (a) $V(x) = x(20 - 2x)(20 - 2x) = 400x - 80x^2 + 4x^3$

(b) The graph shows $y = -500 + 400x - 80x^2 + 4x^3$. The graph crosses the x-axis at $x = 5$ or $x \approx 1.9$, which are values of x that produce a volume of 500 in^3.

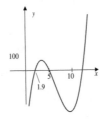

2.3 Finding Factors and Zeros of Polynomials

A polynomial $D(x)$ is a *factor* of another polynomial $P(x)$, provided there is a third polynomial $Q(x)$ with

$$P(x) = D(x) \cdot Q(x).$$

In general, $D(x)$ may not divide evenly into $P(x)$, in which case there is a remainder term $R(x)$, and

$$P(x) = D(x) \cdot Q(x) + R(x).$$

For this case, we can also write

$$\frac{P(x)}{D(x)} = Q(x) + \frac{R(x)}{D(x)}.$$

This is similar to the situation of the long division of positive real numbers. If $0 < d \le a$, we can divide d into a to obtain

$$a = dq + r, \quad \text{or} \quad \frac{a}{d} = q + \frac{r}{d},$$

for some number $0 < r < d$. For example, if 7 is divided by 2, then

$$7 = 2 \cdot 3 + 1, \quad \text{or} \quad \frac{7}{2} = 3 + \frac{1}{2}.$$

The standard terminology used is

	Divisor	Dividend	Quotient	Remainder
Real Numbers a/d	d	a	q	r
Polynomials $P(x)/D(x)$	$D(x)$	$P(x)$	$Q(x)$	$R(x)$

DIVISION OF POLYNOMIALS

EXAMPLE 2.3.1   Divide the polynomial $P(x) = x^3 - 2x^2 + x + 1$ by $x - 1$, and specify the quotient and remainder.

SOLUTION:

$$
\begin{array}{r}
x^2 - x \\
x - 1 \overline{\big)\, \boxed{x^3} - 2x^2 + x + 1} \\
\underline{x^3 - x^2} \\
\boxed{-x^2} + x + 1 \\
\underline{-x^2 + x} \\
1
\end{array}
$$

The quotient is

$$Q(x) = x^2 - x,$$

and the remainder is

$$R(x) = 1,$$

with

$$P(x) = x^3 - 2x^2 + x + 1 = (x - 1)(x^2 - x) + 1.$$

The division process ends when the degree of the remainder is less than the degree of the divisor. The key terms in the division process are highlighted in boxes. At each step, we determine what has to be multiplied by the first term in the divisor to yield the term in the box. So, for the first step, $x \cdot x^2 = x^3$, and at the second step, $x \cdot (-x) = -x^2$. Then we multiply each term of the divisor by the same amount and subtract the result. So for example, after the first step, perform the subtraction

$$x^3 - 2x^2$$
$$- \ \underline{x^3 - x^2}.$$

The result is

$$x^3 - 2x^2 - (x^3 - x^2) = x^3 - 2x^2 - x^3 + x^2 = -x^2.$$

◇

EXAMPLE 2.3.2   Let $P(x) = 5x^4 + 4x^3 - 2x^2 + x + 3$, and $D(x) = x^2 + x - 2$. Find the quotient $Q(x)$ and remainder $R(x)$ when $P(x)$ is divided by $D(x)$.

SOLUTION:

$$
\begin{array}{r}
5x^2 - x + 9 \\
x^2 + x - 2 \overline{\smash{\big)}\ 5x^4 + 4x^3 - 2x^2 + x + 3} \\
\underline{5x^4 + 5x^3 - 10x^2} \\
-x^3 + 8x^2 + x + 3 \\
\underline{-x^3 - x^2 + 2x} \\
9x^2 - x + 3 \\
\underline{9x^2 + 9x - 18} \\
-10x + 21
\end{array}
$$

The quotient is

$$Q(x) = 5x^2 - x + 9,$$

the remainder is

$$R(x) = -10x + 21,$$

and

$$5x^4 + 4x^3 - 2x^2 + x + 3 = (5x^2 - x + 9)(x^2 + x - 2) - 10x + 21.$$

◇

THE FACTOR THEOREM

The Factor Theorem states that if a polynomial $P(x)$ is divided by $x - c$, then the remainder is $P(c)$. In the special case when $P(c) = 0$, we have $x - c$ is a factor of $P(x)$ if and only if c is a zero of $P(x)$.

EXAMPLE 2.3.3   Let $P(x) = 3x^4 - 14x^3 + 19x^2 - 4x - 4$, $c = 1, c = 2$. Use the Factor Theorem to show that $x - c$ is a factor for the given values of c, and factor $P(x)$ completely.

SOLUTION: It is not enough to simply verify that $x = 1$ and $x = 2$ are zeros, that is $P(1) = 0$ and $P(2) = 0$. We need to perform long division so we can also factor the polynomial. So

$$3x^3 - 11x^2 + 8x + 4$$

$$x - 1 \overline{)\, 3x^4 - 14x^3 + 19x^2 - 4x - 4}$$

$$\underline{3x^4 - 3x^3}$$

$$-11x^3 + 19x^2 - 4x - 4$$

$$\underline{-11x^3 + 11x^2}$$

$$8x^2 - 4x - 4$$

$$\underline{8x^2 - 8x}$$

$$4x - 4$$

$$\underline{4x - 4}$$

$$0$$

Since the remainder is 0,

$$P(x) = (x - 1)(3x^3 - 11x^2 + 8x + 4).$$

To show that 2 is a zero of $P(x)$, it is enough to show it is a zero of $3x^3 - 11x^2 + 8x + 4$. After dividing the factor $(x - 2)$ into $3x^3 - 11x^2 + 8x + 4$, we have

$$P(x) = (x - 1)(3x^3 - 11x^2 + 8x + 4)$$

$$= (x - 1)(x - 2)(3x^2 - 5x - 2)$$

$$= (x - 1)(x - 2)(3x + 1)(x - 2).$$

SYNTHETIC DIVISION

Synthetic division is a process which simplifies the long division of polynomials in cases where the divisor has the form $x - c$. It can be used to find quotients and remainders, solving polynomial equations $P(x) = 0$, and for evaluating polynomials at specific values.

To introduce the technique of synthetic division, let's begin with another example of long division. We will divide $x - 2$ into $x^5 - 3x^4 + 3x^3 - x^2 + x - 4$. The long division is

$$
\begin{array}{r}
x^4 - x^3 + x^2 + x + 3 \\
x - 2\overline{\smash{\big)}\ x^5 - 3x^4 + 3x^3 - x^2 + x - 4} \\
\underline{x^5 - 2x^4} \\
-x^4 + 3x^3 - x^2 + x - 4 \\
\underline{-x^4 + 2x^3} \\
x^3 - x^2 + x - 4 \\
\underline{x^3 - 2x^2} \\
x^2 + x - 4 \\
\underline{x^2 - 2x} \\
3x - 4 \\
\underline{3x - 6} \\
2
\end{array}
$$

The quotient is

$$Q(x) = x^4 - x^3 + x^2 + x + 3,$$

and the remainder is

$$R(x) = 2.$$

Recall that by the Factor Theorem, if a polynomial $P(x)$ is divided by the linear term $x - c$, then the remainder is $P(c)$. So, $P(2) = 2$.

At each step of the long division process, the first columns always cancel on subtraction, so only the second columns are important for the result. If terms for every power of x are included, 0 for any missing powers, only the coefficients are needed and the variable terms can be dropped. The column will indicate the power of x.

What is recorded in or first step in the synthetic division process are the coefficients of the terms of the dividend and the constant term, $-c$, of the divisor, $x - c$. In the long division example, we start with the form

$$\underline{-2\,|\,1 \quad -3 \quad 3 \quad -1 \quad 1 \quad -4}$$

$$1$$

which records the coefficients of the dividend $x^5 - 3x^4 + 3x^3 - x^2 + x - 4$, and the constant term $-c = -2$ from the divisor, $x - 2$. To start the process, drop the first coefficient 1 down to the third row.

Now multiply the first term in row 3 by -2, and place it in column 2 of row 2. Subtract the terms in column 2 recording the result in column 2 of row 3. We now have the form

$$\begin{array}{r} -2\,|\,1 \quad -3 \quad 3 \quad -1 \quad 1 \quad -4 \\ \underline{\quad\quad -2 \quad\quad\quad\quad\quad\quad\quad} \\ 1 \quad -1 \end{array}$$

Repeat the process with each successive column to obtain

$$
\begin{array}{r|rrrrrr}
-2 & 1 & -3 & 3 & -1 & 1 & -4 \\
 & & -2 & 2 & -2 & -2 & -6 \\
\hline
 & 1 & -1 & 1 & 1 & 3 & 2
\end{array}
$$

The coefficients of the quotient and remainder are read from row 3. The remainder is the constant in the far right column, so $R(x) = 2$ and the quotient, in decreasing powers of x is, $Q(x) = x^4 - x^3 + x^2 + x + 3$.

In the procedure as outlined thusfar, the second term in each column is subtracted from the one above. To avoid possible subtraction errors, we make the final modification in the synthetic division process. We change the -2 in the divisor to 2, and **add**, rather than subtract, the columns. The results are the same, as shown below.

$$
\begin{array}{r|rrrrrr}
2 & 1 & -3 & 3 & -1 & 1 & -4 \\
 & & 2 & -2 & 2 & 2 & 6 \\
\hline
 & 1 & -1 & 1 & 1 & 3 & 2
\end{array}
$$

EXAMPLE 2.3.4   Find the quotient and remainder when $x^4 + 2x^3 - 3x^2 + 2x - 5$ is divided by $x + 3$.

SOLUTION: First note the divisor has the form

$$x + 3 = x - (-3).$$

When setting up synthetic division, in order to be able to add columns, use (-3) for the multiplier.

$$
\begin{array}{r|rrrrr}
-3 & 1 & 2 & -3 & 2 & -5 \\
   &   & -3 & 3 & 0 & -6 \\
\hline
   & 1 & -1 & 0 & 2 & -11
\end{array}
$$

The quotient is

$$Q(x) = x^3 - x^2 + 2,$$

and the remainder is

$$R(x) = -11.$$

◇

EXAMPLE 2.3.5   Let $P(x) = x^6 - 2x^4 + x^3 - 3x + 5$. Find $P(3)$.

SOLUTION: To find $P(3)$, we could simply substitute 3 for x in the expression for $P(x)$ and do the arithmetic. But since the remainder on division of $P(x)$ by $x - 3$ is also $P(3)$, synthetic division can be used to organize the work involved.

The polynomial is missing an x^5 and an x^2 term, so remember to use zeros for the coefficients of these terms in the synthetic division process. The synthetic division takes the form

$$
\begin{array}{r|rrrrrrr}
3 & 1 & 0 & -2 & 1 & 0 & -3 & 5 \\
  &   & 3 & 9 & 21 & 66 & 198 & 585 \\
\hline
  & 1 & 3 & 7 & 22 & 66 & 195 & 590
\end{array}
$$

and we have $P(3) = 590$.   ◇

EXAMPLE 2.3.6   Is $x - 3$ a factor of the polynomial $P(x) = x^5 - x^4 - 6x^3 + 4x - 12$?

SOLUTION: The linear factor $x - 3$ is a factor of the polynomial $P(x)$ if when divided into $P(x)$, the remainder is 0.

$$
\begin{array}{r|rrrrrr}
3 & 1 & -1 & -6 & 0 & 4 & -12 \\
  &   & 3 & 6 & 0 & 0 & 12 \\
\hline
  & 1 & 2 & 0 & 0 & 4 & 0
\end{array}
$$

So the remainder is $0, x - 3$ is a factor, and

$$P(x) = (x - 3)(x^4 + 2x^3 + 4).$$

◇

THE RATIONAL ZERO TEST

The rational numbers that are possible zeros of the polynomial

$$P(x) = a_n x^n + a_{n-1} x^{n-1} + \cdots + a_0$$

must be of the form

$$\frac{\pm \text{ factors of } a_0}{\pm \text{ factors of } a_n}.$$

EXAMPLE 2.3.7   Let

$$P(x) = 6x^5 - 8x^3 + 4x^2 + 5x - 4.$$

Determine all possibilities for rational zeros of $P(x)$.

SOLUTION: The factors of the constant term are $1, 2$, and 4, and the factors of the leading coefficient are $1, 2, 3$, and 6. So the possible rational zeros are

$$\frac{\pm 1, \pm 2, \pm 4}{\pm 1, \pm 2, \pm 3, \pm 6} = \pm 1, \pm \frac{1}{2}, \pm \frac{1}{3}, \pm \frac{1}{6}, \pm 2, \pm \frac{2}{3}, \pm 4, \pm \frac{4}{3}.$$

$\diamond$

Graphing devices are very useful here in eliminating possibilities from the list of possible rational zeros, which can often be a very extensive list.

EXAMPLE 2.3.8   Find all rational and irrational zeros of the polynomial $P(x) = x^3 - 2x^2 - 5x + 6$, and factor the polynomial completely.

SOLUTION: The possible rational zeros are

$$\frac{\pm \text{ factors of } 6}{\pm \text{ factors of } 1} = \pm 1, \pm 2, \pm 3, \pm 6.$$

By inspection we can see that $P(1) = 1 - 2 - 5 + 6 = 0$, so $x - 1$ is a factor of the polynomial. This can also be seen from the graph shown in the figure. The graph also indicates there are zeros at $x = -2$ and $x = 3$. Dividing the polynomial by $x - 1$, we get

$$
\require{enclose}
\begin{array}{r}
x^2 - x - 6 \\
x - 1 \enclose{longdiv}{x^3 - 2x^2 - 5x + 6} \\
\underline{x^3 - x^2} \\
-x^2 - 5x + 6 \\
\underline{-x^2 + x} \\
-6x + 6 \\
\underline{-6x + 6} \\
0
\end{array}
$$

or by using synthetic division

$$\begin{array}{r|rrrr}
1 & 1 & -2 & -5 & 6 \\
 & & 1 & -1 & -6 \\
\hline
 & 1 & -1 & -6 & 0.
\end{array}$$

So

$$P(x) = (x-1)(x^2 - x - 6) = (x-1)(x-3)(x+2),$$

and the zeros are $x = 1, x = 3$, and $x = -2$. There are no irrational zeros for this polynomial.

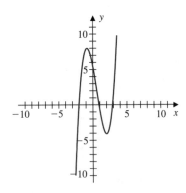

Figure 2.22

$\diamond$

EXAMPLE 2.3.9 Find all rational and irrational zeros of the polynomial $P(x) = x^5 - 2x^3 + 2x^2 - 3x + 2$, and factor the polynomial completely.

SOLUTION: The possible rational zeros are $\pm 1, \pm 2$. The graph indicates a zero at $x = -2$ and another zero, of multiplicity at least 2, at $x = 1$. If we divide the polynomial by $(x-1)^2$, we get

$$P(x) = (x-1)^2(x^3 + 2x^2 + x + 2).$$

Here we can use synthetic division to quickly show that $x = 1$ is a zero of multiplicity 2. First divide $P(x)$ by $x - 1$.

$$
\begin{array}{r|rrrrrr}
\underline{1} & 1 & 0 & -2 & 2 & -3 & 2 \\
 &   & 1 & 1 & -1 & 1 & -2 \\
\hline
 & 1 & 1 & -1 & 1 & -2 & 0
\end{array}
$$

Now use synthetic division to divide the quotient $Q(x) = x^4 + x^3 - x^2 + x - 2$ by $x - 1$.

$$
\begin{array}{r|rrrrr}
\underline{1} & 1 & 1 & -1 & 1 & -2 \\
 &   & 1 & 2 & 1 & 2 \\
\hline
 & 1 & 2 & 1 & 2 & 0
\end{array}
$$

So $P(x) = (x-1)^2(x^3 + 2x^2 + x + 2)$. Next divide $x^3 + 2x^2 + x + 2$ by $x + 2$.

$$
\begin{array}{r|rrrr}
\underline{-2} & 1 & 2 & 1 & 2 \\
 &   & -2 & 0 & -2 \\
\hline
 & 1 & 0 & 1 & 0
\end{array}
$$

Since the remainder is 0, $x + 2$ is a factor of $x^3 + 2x^2 + x + 2$ and consequently of $P(x)$, with

$$P(x) = (x-1)^2(x^3 + 2x^2 + x + 2) = (x-1)^2(x+2)(x^2 + 1).$$

The polynomial is now in complete factored form since $x^2 + 1$ has no real factors.

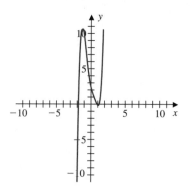

Figure 2.23

◇

EXAMPLE 2.3.10   Find all rational and irrational zeros of the polynomial $P(x) = 2x^5 - 10x^4 + 7x^3 + 28x^2 - 45x + 18$, and factor the polynomial completely.

SOLUTION: The possible rational zeros are

$$\pm 1, \pm 2, \pm 3, \pm 6, \pm 9, \pm 18, \pm \frac{1}{2}, \pm \frac{3}{2}, \pm \frac{9}{2}.$$

The list is extensive, but we could simply substitute each value into the polynomial until we hopefully find a zero, perform long division and continue until the last factor is a quadratic which can be handled. A graphing device will reduce the process in this example very quickly.

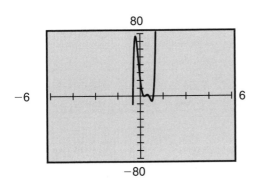

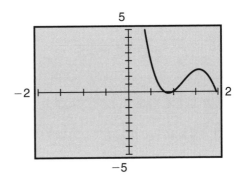

Figure 2.24   Figure 2.25

The graph in figure (a) indicates that there is a zero between -2 and -1, a zero of multiplicity 2 at $x = 1$ and zeros at $x = 2$ and $x = 3$. The close up view near $x = 1$ in figure (b) indicates $x = 1$ is *not* a root of multiplicity 2! Then

$$(x - 1)(x - 2)(x - 3) = x^3 - 6x^2 + 11x - 6,$$

which when divided into $P(x)$ gives

$$P(x) = (x^3 - 6x^2 + 11x - 6)(2x^2 + 2x - 3).$$

To factor the quadratic, we use the quadratic formula. The zeros are

$$x = \frac{-2 \pm \sqrt{4 - 4(2)(-3)}}{4} = \frac{-2 \pm \sqrt{28}}{4}$$
$$= \frac{-2 \pm 2\sqrt{7}}{4} = \frac{-1 \pm \sqrt{7}}{2}.$$

The final factorization is

$$P(x) = (x^3 - 6x^2 + 11x - 6)(2x^2 + 2x - 3)$$
$$= (x - 1)(x - 2)(x - 3)\left(x - \left(\frac{-1 + \sqrt{7}}{2}\right)\right)\left(x - \left(\frac{-1 - \sqrt{7}}{2}\right)\right).$$

Note that the Rational Zero test is no help in finding the last two zeros since they are irrational numbers.   ◇

EXAMPLE 2.3.11   Sketch the graphs of $f(x) = x^3 - x^2$ and $g(x) = x^2 + x - 2$, and determine all the points of intersection.

SOLUTION: To graph the cubic $y = f(x)$, we factor to obtain

$$f(x) = x^2(x - 1),$$

so f has a root of multiplicity 2 at $x = 0$ and a root of multiplicity 1 at $x = 1$. To graph $y = g(x)$, place the quadratic in standard form,

$$
\begin{aligned}
g(x) &= x^2 + x - 2 \\
&= \left(x^2 + x\right) - 2 \\
&= \left(x^2 + x + \frac{1}{4} - \frac{1}{4}\right) - 2 \\
&= \left(x + \frac{1}{2}\right)^2 - 2 - \frac{1}{4} \\
&= \left(x + \frac{1}{2}\right)^2 - \frac{9}{4}.
\end{aligned}
$$

The graphs of f and g are shown below.

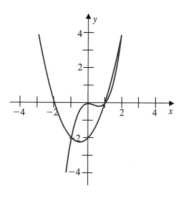

Figure 2.26

To find the points of intersection solve

$$x^3 - x^2 = x^2 + x - 2,$$

or

$$x^3 - 2x^2 - x + 2 = 0.$$

Possible rational zeros: $\pm 1, \pm 2$

The figure implies $x = 1$ is a zero.

$$
\require{enclose}
\begin{array}{r}
x^2 - \ x - 2 \\
x - 1 \enclose{longdiv}{x^3 - 2x^2 - x + 2} \\
\underline{x^3 - x^2} \\
-x^2 - x + 2 \\
\underline{-x^2 + x} \\
-2x + 2 \\
\underline{-2x + 2} \\
0
\end{array}
$$

or by synthetic division,

$$
\begin{array}{r|rrrr}
1 & 1 & -2 & -1 & 2 \\
  &   & 1 & -1 & -2 \\
\hline
  & 1 & -1 & -2 & 0.
\end{array}
$$

So

$$x^3 - 2x^2 - x + 2 = (x - 1)(x^2 - x - 2) = (x - 1)(x - 2)(x + 1),$$

and the points of intersection are

$$(1,0), (2,4), (-1,-2).$$

◇

EXAMPLE 2.3.12 Find a third degree polynomial $P(x)$ that has zeros at $x = -2$, $x = -1$, and $x = 2$, and whose x^2-term has coefficient 3.

SOLUTION: A polynomial with the desired zeros is

$$Q(x) = (x+2)(x+1)(x-2) = x^3 + x^2 - 4x - 4.$$

Multiplying a polynomial by a constant does not change the zeros of the polynomial. So to make the coefficient of the x^2-term 3, multiply $Q(x)$ by 3. Therefore,

$$P(x) = 3(x^3 + x^2 - 4x - 4) = 3x^3 + 3x^2 - 12x - 12.$$

◇

DESCARTE'S RULE OF SIGNS

This is a simple test that provides information about the possible *number* of positive real zeros and negative real zeros. If $P(x)$ is a polynomial with real coefficients,

(i) the number of positive zeros is either the number of variations in sign of $P(x)$ or less than this by an even number;

(ii) the number of negative zeros is either the number of variations in sign of $P(-x)$ or less than this by an even number.

EXAMPLE 2.3.13   Use Descarte's Rule of Signs to determine the maximum number of positive and negative zeros of the polynomial $P(x) = 3x^4 - 4x^3 - 7x^2 + x + 4$.

SOLUTION:

Variations in sign of $P(x)$: $P(x) = 3x^4 \underbrace{- 4x^3}_{+\text{ to }-} \underbrace{- 7x^2 + x}_{-\text{ to }+} + 4.$

There are two variations in sign, so the maximum number of positive real zeros is 2.

Variations in sign of $P(-x)$:

$$P(-x) = 3(-x)^4 - 4(-x)^3 - 7(-x)^2 + (-x) + 4$$

$$= 3x^4 + \underbrace{4x^3 - 7x^2}_{+\text{ to }-} \underbrace{- x + 4}_{-\text{ to }+}$$

The maximum number of negative real zeros is also 2.   ◇

Solutions for Exercise Set 2.3

1. We have $Q(x) = 3x + 1$ and $R(x) = 3$ since

$$
\begin{array}{r}
3x + 1 \\
x - 1 \overline{\smash{\big)}\, 3x^2 - 2x + 2} \\
\underline{3x^2 - 3x} \\
x + 2 \\
\underline{x - 1} \\
3
\end{array}
$$

3. We have $Q(x) = x^2 + 2x + 2$ and $R(x) = 0$ since

$$
\begin{array}{r}
x^2 + 2x + 2 \\
x - 1 \overline{\smash{\big)}\, x^3 + x^2 + 0x - 2} \\
\underline{x^3 - x^2} \\
2x^2 + 0x - 2 \\
\underline{2x^2 - 2x} \\
2x - 2 \\
\underline{2x - 2} \\
0
\end{array}
$$

5. We have $Q(x) = 2x^2 + 6$ and $R(x) = 7x + 4$ since

$$
\require{enclose}
\begin{array}{r}
2x^2 + 6 \\
x^2 - x - 1 \enclose{longdiv}{2x^4 - 2x^3 + 4x^2 + x - 2} \\
\underline{2x^4 - 2x^3 - 2x^2} \\
6x^2 + x - 2 \\
\underline{6x^2 - 6x - 6} \\
7x + 4
\end{array}
$$

7. We have $P(x) = x^3 - 5x^2 + 8x - 4$ and $P(1) = 1^3 - 5(1)^2 + 8(1) - 4 = 0$, so $x - 1$ is a factor. Dividing gives

$$
\require{enclose}
\begin{array}{r}
x^2 - 4x + 4 \\
x - 1 \enclose{longdiv}{x^3 - 5x^2 + 8x - 4} \\
\underline{x^3 - x^2} \\
-4x^2 + 8x - 4 \\
\underline{-4x^2 + 4x} \\
4x - 4 \\
\underline{4x - 4} \\
0
\end{array}
$$

and $P(x) = (x-1)(x^2 - 4x + 4) = (x-1)(x-2)(x-2) = (x-1)(x-2)^2$.

9. We have

$$P(x) = 2x^4 + 3x^3 - 12x^2 - 7x + 6 \quad \text{and} \quad P(-1) = 2 - 3 - 12 + 7 + 6 = 0,$$

so $x + 1$ is a factor. Dividing gives

$$
\begin{array}{r}
2x^3 + x^2 - 13x + 6 \\
x + 1\overline{)\,2x^4 + x^3 - 12x^2 - 7x + 6} \\
\underline{2x^4 + 2x^3} \qquad\qquad\qquad \\
x^3 - 12x^2 - 7x + 6 \\
\underline{x^3 + x^2} \qquad\qquad \\
-13x^2 - 7x + 6 \\
\underline{-13x^2 - 13x} \qquad \\
6x + 6 \\
\underline{6x + 6} \\
0
\end{array}
$$

and

$$
P(x) = (x+1)(2x^3 + x^2 - 13x + 6).
$$

If $x = -3$ is a zero of $P(x)$, then it will also be a zero of $Q(x) = 2x^3 + x^2 - 13x + 6$. Since

$$
Q(-3) = 2(-3)^3 + (-3)^2 - 13(-3) + 6 = 0,
$$

$x + 3$ is a factor of $Q(x)$ and also of $P(x)$. Then

$$
\begin{array}{r}
2x^2 - 5x + 2 \\
x + 3{\overline{\smash{\big)}\,2x^3 + x^2 - 13x + 6}} \\
\underline{2x^3 + 6x^2} \\
-5x^2 - 13x + 6 \\
\underline{-5x^2 - 15x} \\
2x + 6 \\
\underline{2x + 6} \\
0
\end{array}
$$

and

$$
P(x) = (x + 1)(x + 3)(2x^2 - 5x + 2) = (x + 1)(x + 3)(2x - 1)(x - 2).
$$

11. We have $P(x) = 3x^3 + x^2 - 8x + 4$ and

$$
P\left(\frac{2}{3}\right) = 3\left(\frac{8}{27}\right) + \frac{4}{9} - \frac{16}{3} + 4 = \frac{8}{9} + \frac{4}{9} - \frac{48}{9} + \frac{36}{9} = 0
$$

so $x - \frac{2}{3}$ is a factor. Dividing gives

$$
\begin{array}{r}
3x^2 + 3x - 6 \\
x - \frac{2}{3}{\overline{\smash{\big)}\,3x^3 + x^2 - 8x + 4}} \\
\underline{3x^3 - 2x^2} \\
3x^2 - 8x + 4 \\
\underline{3x^2 - 2x} \\
-6x + 4 \\
\underline{-6x + 4} \\
0
\end{array}
$$

and

$$P(x) = \left(x - \frac{2}{3}\right)(3x^2 + 3x - 6) = \left(x - \frac{2}{3}\right)3(x^2 + x - 2)$$

$$= 3\left(x - \frac{2}{3}\right)(x + 2)(x - 1) = 3\frac{1}{3}(3x - 2)(x + 2)(x - 1)$$

$$= (3x - 2)(x + 2)(x - 1).$$

13. The Rational Root Test gives

$$\frac{\text{factors of } 6}{\text{factors of } 1} = \frac{\pm 1, \pm 2, \pm 3, \pm 6}{\pm 1} = \pm 1, \pm 2, \pm 3, \pm 6.$$

15. Since the rational zeros will be the same if we divide the equation by 2, consider the equation $5x^5 - 7x^3 + 9x^2 + 3x - 2 = 0$. We have

$$\frac{\text{factors of } 2}{\text{factors of } 5} = \frac{\pm 1, \pm 2}{\pm 1, \pm 2, \pm 5} = \pm 1, \pm 2, \pm \frac{1}{2}, \pm \frac{1}{5}, \pm \frac{2}{5}.$$

17. Since the rational zeros will be the same if we multiply the equation by 3, consider the equation $2x^3 - 19x^2 + 54x - 45 = 0$.

$$\frac{\text{factors of } 45}{\text{factors of } 2} = \frac{\pm 1, \pm 3, \pm 5, \pm 9, \pm 15, \pm 45}{\pm 1, \pm 2}$$

$$= \pm 1, \pm 3, \pm 5, \pm 9, \pm 15, \pm 45, \pm \frac{1}{2}, \pm \frac{3}{2}, \pm \frac{5}{2}, \pm \frac{9}{2}, \pm \frac{15}{2}, \pm \frac{45}{2}$$

19. Since

$$P(x) = 6x^4 \underbrace{+5x^3 - 14x^2 + x}_{+ \text{ to } - \text{ to } +} + 2$$

has two sign changes, the polynomial has a maximum of two positive roots. Also, since

$$P(-x) = \underbrace{6x^4 - 5x^3}_{+ \text{ to } -} - 14x^2 \underbrace{- x + 2}_{- \text{ to } +}$$

has two sign changes, the polynomial has a maximum of two negative roots.

21. Since

$$P(x) = x^5 + \underbrace{2x^4 - x}_{+ \text{ to } -} - 2$$

has one sign change, the polynomial has a maximum of one positive root. Also, since

$$P(-x) = \underbrace{-x^5 + 2x^4}_{- \text{ to } +} + \underbrace{x - 2}_{+ \text{ to } -}$$

has two sign changes, the polynomial has a maximum of two negative roots.

23. For the polynomial $P(x) = x^3 - 3x^2 + 4$ the possible rational zeros are $\pm 1, \pm 2, \pm 4$. From substitution or from a computer generated graph, the polynomial has zeros at $x = -1$, and 2 with the zero at $x = 2$ of multiplicity 2. Dividing $(x + 1)$ into $P(x)$ gives

$$P(x) = (x + 1)(x^2 - 4x + 4) = (x + 1)(x - 2)^2.$$

25. Since

$$P(x) = 2x^4 - 3x^3 - 3x^2 + 2x = x(2x^3 - 3x^2 - 3x + 2),$$

$P(x)$ has a zero at $x = 0$. The possible rational zeros of $Q(x) = 2x^3 - 3x^2 - 3x + 2$ are

$$\frac{\pm 1, \pm 2}{\pm 1, \pm 2} = \pm 1, \pm 2, \pm \frac{1}{2}.$$

From substitution or from a computer generated graph, the polynomial $Q(x)$ has zeros at $x = -1, \frac{1}{2}, 2$. Dividing $(x + 1)$ into $Q(x)$ gives

$$P(x) = x(x + 1)(2x^2 - 5x + 2) = x(x + 1)(2x - 1)(x - 2).$$

27. For the polynomial

$$P(x) = 2x^5 + x^4 - 12x^3 + 10x^2 + 2x - 3$$

the possible rational zeros are

$$\frac{\pm 1, \pm 2, \pm 3}{\pm 1, \pm 2} = \pm 1, \pm 2, \pm 3, \pm \frac{1}{2}, \pm \frac{3}{2}.$$

From substitution or from a computer generated graph, the polynomial has zeros at $x = 1$, of multiplicity 3, and zeros at $x = -3, -\frac{1}{2}$. So

$$P(x) = (x - 1)^3 (2x + 1)(x + 3).$$

29. For the polynomial $P(x) = x^3 + x^2 - 3x + 1$ the possible rational zeros are ± 1. From substitution or from a computer generated graph, the polynomial has a zero at $x = 1$. Dividing $(x - 1)$ into $P(x)$ gives

$$P(x) = (x - 1)(x^2 + 2x - 1).$$

To factor the quadratic requires the quadratic formula to find the roots of $x^2 + 2x - 1 = 0$. That is,

$$
\begin{aligned}
x &= \frac{-2 \pm \sqrt{4 - 4(1)(-1)}}{2} = \frac{-2 \pm \sqrt{8}}{2} \\
&= \frac{-2 \pm 2\sqrt{2}}{2} = -1 \pm \sqrt{2},
\end{aligned}
$$

and

$$P(x) = (x - 1)\left(x - \left(-1 + \sqrt{2}\right)\right)\left(x - \left(-1 - \sqrt{2}\right)\right).$$

31. For the polynomial $P(x) = 4x^3 - 12x^2 + 9x - 1$ the possible rational zeros are $\pm 1, \pm \frac{1}{2}, \pm \frac{1}{4}$. From substitution or from a computer generated graph, the polynomial has a zero at $x = 1$. Dividing $(x - 1)$ into $P(x)$ gives

$$P(x) = (x - 1)(4x^2 - 8x + 1).$$

To factor the quadratic requires the quadratic formula to find the roots of $4x^2 - 8x + 1 = 0$. That is,

$$x = \frac{8 \pm \sqrt{64 - 4(4)(1)}}{8} = \frac{8 \pm \sqrt{48}}{8}$$

$$= \frac{8 \pm 4\sqrt{3}}{8} = 1 \pm \frac{\sqrt{3}}{2},$$

and

$$P(x) = (x - 1)\left(x - \left(1 + \frac{\sqrt{3}}{2}\right)\right)\left(x - \left(1 - \frac{\sqrt{3}}{2}\right)\right).$$

33. The curves $y = f(x) = x^3$ and $y = g(x) = 2x^2 + x - 2$ intersect if and only if $f(x) = g(x) \Leftrightarrow x^3 = 2x^2 + x - 2 \Leftrightarrow x^3 - 2x^2 - x + 2 = 0$. For the polynomial $P(x) = x^3 - 2x^2 - x + 2$ the possible rational zeros are $\pm 1, \pm 2$. From substitution or from a computer generated graph, the polynomial has a zero at $x = 1$. Dividing $(x - 1)$ into $P(x)$ gives

$$P(x) = (x - 1)(x^2 - x - 2)$$

$$= (x - 1)(x + 1)(x - 2).$$

The points of intersection are $(-1, -1), (1, 1)$, and $(2, 8)$. To sketch the parabola the quadratic in standard form is

$$2x^2 + x - 2 = 2\left(x^2 + \frac{1}{2}x\right) - 2$$

$$= 2\left(x^2 + \frac{1}{2}x + \frac{1}{16} - \frac{1}{16}\right) - 2$$

$$= 2\left(x + \frac{1}{4}\right)^2 - \frac{17}{8}.$$

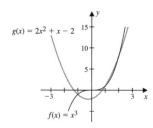

35. The curves $y = f(x) = x^3 - 1$ and $y = g(x) = 6x + 3$ intersect if and only if $f(x) = g(x) \Leftrightarrow x^3 - 1 = 6x + 3 \Leftrightarrow x^3 - 6x - 4 = 0$. For the polynomial $P(x) = x^3 - 6x + 4$ the possible rational zeros are $\pm 1, \pm 2, \pm 4$. From substitution or from a computer generated graph, the polynomial has a zero at $x = -2$. Dividing $(x + 2)$ into $P(x)$ gives

$$P(x) = (x + 2)(x^2 - 2x - 2).$$

To factor the quadratic requires the quadratic formula to find the roots of $x^2 - 2x - 2 = 0$. That is,

$$x = \frac{2 \pm \sqrt{4 - 4(1)(-2)}}{2} = \frac{2 \pm \sqrt{12}}{2}$$

$$= \frac{2 \pm 2\sqrt{3}}{2} = 1 \pm \sqrt{3}.$$

and

$$P(x) = (x + 2)\left(x - \left(1 + \sqrt{3}\right)\right)\left(x - \left(1 - \sqrt{3}\right)\right).$$

The points of intersection are $(-2, -9)$, $\left(1 + \sqrt{3}, \left(1 + \sqrt{3}\right)^3 - 1\right)$, and $\left(1 - \sqrt{3}, \left(1 - \sqrt{3}\right)^3 - 1\right)$.

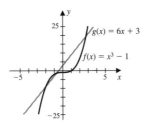

37. Since the polynomial has zeros of multiplicity 1 at $x = -3, x = -1, x = 1$, and $x = 4$, we have

$$P(x) = a(x+3)(x+1)(x-1)(x-4).$$

Since the graph passes through $(2, -3)$, we have

$$-3 = P(2) = a(5)(3)(1)(-2) = -30a \Rightarrow a = \frac{1}{10}.$$

So

$$P(x) = \frac{1}{10}(x+3)(x+1)(x-1)(x-4) = \frac{1}{10}x^4 - \frac{1}{10}x^3 - \frac{13}{10}x^2 + \frac{1}{10}x + \frac{6}{5}.$$

39. A fourth degree polynomial with zeros $x = -2 - 1, 0, 1$ and hence, the factors $(x+2), (x+1), x, (x-1)$, is

$$Q(x) = (x+1)x(x-1)(x+2) = x^4 + 2x^3 - x^2 - 2x.$$

Multiplying a polynomial by a constant does not change the zeros. To have the coefficient of the x^2 term be 5, multiply the polynomial by -5. So

$$P(x) = -5(x^4 + 2x^3 - x^2 - 2x) = -5x^4 - 10x^3 + 5x^2 + 10x.$$

41. We have

$$P(x) = (x-2)^3 \cdot x^2(x-1) = x^6 - 7x^5 + 18x^4 - 20x^3 + 8x^2.$$

43. If n is a positive integer with $n \geq 1$, then $(1)^n - 1 = 1 - 1 = 0$ and hence $x - 1$ is a factor of $x^n - 1$.

2.4 Rational Functions

A *rational function* has the form

$$f(x) = \frac{P(x)}{Q(x)},$$

where $P(x)$ and $Q(x)$ are polynomials. The domain of f is the set of all real numbers with $Q(x) \neq 0$.

DOMAINS AND RANGES OF RATIONAL FUNCTIONS

EXAMPLE 2.4.1   Determine the domain and range of the rational function whose graph is shown in the figure.

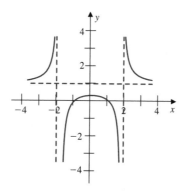

Figure 2.27

SOLUTION:

Domain: Any vertical line, except $x = -2$ and $x = 2$, will cross the graph of the function in one point, so the domain is $(-\infty, -2) \cup (-2, 2) \cup (2, \infty)$.

<u>Range:</u> Horizontal lines between the line $y = \frac{1}{4}$ and the line $y = 1$, including $y = 1$, will miss the graph, and so will be excluded from the range. The range is $\left(-\infty, \frac{1}{4}\right] \cup (1, \infty)$. ◇

EXAMPLE 2.4.2   Find the domain and any x- and y-intercepts of the rational function.

(a) $f(x) = \dfrac{(x-1)(x+2)}{(x+3)(x-4)}$ (b) $f(x) = \dfrac{x^2 - x - 6}{x^2 - 4}$

SOLUTION:

(a) <u>Domain:</u> All real numbers for which the denominator is *not* zero. So,

$$(x+3)(x-4) = 0 \Rightarrow x = -3 \quad \text{or} \quad x = 4,$$

and the domain is $(-\infty, -3) \cup (-3, 4) \cup (4, \infty)$.

<u>x-intercepts:</u> Solve $y = f(x) = 0$. The fraction will be 0 provided the numerator is 0, so

$$(x-1)(x+2) = 0 \Rightarrow x = 1 \quad \text{or} \quad x = -2,$$

and the graph will cross the x-axis at the points $(1, 0)$ and $(-2, 0)$.

<u>y-intercept:</u> Set $x = 0$, so $y = \dfrac{(-1)(2)}{(3)(-4)} = \dfrac{1}{6}$ and the graph crosses the y-axis at $\left(0, \frac{1}{6}\right)$.

(b) First factor the numerator and denominator,

$$f(x) = \frac{x^2 - x - 6}{x^2 - 4} = \frac{(x-3)(x+2)}{(x-2)(x+2)} = \frac{x-3}{x-2}, \quad \text{when} \quad x \neq 2.$$

<u>Domain:</u>

$$x - 2 = 0 \Rightarrow x = 2,$$

so $x = 2$ must be eliminated from the domain. The value $x = -2$ must also be eliminated from the domain since the original equation is not defined at $x = -2$.

The domain is $(-\infty, -2) \cup (-2, 2) \cup (2, \infty)$. The graph will have a hole at the point $\left(-2, \frac{5}{4}\right)$.

<u>x-intercepts:</u> Solve for $y = f(x) = 0$. We have

$$x - 3 = 0, \quad \text{so} \quad x = 3,$$

and the graph will cross the x-axis at the point $(3, 0)$.

<u>y-intercept:</u> Set $x = 0$. Then $y = \frac{3}{2}$, and the graph crosses the y-axis at $\left(0, \frac{3}{2}\right)$.

◇

HORIZONTAL AND VERTICAL ASYMPTOTES

An *asymptote* of a graph is a line that the graph approaches. A horizontal line $y = a$ is a *horizontal asymptote* to the graph of f, if $f(x) \to a$ as $x \to \infty$ or $x \to -\infty$. The vertical line $x = a$ is a *vertical asymptote* to the graph of f, if $f(x) \to \infty$ or $f(x) \to -\infty$ as x approaches a from the left side or the right side.

To find horizontal asymptotes, use the end behavior of the polynomials in the numerator and denominator of the rational function. For example,

$$f(x) = \frac{2x^3 - 2x^2 - 8x + 2}{3x^3 + x^2 - x + 5} \approx \frac{2x^3}{3x^3} = \frac{2}{3}, \quad \text{for } |x| \text{ large.}$$

This says that as x is selected further and further from 0, on the positive or negative x-axis, the values of $f(x)$ get closer to $\frac{2}{3}$. So the curve flattens to the horizontal line $y = \frac{2}{3}$.

Vertical asymptotes can only be vertical lines $x = a$, where the denominator of the rational function is 0 at $x = a$. For example,

$$f(x) = \frac{1}{(x - 1)(x + 2)}$$

will have vertical asymptotes $x = 1$ and $x = -2$.

As x gets close to 1 from the right, $x > 1$, so $x - 1 > 0$, but getting close to zero. The factor $(x + 2)$ approaches 3, so $(x - 1)(x + 2) > 0$ and approaching 0, and $f(x) = \frac{1}{(x-1)(x+1)}$ is getting arbitrarily large. This is written

$$f(x) \to \infty \quad \text{as} \quad x \to 1^{+}.$$

As x gets close to 1 from the left, $x < 1$, so $x - 1 < 0$, but getting close to zero. The factor $(x + 2)$ still approaches 3, so $(x - 1)(x + 2) < 0$ and approaching 0, and $f(x) = \frac{1}{(x-1)(x+1)}$ becomes arbitrarily large in magnitude but negative. This is written

$$f(x) \to -\infty \quad \text{as} \quad x \to 1^{-}.$$

The same analysis near the vertical line $x = -2$, gives

$$f(x) \to -\infty \quad \text{as} \quad x \to -2^{+}$$
$$f(x) \to \infty \quad \text{as} \quad x \to -2^{-}.$$

EXAMPLE 2.4.3   Sketch the graph of

$$f(x) = \frac{x^2 - x - 2}{x - 2},$$

labeling all horizontal and vertical asymptotes and x- and y-intercepts.

SOLUTION: First factor the numerator, so

$$f(x) = \frac{(x - 2)(x + 1)}{x - 2} = x + 1, \quad \text{when} \quad x \neq 2.$$

The value $x = 2$ is not in the domain of f since it makes the denominator 0, and it also cannot be included in the final definition even though the fraction simplified nicely. The graph has no horizontal or vertical asymptotes. The intercepts are $(0,1)$ and $(-1,0)$. The graph is the same as the graph of $y = x + 1$, except the point $(2,3)$ is removed.

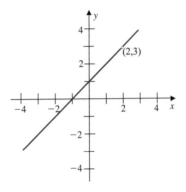

Figure 2.28

Be very careful not to assume a graph has a vertical asymptote whenever the denominator is 0. These are the possibilities, but if the numerator and denominator have a common factor that can be cancelled, a zero of the denominator will not yield a vertical asymptote.   ◇

EXAMPLE 2.4.4 Sketch the graph of

$$f(x) = \frac{x - 1}{x^2 - 2x - 3},$$

labeling any horizontal and vertical asymptotes, and x- and y-intercepts.

SOLUTION: Factoring the denominator we have,

$$f(x) = \frac{x - 1}{x^2 - 2x - 3} = \frac{x - 1}{(x - 3)(x + 1)}.$$

<u>Vertical Asymptotes:</u> Setting the denominator equal to 0,

$$(x - 3)(x + 1) = 0 \Rightarrow x = 3 \quad \text{or} \quad x = -1.$$

The vertical asymptotes are $x = 3$ and $x = -1$. To sketch the graph, analyze the behavior of the graph near the vertical asymptotes. Whether the graph goes to plus or minus infinity depends on the sign of the fraction.

As $x \to 3^+$, the sign of the fraction is $\dfrac{+}{(+)(+)}$, so $f(x) \to \infty$.

As $x \to 3^-$, the sign of the fraction is $\dfrac{+}{(-)(+)}$, so $f(x) \to -\infty$.

As $x \to -1^+$, the sign of the fraction is $\dfrac{-}{(-)(+)}$, so $f(x) \to \infty$.

As $x \to -1^-$, the sign of the fraction is $\dfrac{-}{(-)(-)}$, so $f(x) \to -\infty$.

<u>Horizontal Asymptotes:</u> The end behavior of the rational function is given by

$$f(x) = \frac{x - 1}{x^2 - 2x - 3} \approx \frac{x}{x^2} = \frac{1}{x},$$

so

$$f(x) \to 0 \quad \text{as} \quad x \to \infty$$

$$f(x) \to 0 \quad \text{as} \quad x \to -\infty.$$

The graph has a horizontal asymptote $y = 0$.

<u>x-intercepts:</u> Solve

$$f(x) = \frac{x - 1}{x^2 - 2x - 3} = 0$$

$$x - 1 = 0$$

so

$$x = 1.$$

<u>y-intercept:</u> Set $x = 0$, then $y = \frac{1}{3}$.

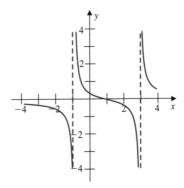

Figure 2.29

◇

EXAMPLE 2.4.5   Sketch the graph of

$$f(x) = \frac{x^2 - 4}{x^2 - 9},$$

labeling all horizontal and vertical asymptotes, and x- and y-intercepts.

SOLUTION:

$$f(x) = \frac{x^2 - 4}{x^2 - 9} = \frac{(x - 2)(x + 2)}{(x - 3)(x + 3)}.$$

<u>Vertical Asymptotes:</u> $x = -3, x = 3$

	Sign of Fraction	Behavior Near Asymptote
$x \to -3^+$	$\dfrac{(-)(-)}{(-)(+)}$	$f(x) \to -\infty$
$x \to -3^-$	$\dfrac{(-)(-)}{(-)(-)}$	$f(x) \to \infty$
$x \to 3^+$	$\dfrac{(+)(+)}{(+)(+)}$	$f(x) \to \infty$
$x \to 3^-$	$\dfrac{(+)(+)}{(-)(+)}$	$f(x) \to -\infty$

Horizontal Asymptote: $y = 1$

$$f(x) = \frac{x^2 - 4}{x^2 - 9} \approx \frac{x^2}{x^2} = 1.$$

x-intercepts: $(2,0), (-2,0)$

$$\frac{(x-2)(x+2)}{(x-3)(x+3)} = 0,$$

implies that

$$(x-2)(x+2) = 0.$$

So

$$x = 2 \quad \text{or} \quad x = -2.$$

y-intercept: $\left(0, \frac{4}{9}\right)$

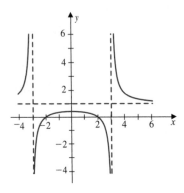

Figure 2.30

◇

SLANT ASYMPTOTES

If the degree of the denominator of a rational function is exactly one greater than the degree of the numerator, then the graph will approach a non-horizontal line as $x \to \infty$ and as $x \to -\infty$. Such a line is called a *slant asymptote* to the graph. The equation of the slant asymptote comes from the fact that in this case

$$f(x) = \frac{P(x)}{Q(x)} = (ax + b) + \frac{R(x)}{Q(x)},$$

where the degree of $R(x)$ is less than the degree of $Q(x)$. The quotient has the form $ax + b$, since the degree of $P(x)$ is exactly one greater than the degree of $Q(x)$.

EXAMPLE 2.4.6   Use a graphing device to sketch the graph of

$$f(x) = \frac{2x^2 + 3x - 1}{x + 2},$$

and show any vertical and slant asymptotes, and x- and y-intercepts.

SOLUTION:

There is a vertical asymptote at $x = -2$, and since the degree of the numerator is one greater than the degree of the denominator, we have a slant asymptote. Long division gives

$$
\begin{array}{r}
2x - 1 \\
x + 2 \overline{\smash{\big)}\ 2x^2 + 3x - 1} \\
\underline{2x^2 + 4x } \\
-x - 1 \\
\underline{-x - 2} \\
1
\end{array}
$$

or, by synthetic division,

$$\begin{array}{r|rrr} -2 & 2 & 3 & -1 \\ & & -4 & 2 \\ \hline & 2 & -1 & 1 \end{array}$$

so

$$f(x) = 2x - 1 + \frac{1}{x+2}.$$

Therefore,

$$\text{as } x \to \infty, \frac{1}{x+2} \to 0, \text{ so } f(x) \to 2x - 1,$$

and

$$\text{as } x \to -\infty, \frac{1}{x+2} \to 0, \text{ so } f(x) \to 2x - 1.$$

The slant asymptote is $y = 2x - 1$.   ◇

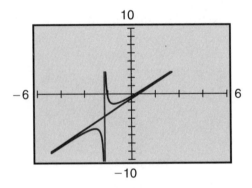

Figure 2.31

EXAMPLE 2.4.7   Determine a rational function that satisfies all of the following conditions:

(i) has the vertical asymptotes $x = 3$ and $x = -2$,

(ii) has the horizontal asymptote $y = 2$,

(iii) has x-intercepts at 2 and 4.

SOLUTION: For the rational function to have vertical asymptotes $x = 3$ and $x = -2$, the simplified form of the function must have a denominator that contains the factors $(x - 3)$ and $(x + 2)$. If the graph is to have the horizontal asymptote $y = 2$, the numerator and the denominator must have the same degree, and the leading coefficient of the numerator must be twice that of the denominator. If the graph is to cross the x-axis at 2 and 4, then the numerator must contain the factors $x - 2$ and $x - 4$. A possible function is then

$$f(x) = \frac{2(x - 2)(x - 4)}{(x - 3)(x + 2)} = \frac{2x^2 - 12x + 16}{x^2 - x - 6}.$$

◇

APPLICATIONS

EXAMPLE 2.4.8   A can in the shape of a right cylinder with radius r and height h is to have a volume of 60 inches^3. The cost of the top and the bottom of the can is 2 cents per square inch, and the cost for the sides is 3 cents per square inch. Express the cost of the box in terms of

(a) the variables r and h; (b) the variable r only; (c) the variable h only.

V=60 in ³

Figure 2.32

SOLUTION: The cost of a piece of material is the area of the piece times the price per square foot. The area of the top and bottom of the can is

$$A_{T,B} = 2(\pi r^2).$$

The area of the side of the can is the product of the height and the circumference of the can,

$$A_S = 2\pi rh.$$

(a) The cost is then

$$C = (2\pi r^2)(0.02) + (2\pi rh)(0.03) = 0.04\pi r^2 + 0.06\pi rh \text{ dollars.}$$

(b) To eliminate the variable h from the equation for cost, we need an equation relating h and r. We have yet to use the information that the volume of the box is 60 inches². The volume is

$$V = (\text{area of the base}) \times (\text{height})$$
$$= \pi r^2 h = 60,$$

so

$$h = \frac{60}{\pi r^2},$$

and

$$C(r) = 0.04\pi r^2 + 0.06\pi \left(\frac{60}{\pi r^2}\right) = 0.04\pi r^2 + \frac{3.6}{r} \text{ dollars.}$$

(c) Using the volume to solve for r in terms of h gives

$$\pi r^2 h = 60, \quad r^2 = \frac{60}{\pi h}, \quad \text{and} \quad r = \sqrt{\frac{60}{\pi h}}.$$

So

$$C(h) = 0.04\pi \left(\sqrt{\frac{60}{\pi h}}\right)^2 + 0.06\pi h \sqrt{\frac{60}{\pi h}} = \frac{2.4}{h} + 0.06\sqrt{60}\sqrt{\pi h} \text{ dollars.}$$

◇

Solutions for Exercise Set 2.4

1. (a) Domain: $\{x \mid x \neq 1\} = (-\infty, 1) \cup (1, \infty)$

(b) Range: $\{x \mid x \neq 0\} = (-\infty, 0) \cup (0, \infty)$

(c) Vertical asymptote: $x = 1$; horizontal asymptote: $y = 0$

3. (a) Domain: $\{x \mid x \neq -1, x \neq 1\} = (-\infty, -1) \cup (-1, 1) \cup (1, \infty)$

(b) Range: $(-\infty, 1) \cup [2, \infty)$

(c) Vertical asymptotes: $x = -1, x = 1$; horizontal asymptote: $y = 1$

5. (a) Domain: $(-\infty, 1)$

(b) Range: $(-\infty, \infty)$

(c) Vertical asymptote: $x = 1$; horizontal asymptote: none

7. For $f(x) = \dfrac{x - 3}{x + 1}$ we have:

(a) Domain: $\{x \mid x \neq -1\} = (-\infty, -1) \cup (-1, \infty)$

(b) x-intercepts: $(3, 0)$

Solve

$$\frac{x - 3}{x + 1} = 0 \Leftrightarrow x - 3 = 0 \Leftrightarrow x = 3.$$

y-intercept: $(0, -3)$

Set

$$x = 0 \Rightarrow y = \frac{0 - 3}{0 + 1} = -3.$$

(c) Vertical asymptote: $x = -1$; horizontal asymptote: $y = 1$, since for x large in magnitude

$$\frac{x - 1}{x + 1} \approx \frac{x}{x} = 1.$$

9. For

$$f(x) = \frac{2x^2 - 5x + 3}{x^2 - 4} = \frac{(2x - 3)(x - 1)}{(x - 2)(x + 2)}$$

we have:

(a) Domain: $\{x \ : \ x \neq -2, x \neq 2\} = (-\infty, -2) \cup (-2, 2) \cup (2, \infty)$

(b) x-intercepts: $\left(\frac{3}{2}, 0\right), (1, 0)$

Solve

$$\frac{(2x - 3)(x - 1)}{(x - 2)(x + 2)} = 0 \Leftrightarrow 2x - 3 = 0, x - 1 = 0 \Leftrightarrow x = \frac{3}{2}, x = 1.$$

y-intercept: $\left(0, -\frac{3}{4}\right)$;

Set

$$x = 0 \Rightarrow y = \frac{(-3)(-1)}{(-2)(2)} = -\frac{3}{4}.$$

(c) Vertical asymptotes: $x = -2, x = 2$; horizontal asymptote: $y = 2$, since for x large in magnitude

$$\frac{(2x - 3)(x - 1)}{(x - 2)(x + 2)} \approx \frac{2x^2}{x^2} = 2.$$

11. For

$$f(x) = \frac{x^3 - 2x^2 - x + 2}{x^2} = \frac{(x + 1)(x - 1)(x - 2)}{x^2}$$

we have:

(a) Domain: $\{x \mid x \neq 0\} = (-\infty, 0) \cup (0, \infty)$

(b) x-intercepts: $(-1, 0), (1, 0), (2, 0)$

Solve

$$\frac{(x + 1)(x - 1)(x - 2)}{x^2} = 0 \Leftrightarrow (x + 1)(x - 1)(x - 2) = 0 \Leftrightarrow x = -1, 1, 2.$$

y-intercept: None, since $x = 0$ is not in the domain of the function.

(c) Vertical asymptotes: $x = 0$; horizontal asymptote: None, since for x large in magnitude

$$\frac{(x + 1)(x - 1)(x - 2)}{x^2} \approx \frac{x^3}{x^2} = x.$$

13. Vertical asymptotes: $x = 1$; horizontal asymptote: $y = 0$; x-intercepts: none; y-intercept: $(0, -2)$

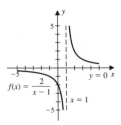

15. Vertical asymptotes: $x = 3$; horizontal asymptote: $y = 1$; x-intercepts: $(0, 0)$; y-intercept: $(0, 0)$

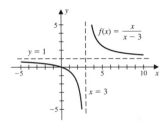

17. Vertical asymptotes: $x = -3, x = \frac{1}{3}$; horizontal asymptote: $y = \frac{2}{3}$; x-intercepts: $\left(-\frac{1}{2}, 0\right), (2, 0)$; y-intercept: $\left(0, \frac{2}{3}\right)$

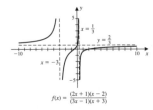

19. For
$$f(x) = \frac{x^2 - 9}{x^2 - 16} = \frac{(x-3)(x+3)}{(x-4)(x+4)}.$$

Vertical asymptotes: $x = 4, x = -4$; horizontal asymptote: $y = 1$;
x-intercepts: $(-3, 0), (3, 0)$; y-intercept: $(0, 9/16)$

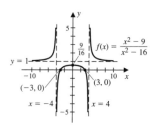

21. For
$$f(x) = \frac{x^2 + 2x - 3}{x - 1} = \frac{(x+3)(x-1)}{x - 1} = x + 3, \text{ for } x \neq 1.$$

Vertical asymptotes: None; horizontal asymptote: None; x-intercept:
$(-3, 0)$; y-intercept: $(0, 3)$

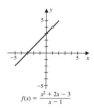

23. For
$$f(x) = \frac{x^2 - x - 2}{x^2 - 2x - 3} = \frac{(x-2)(x+1)}{(x-3)(x+1)} = \frac{x - 2}{x - 3}, \quad \text{for } x \neq -1.$$

Vertical asymptote: $x = 3$; horizontal asymptote: $y = 1$; x-intercepts:
$(2, 0)$; y-intercept: $(0, 2/3)$

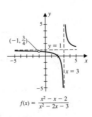

$$f(x) = \frac{x^2 - x - 2}{x^2 - 2x - 3}$$

25. For

$$f(x) = \frac{2x - 3}{x^2 - x - 6} = \frac{2x - 3}{(x - 3)(x + 2)}.$$

Vertical asymptotes: $x = -2, x = 3$; horizontal asymptote: $y = 0$;

x-intercepts: $(3/2, 0)$; y-intercept: $(0, 1/2)$

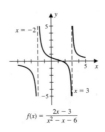

$$f(x) = \frac{2x - 3}{x^2 - x - 6}$$

27. $f(x) = \dfrac{1}{x - 1}$

29. One example is

$$f(x) = \frac{(x - 3)(x + 4)}{(x - 2)(x + 3)}.$$

31. One example is

$$f(x) = \frac{3(x - 2)}{x} = \frac{3x - 6}{x}.$$

33. $f(x) = \dfrac{x^2}{x - 1} = x + 1 + \dfrac{1}{x - 1}$

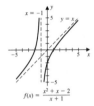

$f(x) = \dfrac{x^2}{x-1}$

35. $f(x) = \dfrac{x^2 + x - 2}{x + 1} = x - \dfrac{2}{x + 1}$

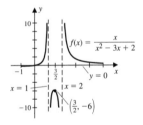

$f(x) = \dfrac{x^2 + x - 2}{x + 1}$

37. $f(x) = \dfrac{x}{x^2 - 3x + 2} = \dfrac{x}{(x - 1)(x - 2)}$

$f(x) = \dfrac{x}{x^2 - 3x + 2}$

39. $f(x) = \dfrac{x^2}{x^2 - 5x + 6} = \dfrac{x^2}{(x - 2)(x - 3)}$

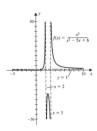

41. We have $A = \pi r^2 + 2\pi rh$ and $2000 = \pi r^2 h \Rightarrow h = \frac{2000}{\pi r^2}$, so

$$A(r) = \pi r^2 + 2\pi r \left(\frac{2000}{\pi r^2}\right) = \pi r^2 + \frac{4000}{r}.$$

From the graph of $y = A(r)$, the minimum point can be estimated to be

$$r \approx 8.6 \Rightarrow h = \frac{2000}{\pi r^2} \approx 8.6.$$

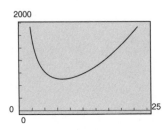

43. The number of bacteria at time t is $n = 10000 \left(\frac{3t^2+1}{t^2+1}\right)$.

(a) For t large,

$$n = 10000 \left(\frac{3t^2 + 1}{t^2 + 1}\right) \approx 10000 \left(\frac{3t^2}{t^2}\right) = 30000$$

so the bacteria colony stabilizes.

(b) The bacteria colony stabilizes to the level 30000.

(c) To find when the number of bacteria exceeds 22000, solve

$$10000 \left(\frac{3t^2 + 1}{t^2 + 1} \right) > 22000$$

$$\left(\frac{3t^2 + 1}{t^2 + 1} \right) > 2.2$$

$$3t^2 + 1 > 2.2t^2 + 2.2$$

$$0.8t^2 > 1.2$$

$$t^2 - 1.5 > 0$$

$$t > \sqrt{1.5} = \frac{\sqrt{6}}{2} \approx 1.2.$$

45. (a) For $P(t) = \frac{at^2}{t^2+1}$, the figure on the left shows $a = 1, 2, 3, \frac{1}{2}, \frac{1}{3}$, and the figure on the right shows $a = -1, -2, -3, -\frac{1}{2}, -\frac{1}{3}$.

(b) As t increases the population stabilizes to the value a.

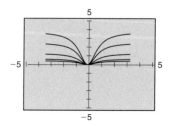
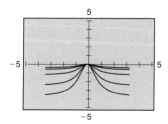

2.5 Other Algebraic Functions

POWER FUNCTIONS

The *rational power functions* have the form

$$f(x) = x^{\frac{m}{n}} = \left(\sqrt[n]{x} \right)^m = \sqrt[n]{x^m},$$

where $\dfrac{m}{n}$ is a rational number, and n is an integer greater than 1.

EXAMPLE 2.5.1   Use the graph of $y = x^{\frac{1}{3}}$, to sketch the graph of the function.

(a) $f(x) = (x - 2)^{\frac{1}{3}} - 1$ (b) $f(x) = -2(x - 1)^{\frac{1}{3}} + 2$

(c) $f(x) = (x + 1)^{\frac{2}{3}}$

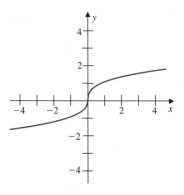

Figure 2.33

SOLUTION:

(a) Shift the graph of $y = x^{\frac{1}{3}}$, to the right 2 units, and downward 1 unit.

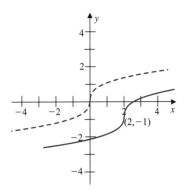

Figure 2.34

(b) Vertically scale the graph of $y = x^{\frac{1}{3}}$ by a factor of 2, so for $x > 0$ the graph of $y = 2x^{\frac{1}{3}}$ is above the graph of $y = x^{\frac{1}{3}}$, and it is below it for $x < 0$. Then reflect the resulting graph about the x-axis and shift it to the right 1 unit and upward 2 units to obtain the graph of $f(x) = -2(x-1)^{\frac{1}{3}} + 2$.

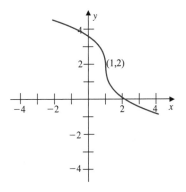

Figure 2.35

(c) To sketch $y = x^{\frac{2}{3}}$, notice that

$$\text{For } 0 < x < 1, \text{ we have } x^{\frac{2}{3}} = \left(x^{\frac{1}{3}}\right)^2 < x^{\frac{1}{3}};$$

$$\text{For } x > 1, \text{ we have } x^{\frac{2}{3}} = \left(x^{\frac{1}{3}}\right)^2 > x^{\frac{1}{3}};$$

$$\text{For } x = 1, \text{ we have } x^{\frac{2}{3}} = x^{\frac{1}{3}}.$$

Since

$$(-x)^{\frac{2}{3}} = \left(\sqrt[3]{-x}\right)^2 = \left(-\sqrt[3]{x}\right)^2 = \left(\sqrt[3]{x}\right)^2 = x^{\frac{2}{3}},$$

the function $g(x) = x^{\frac{2}{3}}$ is an even function and the graph is symmetric with respect to the y-axis. The graph of $f(x) = (x+1)^{\frac{2}{3}}$ is obtained by shifting the graph of $y = x^{\frac{2}{3}}$ to the left 1 unit.

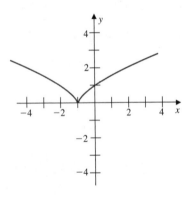

Figure 2.36

◇

EXAMPLE 2.5.2   Determine the domain of the function $\sqrt{\dfrac{2-x}{x+1}}$.

SOLUTION: The domain is the set of all real numbers that make the expression under the radical greater than or equal to 0. So solve

$$\frac{2-x}{x+1} \geq 0,$$

which is greater than 0 when numerator and denominator are both positive, or are both negative. It is zero when $2-x = 0$, or $x = 2$. The linear factors separate the real line into the intervals $(-\infty, -1) \cup (-1, 2) \cup (2, \infty)$, as shown on the line chart.

$2 - x$ $+ + + + + + + + + + + + + + 0 \ - \ - \ - \ - \ - \ - \ -$

$x + 1$ $- \ - \ - \ - \ - \ - \ - \ - \ - \ 0 + + + + + + + + + + + + +$

$\dfrac{(2 - x)}{(x + 1)}$ $- \ - \ - \ - \ - \ - \ - \ - \ - \ \square \ + + + + + 0 \ - \ - \ - \ - \ - \ - \ -$

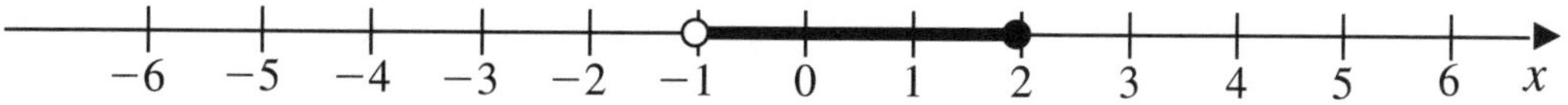

Figure 2.37

The domain is $(-1, 2]$.   ◇

EXAMPLE 2.5.3   Sketch the graph of $f(x) = \sqrt{\dfrac{x - 3}{x + 1}}$, and label any axis intercepts and asymptotes.

SOLUTION:

Domain: $\left\{ x \mid \dfrac{x - 3}{x + 1} \geq 0 \right\} = (-\infty, -1) \cup [3, \infty)$. This can be verified from the sign chart.

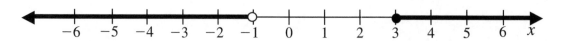

$x - 3$ $------------------\ 0 + + + + +$

$x + 1$ $-----------\ 0 + + + + + + + + + + + +$

$\dfrac{(x - 3)}{(x + 1)}$ $+ + + + + + + + + ⬚ ------- 0 + + + + +$

Figure 2.38

x-intercepts: Solve,

$$\sqrt{\dfrac{x - 3}{x + 1}} = 0$$

$$\dfrac{x - 3}{x + 1} = 0$$

$$x - 3 = 0$$

so

$$x = 3.$$

The graph crosses the x-axis at $(3, 0)$.

y-intercepts: None, since $x = 0$ is not in the domain of the function.

Horizontal asymptote: $y = 1$, since for $|x|$ large,

$$\sqrt{\dfrac{x - 3}{x + 1}} \approx \dfrac{\sqrt{x}}{\sqrt{x}} = 1.$$

Vertical asymptote: $x = -1$. Because of the domain of f, x can only approach -1 from the left, so

$$f(x) \to \infty \quad \text{as} \quad x \to -1^-.$$

◇

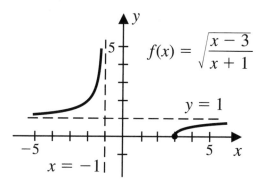

Figure 2.39

Solutions for Exercise Set 2.5

1. The domain of

$$f(x) = \sqrt{x^2 + 4x - 12}$$

consists of all x satisfying

$$x^2 + 4x - 12 \geq 0 \Leftrightarrow (x+6)(x-2) \geq 0 \Leftrightarrow x \leq -6 \quad \text{or} \quad x \geq 2,$$

that is, $(-\infty, -6] \cup [2, \infty)$.

3. The domain of

$$f(x) = \sqrt{(x-1)^2(x+2)}$$

consists of all x satisfying

$$(x-1)^2(x+2) \geq 0 \Leftrightarrow (x+2) \geq 0 \Leftrightarrow x \geq -2,$$

that is, $[-2, \infty)$.

5. The domain of

$$f(x) = \sqrt{(x-1)(x-4)(x+3)}$$

consists of all x satisfying

$$(x-1)(x-4)(x+3) \geq 0.$$

The solution to this inequality can be found from a sign graph. The domain is $[-3, 1] \cup [4, \infty)$.

7. The domain of

$$f(x) = \sqrt{\frac{1-x}{x+3}}$$

consists of all x satisfying

$$\frac{1-x}{x+3} \geq 0 \Leftrightarrow \frac{x-1}{x+3} \leq 0 \Leftrightarrow x > -3 \quad \text{and} \quad x \leq 1,$$

that is, $(-3, 1]$.

9. The domain of

$$f(x) = \sqrt{\frac{x+1}{x^2 + 2x - 3}} = \sqrt{\frac{x+1}{(x+3)(x-1)}}$$

consists of all x satisfying

$$\frac{x+1}{(x+3)(x-1)} \geq 0.$$

The solution to this inequality can be found from a sign graph. The domain is $(-3, -1] \cup (1, \infty)$.

11. (a) iii (b) vi (c) i (d) v (e) iv (f) ii

13. For $f(x) = x^{3/2} - 1$

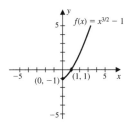

15. For $f(x) = (x - 1)^{1/3} + 1$

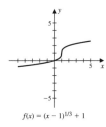

17. For $f(x) = -2(x + 1)^{2/3} - 2$

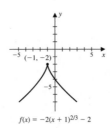

$f(x) = -2(x + 1)^{2/3} - 2$

19. For $f(x) = (1 - x)^{4/3} - 1 = (-(x - 1))^{4/3} - 1 = (x - 1)^{4/3} - 1$

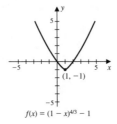

$f(x) = (1 - x)^{4/3} - 1$

21. For $f(x) = x^{-1/2} - 2$

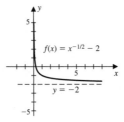

$f(x) = x^{-1/2} - 2$

$y = -2$

23. For $f(x) = \sqrt{\frac{x+2}{x-1}}$

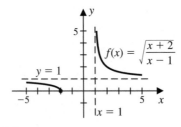

$f(x) = \sqrt{\dfrac{x + 2}{x - 1}}$

$y = 1$

$x = 1$

25. For $f(x) = \sqrt{\dfrac{2-x}{x+2}}$

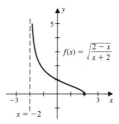

27. For $f(x) = \dfrac{x}{\sqrt{x^2+1}}$

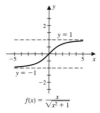

29. For $f(x) = \dfrac{x-1}{\sqrt{(x+1)(x-2)}}$

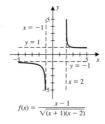

31. For $f(x) = \dfrac{x}{\sqrt{4-x^2}} = \dfrac{x}{\sqrt{(2-x)(2+x)}}$

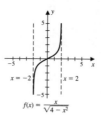

$$f(x) = \frac{x}{\sqrt{4 - x^2}}$$

2.6 Complex Roots of Polynomials

A *complex number* is an expression of the form $a + bi$, where a and b are real numbers, and $i = \sqrt{-1}$, that is, $i^2 = -1$. The number a is called the *real part*, and b is called the *imaginary part*. The *conjugate* of the complex number $a + bi$ is $\overline{a + bi} = a - bi$.

ARITHMETIC OPERATIONS ON COMPLEX NUMBERS

EXAMPLE 2.6.1   Write the complex number in standard form $a + bi$.

(a) $(1 - 3i) + (2 + 4i)$ (b) $(3 - 2i) - (7 + 9i)$

(c) $(2 - 3i) \cdot (3 + 5i)$ (d) $(4 - 3i) \cdot \overline{(7 - 2i)}$

(e) $\dfrac{2}{3 - i}$ (f) $\dfrac{4 - i}{2 + 3i}$

SOLUTION: Addition and subtraction of complex numbers is performed simply by adding and subtracting real and imaginary parts separately.

(a)
$$(1 - 3i) + (2 + 4i) = (1 + 2) + (-3 + 4)i = 3 + i$$

(b)
$$(3 - 2i) - (7 + 9i) = (3 - 7) + (-2 - 9)i = -4 - 11i$$

Multiplication is performed using the distributive law for multiplying real numbers, keeping in mind that $i^2 = -1$.

(c)

$$(2 - 3i) \cdot (3 + 5i) = 2(3 + 5i) - 3i(3 + 5i)$$

$$= 6 + 10i - 9i - 15i^2$$

$$= 6 + i - 15(-1)$$

$$= 21 + i$$

(d)

$$(4 - 3i) \cdot \overline{(7 - 2i)} = (4 - 3i) \cdot (7 + 2i)$$

$$= 28 + 8i - 21i - 6i^2$$

$$= 28 - 13i + 6$$

$$= 34 - 13i$$

Complex division is similar to rationalizing a fraction containing radicals. Here we multiply the numerator and denominator of the complex fraction by the conjugate of the denominator.

(e)

$$\frac{2}{3 - i} = \frac{2}{3 - i} \cdot \frac{\overline{3 - i}}{\overline{3 - i}}$$

$$= \frac{2}{3 - i} \cdot \frac{3 + i}{3 + i}$$

$$= \frac{6 + 2i}{9 - i^2}$$

$$= \frac{6 + 2i}{10} = \frac{3}{5} + \frac{1}{5}i$$

(f)

$$\frac{4-i}{2+3i} = \frac{4-i}{2+3i} \cdot \frac{2-3i}{2-3i}$$

$$= \frac{8-12i-2i+3i^2}{4-9i^2}$$

$$= \frac{5-14i}{13} = \frac{5}{13} - \frac{14}{13}i$$

◇

EXAMPLE 2.6.2   Show that $1 - i$ is a solution to the equation $f(x) = x^2 - 2x + 2 = 0$.

SOLUTION: Substituting $1 - i$ for x gives

$$f(x) = (1-i)^2 - 2(1-i) + 2$$

$$= (1-i)(1-i) - 2 + 2i + 2$$

$$= 1 - 2i + i^2 + 2i$$

$$= 1 + i^2$$

$$= 1 - 1 = 0.$$

◇

COMPLEX ZEROS OF POLYNOMIALS

EXAMPLE 2.6.3   Find the zeros of the quadratic function $f(x) = x^2 - 2x + 4$, and write the function in factored form.

SOLUTION: Factoring directly is not possible, so we use the quadratic formula to first find the zeros. The quadratic equation

$$x^2 - 2x + 4 = 0$$

has

$$a = 1, \ b = -2, \ c = 4,$$

so

$$x = \frac{-b \pm \sqrt{b^2 - 4ac}}{2a}$$

$$= \frac{2 \pm \sqrt{4 - 4(1)(4)}}{2(1)}$$

$$= \frac{2 \pm \sqrt{-12}}{2} = 1 \pm \sqrt{-3}.$$

The discriminant, which is the part under the radical, is negative, so the zeros are complex numbers. Since

$$\sqrt{-3} = \sqrt{3}\sqrt{-1} = \sqrt{3}i,$$

the zeros are

$$x = 1 \pm \sqrt{3}i.$$

The quadratic can then be factored as

$$f(x) = x^2 - 2x + 4$$

$$= \left(x - \left(1 + \sqrt{3}i\right)\right)\left(x - \left(1 - \sqrt{3}i\right)\right).$$

◇

EXAMPLE 2.6.4   Find all solutions of the equation

$$f(x) = x^4 + 2x^2 - 8 = 0.$$

SOLUTION: The expression $x^4 + 2x^2 - 8$ is quadratic in the variable x^2. Let $u = x^2$. Then

$$x^4 + 2x^2 - 8 = u^2 + 2u - 8 = 0,$$

so

$$(u - 2)(u + 4) = 0,$$

and

$$u = 2 \quad \text{or} \quad u = -4.$$

Hence

$$x^2 = 2, x^2 = -4,$$

so

$$x = \pm\sqrt{2}, x = \pm\sqrt{-4} = \pm 2i.$$

$\diamond$

Polynomials with real coefficients have the property that if $a + bi$ is a zero, then so is its conjugate $a - bi$.

EXAMPLE 2.6.5   Show that $x = -1 + 2i$ is a solution of the equation $f(x) = x^3 + x - 10 = 0$, and then find all solutions.

SOLUTION:

$$f(-1 + 2i) = (-1 + 2i)^3 + (-1 + 2i) - 10$$
$$= (-1 + 2i)(-1 + 2i)^2 - 11 + 2i$$
$$= (-1 + 2i)(1 - 4i + 4i^2) - 11 + 2i$$
$$= (-1 + 2i)(-3 - 4i) - 11 + 2i$$
$$= 3 + 4i - 6i - 8i^2 - 11 + 2i$$
$$= 3 + 6i - 6i + 8 - 11 = 0$$

Since $-1 + 2i$ is a zero of the function $f(x)$, its conjugate $-1 - 2i$ is also a zero. To find the last solution, divide the polynomial

$$(x - (-1 + 2i))(x - (-1 - 2i)) = x^2 + 2x + 5,$$

into $x^3 + x - 10$, which must divide $f(x)$ evenly. We have

$$
\begin{array}{r}
x - 2 \\
x^2 + 2x + 5 \overline{)\, x^3 + 0x^2 + x - 10} \\
\underline{x^3 + 2x^2 + 5x} \\
-2x^2 - 4x - 10 \\
\underline{-2x^2 - 4x - 10} \\
0.
\end{array}
$$

Then

$$x^3 + x - 10 = (x - (-1 + 2i))(x - (-1 - 2i))(x - 2),$$

and the solutions to $x^3 + x - 10 = 0$ are

$$x = -1 + 2i, \; x = -1 - 2i, \text{ and } x = 2.$$

◇

EXAMPLE 2.6.6   Find all zeros and factor completely the function $f(x) = x^4 + 2x^3 - 2x^2 - 8x - 8$.

SOLUTION:

<u>Possible rational zeros:</u> $\pm 1, \pm 2, \pm 4, \pm 8$

There is one sign change in $f(x)$, so there is only one possible positive root. Since

$$f(-x) = x^4 - 2x^3 - 2x^2 + 8x - 8,$$

there are either 3 or 1 negative roots, and since $f(1) \neq 0$ and $f(-1) \neq 0$, let us use synthetic division on $f(x)$ with $x = 2$.

$$
\begin{array}{r|rrrrr}
2 & 1 & 2 & -2 & -8 & -8 \\
  &   & 2 & 8 & 12 & 8 \\
\hline
  & 1 & 4 & 6 & 4 & 0
\end{array}
$$

So $x = 2$ is a root, the only positive one. If we now use synthetic division on the resulting polynomial with $x = -2$, we have

$$
\begin{array}{r|rrrr}
-2 & 1 & 4 & 6 & 4 \\
   &   & -2 & -4 & -4 \\
\hline
   & 1 & 2 & 2 & 0.
\end{array}
$$

So

$$f(x) = (x - 2)(x + 2)(x^2 + 2x + 2).$$

The solutions to $x^2 + 2x + 2 = 0$ are

$$x = \frac{-2 \pm \sqrt{4 - 4(1)(2)}}{2} = \frac{-2 \pm \sqrt{-4}}{2} = \frac{-2 \pm 2i}{2} = -1 \pm i.$$

Zeros: $x = 2$, $x = -2$, $x = -1 + i$, $x = -1 - i$

Factored f : $f(x) = (x - 2)(x + 2)(x - (-1 + i))(x - (-1 - i))$

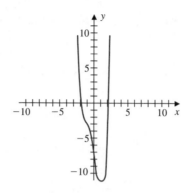

Figure 2.40

◇

EXAMPLE 2.6.7   Find a polynomial with leading coefficient one that has degree 5, zeros i, and $3 - i$, and whose graph passes through the origin.

SOLUTION: Since the polynomial has real coefficients, and i and $3 - i$ are zeros, their conjugates $-i$ and $3 + i$ are also zeros. Since the graph passes through the origin, 0 is another zero. So define the polynomial by

$$f(x) = x(x + i)(x - i)(x - (3 - i))(x - (3 + i))$$
$$= x^5 - 6x^4 + 11x^3 - 6x^2 + 10x.$$

◇

Solutions for Exercise Set 2.6

1.   We have $(2+i)+(-3-2i)=(2-3)+(1-2)i=-1-i$.

3.   We have $(-3+5i)-(2-3i)=(-3-2)+(5+3)i=-5+8i$.

5.   We have $2i\cdot(3+i)=6i+2i^2=-2+6i$.

7.   We have $(2-i)\cdot(3+i)=6+2i-3i-i^2=7-i$.

9.   We have $(6+5i)\cdot(-3-2i)=-18-12i-15i-10i^2=-8-27i$.

11.   We have $(2-3i)\cdot(2+3i)=4+6i-6i-9i^2=13$.

13.   We have $(3-8i)\cdot\overline{(2+i)}=(3-8i)(2-i)=6-3i-16i+8i^2=-2-19i$.

15.   We have $i^5=i^4\cdot i=i^2\cdot i^2\cdot i=(-1)(-1)i=i$.

17.   We have $i^{100}=(i^4)^{25}=1^{25}=1$.

19.   We have $\sqrt{-9}=3\sqrt{-1}=3i$.

21.   We have

$$\left(2+\sqrt{-5}\right)\left(1+\sqrt{-1}\right)=\left(2+\sqrt{5}i\right)(1+i)=2+2i+\sqrt{5}i+\sqrt{5}i^2$$
$$=\left(2-\sqrt{5}\right)+\left(2+\sqrt{5}\right)i.$$

23.   We have $\sqrt{-\frac{16}{9}}=\frac{4}{3}\sqrt{-1}=\frac{4}{3}i$.

25.   We have $\frac{1}{1-i}=\frac{1}{1-i}\cdot\frac{1+i}{1+i}=\frac{1+i}{2}=\frac{1}{2}+\frac{1}{2}i$.

27.   We have $\frac{1}{2+3i}=\frac{1}{2+3i}\cdot\frac{2-3i}{2-3i}=\frac{2-3i}{13}=\frac{2}{13}-\frac{3}{13}i$.

29.   We have $\frac{2-3i}{2+3i}=\frac{2-3i}{2+3i}\cdot\frac{2-3i}{2-3i}=\frac{4-12i+9i^2}{13}=\frac{-5-12i}{13}=-\frac{5}{13}-\frac{12}{13}i$.

31.   (a) Solving for the zeros we have

$$x^2+4=0\Leftrightarrow x^2=-4\Leftrightarrow x=\pm\sqrt{-4}=\pm2i.$$

(b) We have $f(x)=x^2+4=(x-2i)(x+2i)$.

33.   (a) Solving for the zeros we have

$$x^2-2x+2=0\Leftrightarrow x=\frac{2\pm\sqrt{(-2)^2-4(1)(2)}}{2}=\frac{2\pm\sqrt{-4}}{2}=\frac{2\pm2i}{2}=1\pm i.$$

(b) We have $f(x) = (x - (1 - i))(x - (1 + i))$.

35. (a) Solving for the zeros we have

$$2x^2 - x + 2 = 0 \Leftrightarrow x = \frac{1 \pm \sqrt{(-1)^2 - 4(2)(2)}}{2(2)} = \frac{1 \pm \sqrt{-15}}{4} = \frac{1 \pm \sqrt{15}i}{4}.$$

(b) We have

$$f(x) = \left(x - \left(\frac{1 - \sqrt{15}i}{4}\right)\right)\left(x - \left(\frac{1 + \sqrt{15}i}{4}\right)\right).$$

37. We have

$$x^4 - 1 = 0$$

$$(x^2 - 1)(x^2 + 1) = 0$$

$$(x - 1)(x + 1)(x - i)(x + i) = 0$$

and the zeros are $x = -1, 1, -i, i$.

39. We have

$$x^4 - x^2 - 6 = 0$$

$$(x^2 - 3)(x^2 + 2) = 0$$

$$\left(x - \sqrt{3}\right)\left(x + \sqrt{3}\right)\left(x - \sqrt{2}i\right)\left(x + \sqrt{2}i\right) = 0$$

and the zeros are $x = -\sqrt{3}, \sqrt{3}, -\sqrt{2}i, \sqrt{2}i$.

41. By inspection $x = -2$ is a solution to $x^3 + 8 = 0$, so $x + 2$ is a factor. Dividing $x^3 + 1$ by $x + 2$ gives $x^3 + 1 = (x + 2)(x^2 - 2x + 4)$. Then solving for the zeros of $x^2 - 2x + 4 = 0$ gives

$$x = \frac{2 \pm \sqrt{4 - 4(1)(4)}}{2} = \frac{2 \pm \sqrt{-12}}{2} = \frac{2 \pm 2\sqrt{3}i}{2} = 1 \pm \sqrt{3}i.$$

The zeros are $x = -2, 1 + \sqrt{3}i, 1 - \sqrt{3}i$.

43. If $f(x) = x^3 - 2x^2 + 9x - 18$, we have

$$f(3i) = (3i)^3 - 2(3i)^2 + 9(3i) - 18 = -27i + 18 + 27i - 18 = 0.$$

Since $3i$ is a zero, the conjugate $-3i$ is also a zero, so $(x - 3i)(x + 3i) = x^2 + 9$ is a factor. Dividing $x^2 + 9$ into $f(x)$ gives

$$f(x) = (x^2 + 9)(x - 2).$$

The third solution is $x = 2$.

45. If $f(x) = x^4 - 2x^3 - 2x^2 - 2x - 3$, we have

$$f(i) = (i)^4 - 2(i)^3 - 2(i)^2 - 2(i) - 3 = 1 + 2i + 2 - 2i - 3 = 0.$$

Since i is a zero, the conjugate $-i$ is also a zero, so

$$(x - i)(x + i) = x^2 + 1$$

is a factor. Dividing $x^2 + 1$ into $f(x)$ gives

$$f(x) = (x^2 + 1)(x^2 - 2x - 3) = (x^2 + 1)(x - 3)(x + 1).$$

The third and fourth solutions are $x = -1, x = 3$.

47. The possible rational zeros of $f(x) = x^3 - 3x^2 + 9x - 27$ are

$$\pm 1, \pm 3, \pm 9, \pm 27,$$

and from direct substitution or a computer generated graph, $x = 3$ is a zero. Then

$$f(x) = (x - 3)(x^2 + 9) = (x - 3)(x - 3i)(x + 3i).$$

49. Since $f(x) = x^4 - x^2 - 2x + 2$ has a zero of multiplicity 2 at $x = 1$, we have

$$f(x) = (x-1)^2(x^2 + 2x + 2) = (x-1)^2(x - (-1 - i))(x - (-1 + i)).$$

51. Since $2i$ is a zero, the conjugate $-2i$ is also a zero and

$$f(x) = (x-2)(x-2i)(x+2i) = (x-2)(x^2+4)$$
$$= x^3 - 2x^2 + 4x - 8.$$

53. Since $\sqrt{3}i$ and $3i$ are zeros, the conjugates $-\sqrt{3}i$ and $-3i$ are also zeros, so a polynomial with the specified zeros is

$$f(x) = (x - \sqrt{3}i)(x + \sqrt{3}i)(x - 3i)(x + 3i)$$
$$= (x^2 + 3)(x^2 + 9) = x^4 + 12x^2 + 27.$$

55. Since $x = -2$ is a zero of multiplicity 2, $(x+2)^2$, is a factor, and since $1 + 2i$ is a zero, the conjugate $1 - 2i$ is also a zero. There is also a zero at 0, so

$$f(x) = x(x+2)^2(x - (1 + 2i))(x - (1 - 2i))$$
$$= x(x^2 + 4x + 4)(x^2 - 2x + 5)$$
$$= x^5 + 2x^4 + x^3 + 12x^2 + 20x.$$

Solutions for Exercise Set 2 Review

1. (a) (ii) The graph crosses the x-axis at $x = 1, x = 2$, and $x = -1$. Since the zero at $x = -1$ is of multiplicity 2, the curve just touches and turns without crossing the x-axis at $x = -1$.

(b) (iii) The graph crosses the x-axis at $x = 1, x = 2$, and $x = -1$. Since the zero at $x = 2$ is of multiplicity 2, the curve just touches and turns without crossing the x-axis at $x = 2$.

(c) (iv) The graph crosses the x-axis at $x = 1, x = 2$, and $x = -1$. Since the zero at $x = 1$ is of multiplicity 2, the curve just touches and turns without crossing the x-axis at $x = -2$.

(d) (i) The graph crosses the x-axis only at $x = 1$ and $x = -1$ and both zeros are of multiplicity 2.

3. For $f(x) = -2(x-1)^2 + 2$

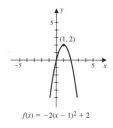

5. For $f(x) = -x^4 - 3$

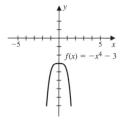

7. For $f(x) = (x+1)(x+2)(x-3)$

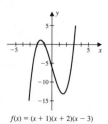

$f(x) = (x + 1)(x + 2)(x - 3)$

9. For $f(x) = \frac{1}{2}(x-1)^3(x+2)$

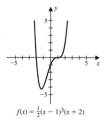

$f(x) = \frac{1}{2}(x-1)^3(x+2)$

11. For $f(x) = x^3 - \frac{1}{2}x^2 - \frac{1}{2}x = \frac{1}{2}x(2x^2 - x - 1) = \frac{1}{2}x(2x+1)(x-1)$

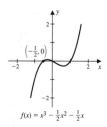

$f(x) = x^3 - \frac{1}{2}x^2 - \frac{1}{2}x$

13. Degree 3, leading coefficient positive. The graph shows a zero of at least multiplicity 2 and one other zero of multiplicity 1, so the degree is at least 3. The end behavior is the same as that for $y = x^3$, so the leading coefficient is positive.

15. Degree 4, leading coefficient negative. Adding or subtracting a positive constant from a polynomial shifts the curve upward or downward but does not change the degree. The curve can be shifted downward so as to appear to have 4 zeros of multiplicity 1, so the degree of the original polynomial is at least 4. Since the end behavior is the opposite of that of $y = x^4$ the leading coefficient is negative.

17. $Q(x) = 4x + 10$ and $R(x) = 19$

19. $Q(x) = 3x^2 - 4x + 6$ and $R(x) = -11$

21. $P(x) = 3x^4 - 9x^3 - 2x^2 + 5x + 3$

(a) We have
$$\frac{\pm 1, \pm 3}{\pm 1, \pm 3} = \pm 1, \pm \frac{1}{3}, \pm 3.$$

(b) The number of sign changes in $P(x)$ is 2, which is the maximum number of positive roots. The number of sign changes in

$$P(-x) = 3(-x)^4 - 9(-x)^3 - 2(-x)^2 + 5(-x) + 3 = 3x^4 + 9x^3 - 2x^2 - 5x + 3$$

is 2, which is the maximum number of negative roots.

(c) Since

$$P(3) = 3(3)^4 - 9(3)^3 - 2(3)^2 + 5(3) + 3 = 3^5 - 3^5 - 18 + 15 + 3 = 0,$$

$x - 3$ is a factor.

(d) $P(x) = (x - 3)(3x^3 - 2x - 1) = (x - 3)(x - 1)(3x^2 + 3x + 1)$

23. $P(x) = x^5 - 3x^4 - 5x^3 + 27x^2 - 32x + 12$

(a) We have
$$\frac{\pm 1, \pm 2, \pm 3, \pm 4, \pm 6, \pm 12}{\pm 1} = \pm 1, \pm 2, \pm 3, \pm 4, \pm 6, \pm 12.$$

(b) The number of sign changes in $P(x)$ is 4, which is the maximum number of positive roots. The number of sign changes in

$$P(-x) = (-x)^5 - 3(-x)^4 - 5(-x)^3 + 27(-x)^2 - 32(-x) + 12$$

$$= -x^5 - 3x^4 + 5x^3 + 27x^2 + 32x + 12$$

is 1, which is the maximum number of negative roots.

(c) $P(-3) = (-3)^5 - 3(-3)^4 - 5(-3)^3 + 27(-3)^2 - 32(-3) + 12 = 0$, so $x + 3$ is a factor.

(d) $P(x) = (x+3)(x^4 - 6x^3 + 13x^2 - 12x + 4) = (x+3)(x-1)^2(x-2)^2$

25. $P(x) = x^4 - 5x^3 + 7x^2 + 3x - 10$

Since $2 + i$ is a zero so is $2 - i$ and

$$(x - (2+i))(x - (2-i)) = x^2 - 4x + 5.$$

Then

$$P(x) = (x^2 - 4x + 5)(x^2 - x - 2)$$

$$= (x - (2+i))(x - (2-i))(x-2)(x+1).$$

27. $P(x) = x^5 - x^4 + 10x^3 - 10x^2 + 9x - 9$

Since i is a zero so is $-i$ and

$$(x - i)(x + i) = x^2 + 1.$$

Then

$$P(x) = (x^2 + 1)(x^3 - x^2 + 9x - 9) = (x^2 + 1)(x - 1)(x^2 + 9)$$

$$= (x - i)(x + i)(x - 1)(x - 3i)(x + 3i).$$

29. For

$$f(x) = \frac{x-4}{x-1}.$$

(a) Domain: $\{x : x \neq 1\} = (-\infty, 1) \cup (1, \infty)$

(b) x-intercepts: Solve

$$\frac{x-4}{x-1} = 0 \Leftrightarrow x - 4 = 0 \Leftrightarrow x = 4;$$

y-intercept: Set $x = 0 \Rightarrow y = 4$.

(c) Vertical asymptotes: Solve $x - 1 = 0 \Leftrightarrow x = 1$; horizontal asymptotes:

For x large in magnitude

$$\frac{x-4}{x-1} \approx \frac{x}{x} = 1,$$

so $y = 1$.

31. For

$$f(x) = \frac{x^2 - 2x + 1}{2x^2 - 18} = \frac{(x-1)^2}{2(x-3)(x+3)}.$$

(a) Domain: $\{x : x \neq -3, x \neq 3\} = (-\infty, -3) \cup (-3, 3) \cup (3, \infty)$

(b) x-intercepts: Solve

$$\frac{(x-1)^2}{2(x-3)(x+3)} = 0 \Leftrightarrow (x-1)^2 = 0 \Leftrightarrow x = 1;$$

y-intercept: Set $x = 0 \Rightarrow y = -\frac{1}{18}$.

(c) Vertical asymptotes: Solve

$$(x-3)(x+3) = 0 \Leftrightarrow x = -3, x = 3;$$

horizontal asymptotes:

$$\frac{(x-1)^2}{2(x-3)(x+3)} \approx \frac{x^2}{2x^2} = \frac{1}{2},$$

so $y = \frac{1}{2}$.

33. For
$$f(x) = \frac{x^3 + 2x^2 - x - 2}{x^3} = \frac{(x+2)(x+1)(x-1)}{x^3}.$$

(a) Domain: $\{x : x \neq 0\} = (-\infty, 0) \cup (0, \infty)$

(b) x-intercepts: Solve
$$\frac{(x+2)(x+1)(x-1)}{x^3} = 0$$

if and only if

$$x + 2 = 0, x + 1 = 0, x - 1 = 0 \Leftrightarrow x = -2, x = -1, x = 1;$$

y-intercept: None, since 0 is not in the domain.

(c) Vertical asymptotes: Solve $x^3 = 0 \Leftrightarrow x = 0$;

horizontal asymptotes: For x large in magnitude

$$\frac{(x+2)(x+1)(x-1)}{x^3} \approx \frac{x^3}{x^3} = 1,$$

so $y = 1$.

35. For $f(x) = \dfrac{3}{x-2}$.

Horizontal asymptotes: $y = 0$; vertical asymptotes: $x = 2$; x-intercepts:

none; y-intercept: $(0, -3/2)$

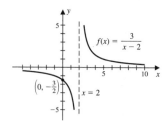

37. For $f(x) = \dfrac{-2}{(x-1)(x-2)}$.

Horizontal asymptotes: $y = 0$; vertical asymptotes: $x = 1, x = 2$; x-intercepts: none; y-intercept: $(0, -1)$

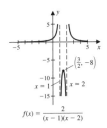

39. For $f(x) = \dfrac{4}{x^2 - 4} = \dfrac{4}{(x - 2)(x + 2)}$.

Horizontal asymptotes: $y = 0$; vertical asymptotes: $x = -2, x = 2$; x-intercepts: none; y-intercept: $(0, -1)$

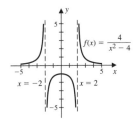

41. For $f(x) = \dfrac{x^2 - 1}{x^2 + 2x} = \dfrac{(x - 1)(x + 1)}{x(x + 2)}$.

Horizontal asymptotes: $y = 1$; vertical asymptotes: $x = -2, x = 0$; x-intercepts: $(1, 0), (-1, 0)$; y-intercept: none

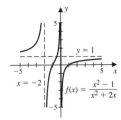

43. For $f(x) = \dfrac{x^2 - 2x + 1}{x + 1} = x - 3 + \dfrac{4}{x + 1}$ we have the following graph.

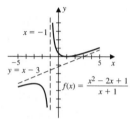

45. For $f(x) = \sqrt{\dfrac{x - 2}{x + 1}}$.

(a) Domain: $(-\infty, -1) \cup [2, \infty)$

(b)

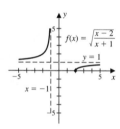

(c) Horizontal asymptotes: $y = 1$; vertical asymptotes: $x = -1$; x-intercepts: $(2, 0)$; y-intercept: none

47. For $f(x) = \sqrt{\dfrac{4 - x}{x + 4}}$.

(a) Domain: $(-4, 4]$

(b)

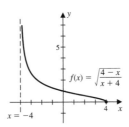

(c) Horizontal asymptotes: none; vertical asymptotes: $x = -4$; x-intercepts: $(4, 0)$; y-intercept: $(0, 1)$

49. For $f(x) = \dfrac{x^2}{\sqrt{4 - x^2}} = \dfrac{x^2}{\sqrt{(2 - x)(2 + x)}}$.

(a) Domain: $(-2, 2)$

(b)

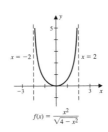

(c) Horizontal asymptotes: none; vertical asymptotes: $x = 2, x = -2$; x-intercept: $(0, 0)$; y-intercept: $(0, 0)$

51. $f(x) = x^3 - 2x^2 - x + 2$

(a) increasing: $(-\infty, -0.2) \cup (1.5, \infty)$; decreasing: $(-0.2, 1.5)$

(b) local maximum: $(-0.2, 2.1)$; local minimum: $(1.5, -0.6)$

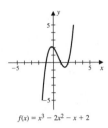

$f(x) = x^3 - 2x^2 - x + 2$

53. $f(x) = \frac{2}{3}x^3 + \frac{7}{2}x^2 - 12x$

   (a) increasing: $(-\infty, -4.7) \cup (1.3, \infty)$; decreasing: $(-4.7, 1.3)$

   (b) local maximum: $(-4.7, 64.5)$; local minimum: $(1.3, -8.2)$

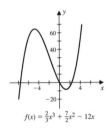

$f(x) = \frac{2}{3}x^3 + \frac{7}{2}x^2 - 12x$

55. We have

$$\frac{2 + i\sqrt{2}}{4} = \frac{1}{2} + \frac{\sqrt{2}}{4}i.$$

57. We have $(3 - i) - (2 - 3i) = (3 - 2) + (-1 + 3)i = 1 + 2i.$

59. We have $(2 - i) \cdot \overline{(2 + i)} = (2 - i) \cdot (2 - i) = 4 - 4i + i^2 = 3 - 4i.$

61. We have $i^{20} = (i^4)^5 = 1^5 = 1.$

63. We have

$$\frac{2 + 3i}{4 - 7i} = \frac{2 + 3i}{4 - 7i} \cdot \frac{4 + 7i}{4 + 7i} = \frac{8 + 14i + 12i + 21i^2}{16 + 49} = \frac{-13 + 26i}{65} = -\frac{1}{5} + \frac{2}{5}i.$$

65. (a) $y = f(x - 1)$ (b) $y = f(x - 1) - 1$

   (c) $y = -f(x + 1) + 1$ (d) $y = |f(x)|$

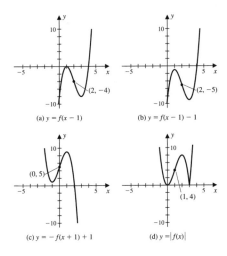

(a) $y = f(x - 1)$

(b) $y = f(x - 1) - 1$

(c) $y = -f(x + 1) + 1$

(d) $y = |f(x)|$

67. A possible graph is as follows.

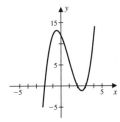

69. First we have

$$P(x) = a(x - 1)(x - 3)(x + 1) = a(x^3 - 3x^2 - x + 3).$$

If $P(0) = 1$, then

$$1 = a((-1)(-3)(1)) \Rightarrow 3a = 1 \Rightarrow a = \frac{1}{3}.$$

So

$$P(x) = \frac{1}{3}x^3 - x^2 - \frac{1}{3}x + 1.$$

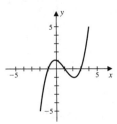

71. (a)

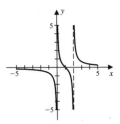

(b) The graph will have vertical asymptotes $x = 0$ and $x = 2$, if the denominator of the function has factors x and $x-2$. To have a horizontal asymptote $y = 0$ the degree of the numerator must be less than the degree of the denominator. If the numerator is $x - 1$, then $f(0) = 1$ and a possible definition for the function is $f(x) = \frac{x-1}{x(x-2)}$.

73. To find the points of intersection of the curves

$$f(x) = x^3 \quad \text{and} \quad g(x) = 2x^2 + x - 2,$$

solve

$$x^3 = 2x^2 + x - 2 \Leftrightarrow x^3 - 2x^2 - x + 2 = 0 \Leftrightarrow (x - 1)(x - 2)(x + 1) = 0$$

$$\Leftrightarrow x = -1, x = 1, x = 2.$$

The points of intersection are $(-1, -1), (1, 1), (2, 8)$.

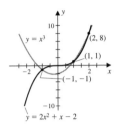

75. If the polynomial has integer coefficients and a zero $-i$, then the conjugate i is also a zero. The polynomial is of degree three with zeros, $1, i, -i$, so

$$P(x) = (x-1)(x-i)(x+i) = (x-1)(x^2+1) = x^3 - x^2 + x - 1.$$

77. If the polynomial has integer coefficients and zeros at $\sqrt{2}i$ and $2i$, then the conjugates $-\sqrt{2}i$ and $-2i$ are also zeros. The polynomial has the form

$$\begin{aligned}
P(x) &= a\left(x - \sqrt{2}i\right)\left(x + \sqrt{2}i\right)(x - 2i)(x + 2i) \\
&= a(x^2 + 2)(x^2 + 4) \\
&= a(x^4 + 6x^2 + 8),
\end{aligned}$$

and if the constant term is to be 8, then $a = 1$, so

$$P(x) = x^4 + 6x^2 + 8.$$

79. If the length and width of the rectangle are l and w, respectively, then the perimeter is

$$2l + 2w = 20 \Rightarrow l + w = 10 \Rightarrow w = 10 - l.$$

The area is

$$A = lw = l(10 - l) = 10l - l^2.$$

To find l so the area is a maximum find the vertex of the parabola. Completing the square gives

$$-l^2 + 10l = -(l^2 - 10l + 25 - 25)$$
$$= -(l - 5)^2 + 25$$

and the vertex is $(5, 25)$. So $l = 5 \Rightarrow w = 10 - 5 = 5$, and the rectangle with maximum area is a square of side 5.

Solutions for Exercise Set 2 Calculus

1. (a) First we have $P(x) = 2x^3 - 3x^2 = x^2(2x - 3)$.

(b) Adding a constant, C, to $P(x)$ shifts the graph upward C units, if $C > 0$, and downward $|C|$ units, if $C < 0$. If $C < 0$, $Q(x) = P(x) + C$ has 1 zero. If $0 < C < 1$, $Q(x) = P(x) + C$ has 3 zeros. If $C = 1$, $Q(x) = P(x) + C$ has 2 zeros. If $C > 1$, $Q(x) = P(x) + C$ has 1 zero. If $C = 0$, $Q(x) = P(x) + C$ has two real zeros.

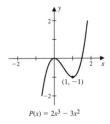

$P(x) = 2x^3 - 3x^2$

3. (a) Let $P(x) = mx + b$. Then

$$(P \circ P)(x) = m(mx + b) + b = m^2 x + b(m + 1),$$

which is a linear polynomial with positive slope since the slope is $m^2 > 0$.

(b) Let $P(x) = ax^2 + bx + c$, so

$$(P \circ P)(x) = a(ax^2 + bx + c)^2 + b(ax^2 + bx + c) + c$$
$$= a^3 x^4 + 2a^2 b x^3 + (2a^2 c + ab^2 + ab)x^2 + (2abc + b^2)x + (ac^2 + bc + c)$$

which is a polynomial of degree 4.

5. The possible rational roots of $f(x) = x^3 + x^2 + kx - 3$ are ± 1 and ± 3. Then

$$f(1) = 0 \Rightarrow k = 1 \qquad f(-1) = 0 \Rightarrow k = -3$$
$$f(3) = 0 \Rightarrow k = -11 \qquad f(-3) = 0 \Rightarrow k = -7.$$

For $k = 1$:

$$f(x) = (x - 1)(x^2 + 2x + 3).$$

For $k = -3$:

$$f(x) = (x + 1)(x^2 - 3) = (x + 1)\left(x - \sqrt{3}\right)\left(x + \sqrt{3}\right).$$

For $k = -11$:

$$f(x) = (x - 3)(x^2 + 4x + 1) = (x - 3)\left(x - \left(-2 + \sqrt{3}\right)\right)\left(x - \left(-2 - \sqrt{3}\right)\right).$$

For $k = -7$:

$$f(x) = (x + 3)(x^2 - 2x - 1) = (x + 3)\left(x - \left(1 + \sqrt{2}\right)\right)\left(x - \left(1 - \sqrt{2}\right)\right).$$

So $f(x)$ has at least one rational zero for $k = 1, -3, -11, -7$.

7. (a) Since

$$P(x_1) = \frac{x_1 - x_2}{x_1 - x_2}y_1 + \frac{x_1 - x_1}{x_2 - x_1}y_2 = y_1,$$

the point (x_1, y_1) is on the line. Similarly,

$$P(x_2) = \frac{x_2 - x_2}{x_1 - x_2}y_1 + \frac{x_2 - x_1}{x_2 - x_1}y_2 = y_2,$$

so (x_2, y_2) is on the line.

(b)

$$y = \frac{x - 1}{-1 - 1}(6) + \frac{x - (-1)}{1 - (-1)}(-2) = -3(x - 1) - (x + 1) = -4x + 2$$

9. The height of an object above the ground is given by $s(t) = v_0 t - \frac{g}{2}t^2$.

(a) The initial height of the object occurs when $t = 0$, and equals $s(0) = 0$.

(b) To determine how long the object is in the air we need to find the time when the object strikes the ground. Solving

$$s(t) = 0 \Leftrightarrow$$

$$v_0 t - \frac{g}{2}t^2 = 0 \Leftrightarrow t\left(v_0 - \frac{g}{2}t\right) = 0 \Leftrightarrow$$

$$t = 0 \quad \text{or} \quad t = \frac{2v_0}{g}$$

and we have $t = \frac{2v_0}{g}$ is the amount of time the object is in the air.

(c) To determine the maximum height reached by the object find the vertex of the parabola. Completing the square we have

$$-\frac{g}{2}t^2 + v_0 t = -\frac{g}{2}\left(t^2 - \frac{2v_0}{g}t\right)$$

$$= -\frac{g}{2}\left(t^2 - \frac{2v_0}{g}t + \frac{v_0^2}{g^2} - \frac{v_0^2}{g^2}\right)$$

$$= -\frac{g}{2}\left(t - \frac{v_0}{g}\right)^2 + \frac{v_0^2}{2g^2}.$$

So the vertex is $\left(\frac{v_0}{g}, \frac{v_0^2}{2g}\right)$ and the maximum height reached is $\frac{v_0^2}{2g}$.

(d) The object reaches the maximum height after $t = \frac{v_0}{g}$.

11. The dimensions on the figure imply that

$$V(x) = x(10 - x)(50 - 3x).$$

CHAPTER 3
TRIGONOMETRIC FUNCTIONS

3.1 Introduction

The *trigonometric functions* $f(t) = \cos t$, and $g(t) = \sin t$ are defined as the x- and y-coordinates of points $P(t) = (\cos t, \sin t)$ on the unit circle. The point $P(t)$ is called the terminal point, and t is the length of the arc along the circle measured counterclockwise from the starting point $(1, 0)$. The measure of t is called radian measure. The sine and cosine functions are the basis for the other trigonometric functions. Master these and you have mastered most of trigonometry.

3.2 The Sine and Cosine Functions

THE UNIT CIRCLE

The *unit circle* is the circle with center at the origin $(0, 0)$ and radius 1. The equation of the unit circle is

$$x^2 + y^2 = 1.$$

EXAMPLE 3.2.1   Verify that the point is on the unit circle.

(a) $\left(\frac{\sqrt{3}}{2}, \frac{1}{2} \right)$ (b) $\left(-\frac{1}{3}, \frac{2\sqrt{2}}{3} \right)$

SOLUTION:

(a) Substitute the coordinates into the equation for the unit circle, and if the equation is satisfied, the point will lie on the unit circle. Substituting $x = \frac{\sqrt{3}}{2}$, and $y = \frac{1}{2}$,

$$\left(\frac{\sqrt{3}}{2}\right)^2 + \left(\frac{1}{2}\right)^2 = \frac{(\sqrt{3})^2}{2^2} + \frac{1}{2^2}$$

$$= \frac{3}{4} + \frac{1}{4} = 1,$$

and the point lies on the unit circle.

(b) Likewise,

$$\left(-\frac{1}{3}\right)^2 + \left(\frac{2\sqrt{2}}{3}\right)^2 = \frac{1}{9} + \frac{2^2(\sqrt{2})^2}{9}$$

$$= \frac{1}{9} + \frac{8}{9} = 1.$$

◇

EXAMPLE 3.2.2   Find the point on the unit circle that satisfies the given conditions.

(a) x-coordinate $\frac{4}{5}$ and y-coordinate negative.

(b) y-coordinate $-\frac{5}{6}$ and in quadrant IV.

(c) x-coordinate $-\frac{\sqrt{2}}{2}$ and in quadrant II.

SOLUTION:

(a) Using the fact that the point lies on the unit circle, and so satisfies the equation $x^2 + y^2 = 1$,

$$\left(\frac{4}{5}\right)^2 + y^2 = 1$$

$$y^2 = 1 - \frac{16}{25} = \frac{9}{25}$$

$$y = \pm\sqrt{\frac{9}{25}} = \pm\frac{3}{5}.$$

Since the y-coordinate is negative, we have the point $\left(\frac{4}{5}, -\frac{3}{5}\right)$.

(b) Solving for the x-coordinate,

$$x^2 + \left(-\frac{5}{6}\right)^2 = 1$$

$$x^2 = 1 - \frac{25}{36} = \frac{11}{36}$$

$$x = \pm\frac{\sqrt{11}}{6}.$$

The quadrants of the plane are labeled I, II, III and IV counterclockwise starting in the upper right. In quadrant IV, the x-coordinate is positive and y-coordinate is negative, so

$$x = \frac{\sqrt{11}}{6},$$

and the point is $\left(\frac{\sqrt{11}}{6}, -\frac{5}{6}\right)$.

(c)

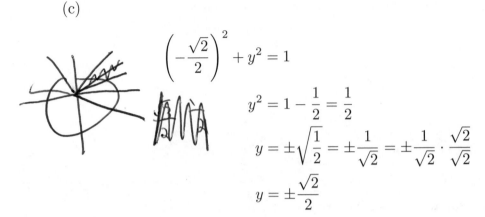

$$\left(-\frac{\sqrt{2}}{2}\right)^2 + y^2 = 1$$

$$y^2 = 1 - \frac{1}{2} = \frac{1}{2}$$

$$y = \pm\sqrt{\frac{1}{2}} = \pm\frac{1}{\sqrt{2}} = \pm\frac{1}{\sqrt{2}} \cdot \frac{\sqrt{2}}{\sqrt{2}}$$

$$y = \pm\frac{\sqrt{2}}{2}$$

In quadrant II, the y-coordinate is positive, so we have the point $\left(-\frac{\sqrt{2}}{2}, \frac{\sqrt{2}}{2}\right)$.

◇

TERMINAL POINTS

If $t > 0$, then $P(t)$ is the point on the unit circle found by measuring t units counterclockwise along the arc of the circle starting at $(1,0)$. If $t < 0$, then measure clockwise around the circle starting at $(1,0)$.

EXAMPLE 3.2.3   Find the location of the terminal point $P(t)$ on the unit circle for the given value of t.

(a) $t = \frac{\pi}{4}$ (b) $t = \frac{7\pi}{6}$ (c) $t = 7\pi$ (d) $t = -\frac{3\pi}{2}$ (e) $t = -\frac{4\pi}{3}$

SOLUTION:

(a) The circumference of the unit circle is 2π, so $\frac{\pi}{2}$ is one fourth the way around. Since $\frac{\pi}{4} = \frac{1}{2}(\frac{\pi}{2})$, and $\frac{\pi}{2}$ is one fourth around the circle, $\frac{\pi}{4}$ is one eighth the way around the circle.

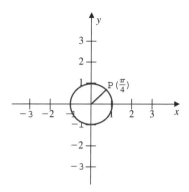

Figure 3.1

(b) Since $\frac{\pi}{6} = \frac{1}{3}(\frac{\pi}{2})$, an arc of length $\frac{\pi}{6}$ is one third of the arc from $(1,0)$ to $(0,1)$. Also, $P(\pi) = (-1,0)$, since an arc of π units is half the unit circle. So $P\left(\frac{7\pi}{6}\right)$ is one third of the way between $(-1,0)$ and $(0,-1)$.

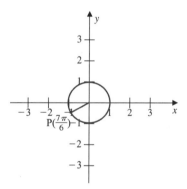

Figure 3.2

(c) An arc of length π is half the length of the unit circle, or one half a complete revolution. Since 2π is one revolution around the circle, 4π is two revolutions, 6π three revolutions, 7π three and one half revolutions, and $P(7\pi)$ is located at the point $(-1,0)$.

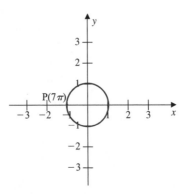

Figure 3.3

(d) $P\left(-\frac{3\pi}{2}\right)$ is three fourths of the way around the circle in the clockwise direction and is located at $(0,1)$.

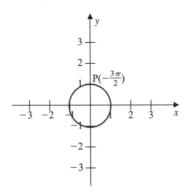

Figure 3.4

(e) Since $-\frac{\pi}{3} = \frac{2}{3}\left(-\frac{\pi}{2}\right)$, the point $P\left(-\frac{\pi}{3}\right)$ is two thirds of the way from $(1,0)$ to $(0,-1)$. Also $-\pi = 3\left(-\frac{\pi}{3}\right)$, so $P\left(-\frac{4\pi}{3}\right)$ is two thirds of the way between $(-1,0)$ and $(0,1)$.   ◇

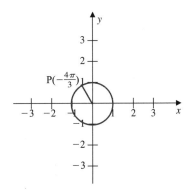

Figure 3.5

REFERENCE NUMBER

For any real number t, the *reference number r* associated with t is the shortest distance along the unit circle from t to the x-axis. For any t, the reference number satisfies $0 \leq r \leq \frac{\pi}{2}$.

EXAMPLE 3.2.4   Find the reference number r for the given value of t, and show $P(t)$ and $P(r)$ on the unit circle.

(a) $t = \frac{5\pi}{6}$ (b) $t = -\frac{2\pi}{3}$ (c) $t = \frac{5\pi}{4}$

SOLUTION:

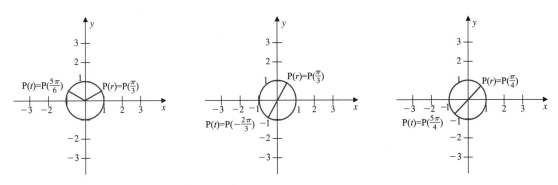

Figure 3.6   Figure 3.7   Figure 3.8

◇

EXAMPLE 3.2.5   If $P(t)$ has coordinates $\left(\frac{12}{13}, \frac{5}{13}\right)$, find the coordinates of

(a) $P(t+\pi)$ (b) $P(-t)$ (c) $P(t-\pi)$ (d) $P(-t-\pi)$

SOLUTION: The reference number for each of the points is t, so the coordinates of the points differ from those of $P(t)$ in one or two signs. For example, $P(t)$ lies in quadrant I, and $P(t+\pi)$ lies in quadrant III, where both coordinates are negative. So

$$P(t+\pi) = \left(-\frac{12}{13}, -\frac{5}{13}\right).$$

The four points are shown in the figure.   ◇

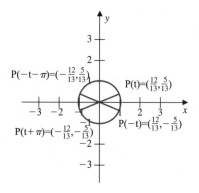

Figure 3.9

SINE AND COSINE

If a point on the unit circle has coordinates $P(t) = (x, y)$, then

$$\sin t = y, \quad \cos t = x.$$

These definitions are also used for the sine and cosine of an angle with radian measure t. The information in the following table is basic.

t	0	$\frac{\pi}{6}$	$\frac{\pi}{4}$	$\frac{\pi}{3}$	$\frac{\pi}{2}$	π	$\frac{3\pi}{2}$	2π
$P(t)$	$(1,0)$	$\left(\frac{\sqrt{3}}{2},\frac{1}{2}\right)$	$\left(\frac{\sqrt{2}}{2},\frac{\sqrt{2}}{2}\right)$	$\left(\frac{1}{2},\frac{\sqrt{3}}{2}\right)$	$(0,1)$	$(-1,0)$	$(0,-1)$	$(1,0)$
$\sin t$	0	$\frac{1}{2}$	$\frac{\sqrt{2}}{2}$	$\frac{\sqrt{3}}{2}$	1	0	-1	0
$\cos t$	1	$\frac{\sqrt{3}}{2}$	$\frac{\sqrt{2}}{2}$	$\frac{1}{2}$	0	-1	0	1

EXAMPLE 3.2.6   Find $\sin t$ and $\cos t$ for the given value of t.

(a) $t = \frac{3\pi}{4}$ (b) $t = \frac{7\pi}{6}$ (c) $t = \frac{5\pi}{3}$ (d) $t = -\frac{7\pi}{4}$

SOLUTION:

(a) The reference number for t is $r = \frac{\pi}{4}$. To find the sine and cosine of t, use the sine and cosine of r and determine the appropriate sign for each based on the quadrant where $P(t)$ lies. So

$$\sin\frac{\pi}{4} = \frac{\sqrt{2}}{2}, \quad \cos\frac{\pi}{4} = \frac{\sqrt{2}}{2}$$

and since $P\left(\frac{3\pi}{4}\right)$ lies in quadrant II, $\sin\frac{3\pi}{4} > 0$ and $\cos\frac{3\pi}{4} < 0$, and

$$\sin\frac{3\pi}{4} = \frac{\sqrt{2}}{2}, \cos\frac{3\pi}{4} = -\frac{\sqrt{2}}{2}.$$

(b)

Reference number: $r = \frac{\pi}{6}$

Quadrant of $P(t)$: III, where $\sin t < 0$ and $\cos t < 0$.

So

$$\sin\frac{7\pi}{6} = -\sin\frac{\pi}{6} = -\frac{1}{2}, \cos\frac{7\pi}{6} = -\cos\frac{\pi}{6} = -\frac{\sqrt{3}}{2}.$$

(c)

Reference number: $r = \frac{\pi}{3}$

Quadrant of $P(t)$: IV, where $\sin t < 0$ and $\cos t > 0$.

So

$$\sin \frac{5\pi}{3} = -\sin \frac{\pi}{3} = -\frac{\sqrt{3}}{2}, \cos \frac{5\pi}{3} = \cos \frac{\pi}{3} = \frac{1}{2}.$$

(d)

<u>Reference number:</u> $r = \frac{\pi}{4}$

<u>Quadrant of</u> $P(t)$: I, where $\sin t > 0$ and $\cos t > 0$.

So

$$\sin \left(-\frac{7\pi}{4} \right) = \sin \frac{\pi}{4} = \frac{\sqrt{2}}{2}, \cos \left(-\frac{7\pi}{4} \right) = \cos \frac{\pi}{4} = \frac{\sqrt{2}}{2}.$$

◇

EXAMPLE 3.2.7   (a) Find all values of t in the interval $[0, 2\pi]$ that satisfy the equation $\cos t = -\frac{1}{2}$.

(b) Find all values of t in the interval $[0, 2\pi]$ that satisfy the equation $\sin 3t = \frac{\sqrt{2}}{2}$.

SOLUTION:

(a) The $\cos t$ is negative in quadrants II and III. We want a reference angle r with $0 < r < \frac{\pi}{2}$ so that

$$\cos r = \frac{1}{2} \implies r = \frac{\pi}{3}.$$

The angle in quadrant II with reference angle $\frac{\pi}{3}$ is $\frac{2\pi}{3}$, and the angle in quadrant III is $\frac{4\pi}{3}$. So

$$\cos t = -\frac{1}{2}$$
$$t = \frac{2\pi}{3}, \frac{4\pi}{3}.$$

(b) First note that if $0 \leq t \leq 2\pi$, then $0 \leq 3t \leq 6\pi$. So, we need to consider values of $3t$ between 0 and 6π. Since

$$\sin 3t = \frac{\sqrt{2}}{2}, \text{ for } 3t = \frac{\pi}{4}, \frac{3\pi}{4}, \frac{9\pi}{4}, \frac{11\pi}{4}, \frac{17\pi}{4}, \frac{19\pi}{4},$$

we have

$$t = \frac{\pi}{12}, \frac{3\pi}{12}, \frac{9\pi}{12}, \frac{11\pi}{12}, \frac{17\pi}{12}, \frac{19\pi}{12}.$$

◇

EXAMPLE 3.2.8   Find all t in the interval $[0, 2\pi]$, satisfying $(\cos t)^2 - \cos t - 2 = 0$.

SOLUTION: If we set $x = \cos t$, then the equation is the same as

$$x^2 - x - 2 = 0$$

$$(x + 1)(x - 2) = 0$$

$$x = -1, x = 2.$$

So solve

$$\cos t = -1, \cos t = 2.$$

The equation $\cos t = 2$ has no solutions since $-1 \leq \cos t \leq 1$ (recall the same holds for $\sin t$, that is $-1 \leq \sin t \leq 1$). The solutions are,

$$\cos t = -1 \Rightarrow$$

$$t = \pi.$$

◇

Solutions for Exercise Set 3.2

1. (a) $P\left(\dfrac{\pi}{2}\right)$ (b) $P(\pi)$ (c) $P(2\pi)$ (d) $P\left(\dfrac{3\pi}{2}\right)$

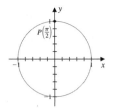 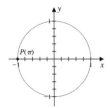 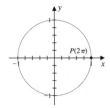 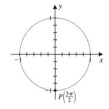

3. (a) $P\left(\dfrac{5\pi}{6}\right)$ (b) $P\left(\dfrac{3\pi}{4}\right)$ (c) $P\left(\dfrac{4\pi}{3}\right)$ (d) $P\left(\dfrac{7\pi}{6}\right)$

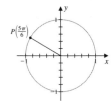 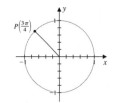 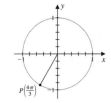 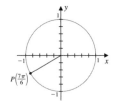

5. (a) $P\left(-\dfrac{\pi}{4}\right)$ (b) $P\left(-\dfrac{4\pi}{3}\right)$ (c) $P\left(-\dfrac{37\pi}{6}\right)$ (d) $P\left(-\dfrac{7\pi}{4}\right)$

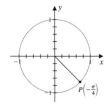 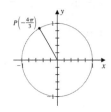 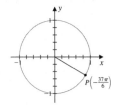 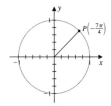

7. (a) $r = \dfrac{\pi}{4}$ (b) $r = \dfrac{\pi}{3}$ (c) $r = \dfrac{\pi}{3}$ (d) $r = \dfrac{\pi}{4}$

9. (a) $r = \dfrac{\pi}{3}$ (b) $r = \dfrac{\pi}{6}$ (c) $r = \dfrac{\pi}{3}$ (d) $r = \dfrac{\pi}{4}$

11. $\sin\dfrac{\pi}{6} = \dfrac{1}{2}$; $\cos\dfrac{\pi}{6} = \dfrac{\sqrt{3}}{2}$

13. $\sin\dfrac{3\pi}{4} = \sin\dfrac{\pi}{4} = \dfrac{\sqrt{2}}{2}$; $\cos\dfrac{3\pi}{4} = -\cos\dfrac{\pi}{4} = -\dfrac{\sqrt{2}}{2}$

15. $\sin\dfrac{4\pi}{3} = -\sin\dfrac{\pi}{3} = -\dfrac{\sqrt{3}}{2}$; $\cos\dfrac{4\pi}{3} = -\cos\dfrac{\pi}{3} = -\dfrac{1}{2}$

17. $\sin\dfrac{11\pi}{6} = -\sin\dfrac{\pi}{6} = -\dfrac{1}{2}$; $\cos\dfrac{11\pi}{6} = \cos\dfrac{\pi}{6} = \dfrac{\sqrt{3}}{2}$

19. $\sin\left(-\dfrac{\pi}{3}\right) = -\sin\dfrac{\pi}{3} = -\dfrac{\sqrt{3}}{2}$; $\cos\left(-\dfrac{\pi}{3}\right) = \cos\dfrac{\pi}{3} = \dfrac{1}{2}$

21. $\sin\left(-\dfrac{5\pi}{6}\right) = -\sin\dfrac{\pi}{6} = -\dfrac{1}{2}$; $\cos\left(-\dfrac{5\pi}{6}\right) = -\cos\dfrac{\pi}{6} = -\dfrac{\sqrt{3}}{2}$

23. $\sin\left(-\dfrac{5\pi}{4}\right) = \sin\dfrac{\pi}{4} = \dfrac{\sqrt{2}}{2}$; $\cos\left(-\dfrac{5\pi}{4}\right) = -\cos\dfrac{\pi}{4} = -\dfrac{\sqrt{2}}{2}$

25. $\sin\left(-\dfrac{5\pi}{2}\right) = \sin\dfrac{3\pi}{2} = -1$; $\cos\left(-\dfrac{5\pi}{2}\right) = \cos\dfrac{3\pi}{2} = 0$

27. $\sin(-7\pi) = \sin(\pi) = 0; \cos(-7\pi) = \cos(\pi) = -1$

29. $\cos t = \dfrac{\sqrt{2}}{2} \Leftrightarrow t = \dfrac{\pi}{4}, \dfrac{7\pi}{4}$

31. $\sin t = -\dfrac{1}{2} \Leftrightarrow t = \dfrac{7\pi}{6}, \dfrac{11\pi}{6}$

33. $\cos t = 1 \Leftrightarrow t = 0, 2\pi$

35. If $0 \le t \le 2\pi$, then $0 \le \dfrac{t}{2} \le \pi$ and $\cos\dfrac{t}{2} = \dfrac{1}{2} \Leftrightarrow \dfrac{t}{2} = \dfrac{\pi}{3} \Leftrightarrow t = \dfrac{2\pi}{3}$.

37. The function $f(x) = (\cos x)^2$ is even since

$$f(-x) = (\cos(-x))^2 = (\cos x)^2 = f(x).$$

39. The function $f(x) = |x|\sin x$, is odd since

$$f(-x) = |-x|\sin(-x) = |x|(-\sin x) = -|x|\sin x = -f(x).$$

41. Let $P(t) = \left(\dfrac{3}{5}, \dfrac{4}{5}\right)$.

(a) Since $P(t)$ is in quadrant I, adding π to t moves the point to quadrant III, so $P(t + \pi) = \left(-\dfrac{3}{5}, -\dfrac{4}{5}\right)$.

(b) Since $P(t)$ is in quadrant I, $P(-t)$ is in quadrant IV, so $P(-t) = \left(\dfrac{3}{5}, -\dfrac{4}{5}\right)$.

(c) Since $P(t)$ is in quadrant I, subtracting π from t moves the point to quadrant III, so $P(t - \pi) = \left(-\dfrac{3}{5}, -\dfrac{4}{5}\right)$.

(d) Since $P(t)$ is in quadrant I, $P(-t)$ is in quadrant IV so subtracting π from $-t$ moves the point to quadrant II, so $P(-t - \pi) = \left(-\dfrac{3}{5}, \dfrac{4}{5}\right)$.

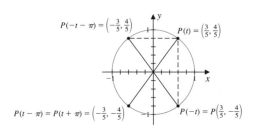

43. We have

$$(\cos t)^2 + \cos t - 2 = 0 \Leftrightarrow (\cos t + 2)(\cos t - 1) = 0$$

which holds when $\cos t + 2 = 0$ or $\cos t - 1 = 0$. Hence $\cos t = -2$, which never holds, or $\cos t = 1 \Leftrightarrow t = 0, 2\pi$.

45. We have

$$0 = \sin t \cos t - \sin t - \cos t + 1 = \sin t(\cos t - 1) - (\cos t - 1)$$
$$= (\sin t - 1)(\cos t - 1).$$

So $\sin t - 1 = 0$ or $\cos t - 1 = 0$, which implies that $\sin t = 1$ or $\cos t = 1$. Hence $t = 0, \dfrac{\pi}{2}, 2\pi$.

47. (a)

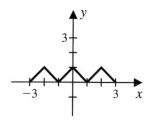

(b) We have $f(x) = 0$ for $x = \pm 1, \pm 3, \pm 5, \ldots$ and $f(x) = 1$ for $x = 0, \pm 2, \pm 4, \pm 6, \ldots$. The range of the function is the interval $[0, 1]$.

3.3 Graphs of the Sine and Cosine Functions

A function f is *periodic* if a positive number T exists with $f(t + T) = f(t)$, for all t in the domain of f. The smallest such T is called the *period* of the function. Since the circumference of the unit circle is 2π,

$$\sin(t + 2\pi) = \sin t \text{ and } \cos(t + 2\pi) = \cos t,$$

and the period of the sine and cosine functions is 2π.

Three important features of the graphs of

$$y = A\sin(Bx + C), \text{ and } y = A\cos(Bx + C)$$

where $A \neq 0$ and $B > 0$ are:

(1) <u>Amplitude:</u> The *amplitude* is the height of the sine or cosine wave which equals $|A|$.

(2) <u>Period:</u> The period of the sine or cosine function is $\frac{2\pi}{B}$.

(3) <u>Phase Shift:</u> To determine the amount the sine or cosine is horizontally shifted first write the function in the form

$$y = A\sin B\left(x + \frac{C}{B}\right) \text{ or } y = A\cos B\left(x + \frac{C}{B}\right).$$

The graph is horizontally shifted $\frac{C}{B}$ units to the left if $\frac{C}{B} > 0$, and horizontally shifted $\left|\frac{C}{B}\right|$ to the right if $\frac{C}{B} < 0$.

EXAMPLE 3.3.1   (i) Use the graph of the sine to sketch one period of the graph. (ii) Specify the amplitude and (iii) the period of the graph.

(a) $y = \frac{1}{2}\cos x$ (b) $y = -2\cos 3x$ (c) $y = 2\sin 2x$ (d) $y = -\frac{1}{2}\sin \frac{1}{2}x$

SOLUTION:

(a)

<u>Amplitude:</u> $A = \frac{1}{2}$

<u>Period:</u> 2π

<u>x-intercepts on $[0, 2\pi]$:</u>

$$\frac{1}{2}\cos x = 0$$

$$\cos x = 0$$

$$x = \frac{\pi}{2}, \frac{3\pi}{2}$$

<u>Maximums and minimums:</u>

$$\frac{1}{2}\cos x = \frac{1}{2}, \frac{1}{2}\cos x = -\frac{1}{2}$$

$$\cos x = 1, \cos x = -1$$

$$x = 0, 2\pi, \ x = \pi$$

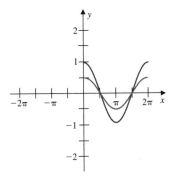

Figure 3.10

(b) $y = -2\cos 3x$ is in the form $y = -A\cos Bx$ with $A = 2$ and $B = 3$.

Amplitude: $A = 2$

Period: $\frac{2\pi}{B} = \frac{2\pi}{3}$

Since the period is $\frac{2\pi}{3}$, one complete wave occurs on the interval $\left[0, \frac{2\pi}{3}\right]$.

<u>x-intercepts on</u> $\left[0, \frac{2\pi}{3}\right]$: Check all values of x so that $0 \le x \le \frac{2\pi}{3} \Rightarrow 0 \le 3x \le 2\pi$. This gives

$$-2\cos 3x = 0$$

$$\cos 3x = 0$$

$$3x = \frac{\pi}{2}, \frac{3\pi}{2}$$

$$x = \frac{\pi}{6}, \frac{\pi}{2}.$$

<u>Maximums and minimums on</u> $\left[0, \frac{2\pi}{3}\right]$:

$$-2\cos 3x = 2, \, -2\cos 3x = -2$$

$$\cos 3x = -1, \, \cos 3x = 1$$

$$3x = \pi, \, 3x = 0, 2\pi$$

$$x = \frac{\pi}{3}, \, x = 0, \frac{2\pi}{3}$$

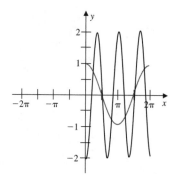

Figure 3.11

(c) $y = 2\sin 2x$ is in the form $y = A\sin Bx$ where $A = 2$, and $B = 2$.

<u>Amplitude:</u> $A = 2$

<u>Period:</u> $\frac{2\pi}{B} = \frac{2\pi}{2} = \pi$

Since the period is π, one complete wave occurs on the interval $[0, \pi]$.

<u>x-intercepts on</u> $[0, \pi]$: Check all values of x so that $0 \le x \le \pi \Rightarrow 0 \le 2x \le 2\pi$.
This gives

$$2\sin 2x = 0$$

$$2x = 0, \pi, 2\pi$$

$$x = 0, \frac{\pi}{2}, \pi.$$

<u>Maximums and minimums on</u> $[0, \pi]$:

$$2\sin 2x = 2, 2\sin 2x = -2$$

$$\sin 2x = 1, \sin 2x = -1$$

$$2x = \frac{\pi}{2}, 2x = \frac{3\pi}{2}$$

$$x = \frac{\pi}{4}, x = \frac{3\pi}{4}$$

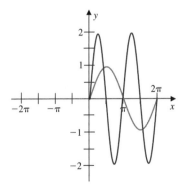

Figure 3.12

(d)$y = -\dfrac{1}{2}\sin\dfrac{1}{2}x$ is in the form $y = A\sin Bx$ where $A = \dfrac{1}{2}$ and $B = \dfrac{1}{2}$.

The graph of $y = -\dfrac{1}{2}\sin\dfrac{1}{2}x$ is the reflection about the x-axis of the graph of $y = \dfrac{1}{2}\sin\dfrac{1}{2}x$.

Amplitude: $A = \frac{1}{2}$

Period: $\frac{2\pi}{B} = \frac{2\pi}{1/2} = 4\pi$

x-intercepts: Checking all x so that $0 \le x \le 4\pi \Rightarrow 0 \le \frac{1}{2}x \le 2\pi$ gives

$$-\frac{1}{2}\sin\frac{1}{2}x = 0$$
$$\frac{1}{2}x = 0, \pi, 2\pi$$
$$x = 0, 2\pi, 4\pi.$$

Maximums and minimums:

$$-\frac{1}{2}\sin\frac{1}{2}x = \frac{1}{2}, -\frac{1}{2}\sin\frac{1}{2}x = -\frac{1}{2}$$
$$\sin\frac{1}{2}x = -1, \sin\frac{1}{2}x = 1$$
$$\frac{1}{2}x = \frac{3\pi}{2}, \frac{1}{2}x = \frac{\pi}{2}$$
$$x = 3\pi, x = \pi$$

◇

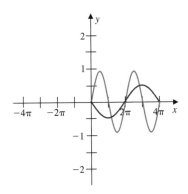

Figure 3.13

EXAMPLE 3.3.2   Use the graph of the sine and cosine to sketch one period of the graph of the given function.

(a) $y = \cos\left(x - \frac{\pi}{4}\right)$ (b) $y = -1 + 2\sin\left(3x - \frac{\pi}{2}\right)$

SOLUTION:

(a)

Amplitude: 1

Period: Same as $y = \cos x$, so 2π.

Phase Shift: $\frac{\pi}{4}$ units to the right.

So the graph is obtained by shifting the graph of $y = \cos x$, to the right $\frac{\pi}{4}$ units.

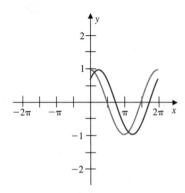

Figure 3.14

(b) First rewrite the function as

$$y = -1 + 2\sin\left(3x - \frac{\pi}{2}\right)$$
$$= -1 + 2\sin 3\left(x - \frac{\pi}{6}\right).$$

Amplitude: 2

Period:

$$\frac{2\pi}{B} = \frac{2\pi}{3}$$

The graph is horizontally compressed and completes one entire wave on an interval of length $\frac{2\pi}{3}$.

Phase Shift: $\frac{\pi}{6}$ units to the right.

x-intercepts of $y = 2\sin\left(3x - \frac{\pi}{2}\right)$:

$$2\sin\left(3x - \frac{\pi}{2}\right) = 0$$

$$\sin\left(3x - \frac{\pi}{2}\right) = 0$$

$$3x - \frac{\pi}{2} = 0, \pi, 2\pi$$

$$3x = \frac{\pi}{2}, \frac{3\pi}{2}, \frac{5\pi}{2}$$

$$x = \frac{\pi}{6}, \frac{\pi}{2}, \frac{5\pi}{6}$$

Maximums and Minimums of $y = 2\sin\left(3x - \frac{\pi}{2}\right)$:

$$2\sin\left(3x - \frac{\pi}{2}\right) = 2 \text{ or } 2\sin\left(3x - \frac{\pi}{2}\right) = -2$$

$$\sin\left(3x - \frac{\pi}{2}\right) = 1 \text{ or } \sin\left(3x - \frac{\pi}{2}\right) = -1$$

$$3x - \frac{\pi}{2} = \frac{\pi}{2} \text{ or } 3x - \frac{\pi}{2} = \frac{3\pi}{2}$$

$$x = \frac{\pi}{3}, x = \frac{2\pi}{3}$$

Notice that $\frac{5\pi}{6} - \frac{\pi}{6} = \frac{4\pi}{6} = \frac{2\pi}{3}$, the period of the function. Also, $\frac{\pi}{3}$ is midway between $\frac{\pi}{6}$ and $\frac{\pi}{2}$, and $\frac{2\pi}{3}$ is midway between $\frac{\pi}{2}$ and $\frac{5\pi}{6}$.

To obtain the graph, start with the graph of $y = \sin x$, and vertically stretch it by a factor of 2. Then horizontally compress the resulting graph so one complete wave occurs on the interval $\left[0, \frac{2\pi}{3}\right]$. Shift this graph to the right $\frac{\pi}{6}$ units, and because of the -1, downward 1 unit.   ◇

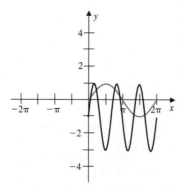

Figure 3.15

EXAMPLE 3.3.3   Find a sine or cosine function whose graph matches the curve.

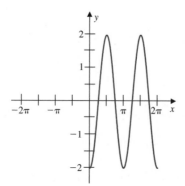

Figure 3.16

SOLUTION: The curve has the appearance of a sine curve of the form $y = A \sin B \left(x + \frac{C}{B} \right)$, which has also been reflected about the x-axis.

Amplitude: The height of the wave is 2, so $|A| = 2$.

<u>Period:</u> One complete wave occurs on the interval $\left[\frac{\pi}{4}, \frac{5\pi}{4}\right]$, so the period is $\frac{5\pi}{4} - \frac{\pi}{4} = \pi$, and

$$\frac{2\pi}{B} = \pi$$
$$B = 2.$$

<u>Phase Shift:</u> The curve is shifted to the left $\frac{\pi}{4}$ units.

The equation of the curve is

$$y = -2\sin 2\left(x + \frac{\pi}{4}\right).$$

The minus sign causes the reflection about the x-axis.

A cosine that gives the same wave is

$$y = -2\cos 2x.$$

◇

EXAMPLE 3.3.4   Match the equation with the graph in the figure,

(a) $y = -\cos 2\left(x - \frac{\pi}{2}\right)$ (b) $y = \cos \frac{1}{2}\left(x + \frac{\pi}{2}\right)$

(c) $y = \sin \frac{1}{2}\left(x - \frac{\pi}{2}\right)$ (d) $y = \sin 2\left(x + \frac{\pi}{2}\right)$

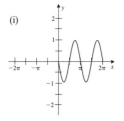

(i)

Figure 3.17

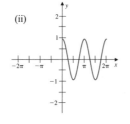

(ii)

Figure 3.18

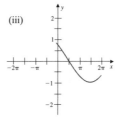

(iii)

Figure 3.19

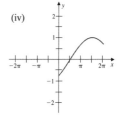

(iv)

Figure 3.20

SOLUTION: The period of the waves in (a) and (d) is $\frac{2\pi}{2} = \pi$, and the period of the waves in (b) and (c) is $\frac{2\pi}{1/2} = 4\pi$. So on an interval of length 4π, we would see one complete wave of the curves described in (b) or (c), but we would see four complete waves of the curves described in (a) or (d). As a consequence, (a) and (d) match (i) or (ii) and (b) and (c) match (iii) or (iv).

(a) If $x = 0$, $y = -\cos 2(0 - \pi/2) = -\cos(-\pi) = 1$, so (a) matches with (ii).

(b) (iii)

(c) If $x = 0$, $y = -\sin \frac{1}{2}(0 - \pi/2) = -\sin(-\pi/4) = \sqrt{2}/2 > 0$ so (c) matches with (iv).

(d) (i)   ◇

EXAMPLE 3.3.5   Find the equation of a cosine wave that is obtained by shifting the graph of $y = \cos x$ to the left 2 units, upward 1 unit, and is horizontally compressed by a factor of 3 when compared with $y = \cos x$.

SOLUTION: The cosine wave has the form

$$y = A \cos B(x + C) + D.$$

Since the graph is not vertically scaled, the amplitude remains one, and $A = 1$. If $y = \cos x$ is horizontally compressed by a factor of 3 ,the period is one third of $y = \cos x$, so

$$\frac{2\pi}{B} = \frac{1}{3}(2\pi) = \frac{2\pi}{3}$$
$$B = 3.$$

To shift the wave to the left 2 units, let $C = -2$, and upward 1 unit, let $D = 1$. So the equation of the wave is

$$y = 1 + \cos 3(x - 2).$$

◇

EXAMPLE 3.3.6   Determine an appropriate viewing rectangle for the function $f(x) = \cos(200x)$, and use it to sketch the graph.

SOLUTION: Since there is no phase shift, the period of the function gives a good viewing rectangle, so compute

$$\frac{2\pi}{200} = \frac{\pi}{100} \approx 0.03.$$

Two complete waves of the function occur on the interval $\left[-\frac{\pi}{100}, \frac{\pi}{100}\right]$, and an appropriate viewing rectangle is $\left[-\frac{\pi}{100}, \frac{\pi}{100}\right] \times [-1, 1]$.   ◇

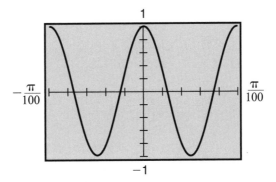

Figure 3.21

EXAMPLE 3.3.7   Use a graphing device to approximate the solutions to the equation $\cos x + x = x^3 + 1$.

SOLUTION: To approximate the solution, use a graphing device to sketch the two curves, $y = \cos x + x$ and $y = x^3 + 1$, and approximate the points of intersection.

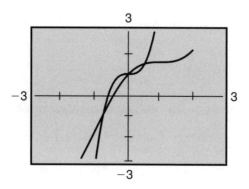

Figure 3.22

The points of intersection are

$$x = 0, \quad x \approx 0.8, \quad \text{and} \quad x \approx -1.25.$$

◇

EXAMPLE 3.3.8   Write each of the following as the composition of two functions $h(x) = f(g(x))$.

(a) $h(x) = \sqrt{\cos x}$ (b) $h(x) = 2\sin(2x + 1)$

SOLUTION:

(a) The inside operation takes the cosine of the number x, and the outside operation takes the square root. So define

$$f(x) = \sqrt{x}, \text{ and } g(x) = \cos x,$$

and

$$f(g(x)) = f(\cos x) = \sqrt{\cos x}.$$

(b) The inside operation is the argument of the sine function. Define

$$f(x) = 2\sin x, \quad \text{and} \quad g(x) = 2x + 1,$$

so

$$f(g(x)) = f(2x + 1) = 2\sin(2x + 1).$$

◇

Solutions for Exercise Set 3.3

1. (a) For $y = 2\cos x$

(i)

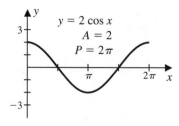

(ii) Amplitude $= 2$ (iii) Period $= 2\pi$

(b) For $y = -3\cos x$

(i)

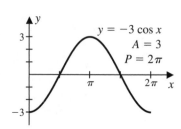

(ii) Amplitude $= 3$; (iii) Period $= 2\pi$

(c) For $y = \frac{1}{2}\cos 2x$

(i)

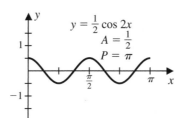

(ii) Amplitude $= \frac{1}{2}$; (iii) Period $= \frac{2\pi}{2} = \pi$

3. (a) For $y = 2 \cos \pi x$,

   (i)

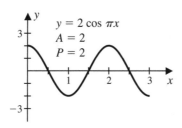

(ii) Amplitude $= 2$; (iii) Period $= \frac{2\pi}{\pi} = 2$

(b) For $y = \cos 2\pi x$

   (i)

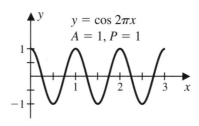

(ii) Amplitude $= 1$; (iii) Period $= \frac{2\pi}{2\pi} = 1$

(c) For $y = -2 \cos \frac{\pi}{2} x$

(i)

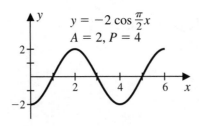

(ii) Amplitude $= 2$;   (iii) Period $= \dfrac{2\pi}{\pi/2} = 4$

5. For $y = 1 + \cos x$

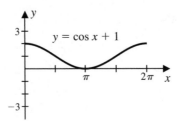

7. For $y = \cos\left(x - \dfrac{\pi}{2}\right)$

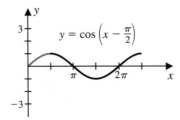

9. For $y = -2 + \cos \pi x$

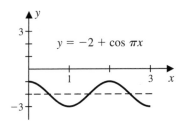

11. For $y = 1 + \cos(2x + \pi) = 1 + \cos 2\left(x + \frac{\pi}{2}\right)$

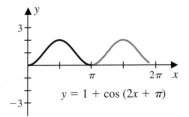

13. For $y = -2 + 3\sin\left(3x - \frac{\pi}{2}\right) = -2 + 3\sin 3\left(x - \frac{\pi}{6}\right)$

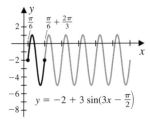

15. For $y = -2\sin(x - 1) + 3$

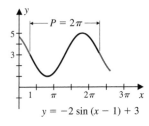

17. For $y = |\cos x|$

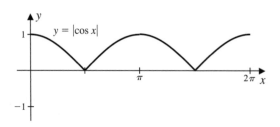

19. (a) The equation of the graph has the form $y = A\cos(Bx)$, where the amplitude is 3 and the period is π. So $A = 3$ and $\frac{2\pi}{B} = \pi \Rightarrow B = 2$ and $y = 3\cos 2x$.

(b) The equation can also be given as a sine curve shifted to the left by $\frac{\pi}{4}$ units. With amplitude 3, period π, and phase shift $-\frac{\pi}{4}$, the sine wave has equation $y = 3\sin 2\left(x + \frac{\pi}{4}\right)$.

21. (a) The equation of the graph has the form

$$y = A\cos(Bx + C) + D = A\cos B\left(x + \frac{C}{B}\right) + D,$$

where the amplitude is 1, the period is 6, the curve has been shifted horizontally to the right $\frac{5}{2}$ units and upward 2 units. So $A = 1$, $\frac{2\pi}{B} = 6 \Rightarrow B = \frac{\pi}{3}$, $\frac{C}{B} = -\frac{5}{2}$, $D = 2$, and $y = \cos\frac{\pi}{3}\left(x - \frac{5}{2}\right) + 2$.

(b) The equation of the graph can also have the form

$$y = A\sin(Bx + C) + D = A\sin B\left(x + \frac{C}{B}\right) + D,$$

where the amplitude is 1, the period is 6, the curve has been shifted horizontally to the right 1 unit, and the curve has been shifted upward 2 units. So $A = 1$, and $\frac{2\pi}{B} = 6 \Rightarrow B = \frac{\pi}{3}$, $\frac{C}{B} = -1$, $D = 2$, and $y = \sin\frac{\pi}{3}(x - 1) + 2$.

23. The inside operation is taking the cosine and the outside operation is raising to the fourth power. Let $f(x) = x^4$ and $g(x) = \cos x$, then $h(x) = f(g(x)) = f(\cos x) = (\cos x)^4$.

25. The argument of the cosine is the inside function, so let $f(x) = 3\cos x$, and $g(x) = 4x - 2$. Then $h(x) = f(g(x)) = f(4x - 2) = 3\cos(4x - 2)$.

27. The amplitude of the waves in (a) and (c) is 2 which could only match (i) and (iv) and the amplitude in (b) and (d) is $\frac{1}{2}$ which could only match (ii) and (iii). Setting $x = 0$ in each of the equations the y-intercepts are $2\sin\left(0 - \frac{\pi}{2}\right) = -2, 2\cos\left(0 - \frac{\pi}{2}\right) = 0, \frac{1}{2}\sin\left(0 + \frac{\pi}{2}\right) = \frac{1}{2}$, and $\frac{1}{2}\cos\left(0 + \frac{\pi}{2}\right) = 0$. The matching is then (a) iv, (b) i, (c) (iii), and (d) ii.

29. $y = -1 + \frac{1}{2}\sin(x - 2)$

31. The period of $f(x) = \cos(100x)$, is $\frac{2\pi}{100} = \frac{\pi}{50} \approx 0.06$, and the amplitude is 1, so a reasonable viewing rectangle is $[-\pi/50, \pi/50] \times [-1, 1]$.

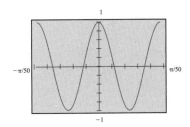

33. The period of $f(x) = \sin\left(\frac{x}{50}\right)$ is $\frac{2\pi}{(1/50)} = 100\pi$, and the amplitude is 1, so a reasonable viewing rectangle is $[-100\pi, 100\pi] \times [-1, 1]$.

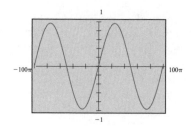

35. $\cos x = x \Rightarrow x \approx 0.7$

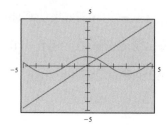

37. $\sin x + \cos x = x \Rightarrow x \approx 1.3$

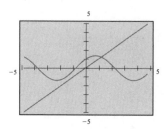

39. For $f(t) = 2\sin\left(3t - \frac{\pi}{4}\right)$:

(a) $f(t) = 0 \Rightarrow \sin\left(3t - \frac{\pi}{4}\right) = 0 \Rightarrow 3t - \frac{\pi}{4} = 0 \Rightarrow t = \frac{\pi}{12}$.

(b) Since the amplitude of the sine curve is 2, we need to find the smallest

t so that $2\sin\left(3t - \frac{\pi}{4}\right) = 2 \Rightarrow \sin\left(3t - \frac{\pi}{4}\right) = 1 \Rightarrow 3t - \frac{\pi}{4} = \frac{\pi}{2} \Rightarrow 3t = \frac{3\pi}{4} \Rightarrow t = \frac{\pi}{4}$.

(c) The minimum value is -2, so $2\sin\left(3t - \frac{\pi}{4}\right) = -2 \Rightarrow \sin\left(3t - \frac{\pi}{4}\right) = -1 \Rightarrow 3t - \frac{\pi}{4} = \frac{3\pi}{2} \Rightarrow 3t = \frac{7\pi}{4} \Rightarrow t = \frac{7\pi}{12}.$

41. The data points exhibit a sine wave pattern centered vertically on the value 1, the amplitude is 2 and the period is 4. So the equation of the graph is of the form $y = 2\sin(Bx) + 1$, where $\frac{2\pi}{B} = 4 \Rightarrow B = \frac{\pi}{2}$. So $y = 2\sin\left(\frac{\pi}{2}x\right) + 1.$

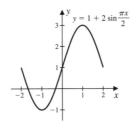

43. In each case the graph of $f(x) = \cos x$ or $f(x) = \sin x$ is rotated $45°$ counterclockwise or clockwise to oscillate about the line $y = x$ or $y = -x.$

45. $y = a\cos(bx + c),$ where $a > 0, b > 0,$ and $c > 0$

(a) If b and c are fixed and a is doubled the amplitude is doubled.

(b) If a and c are fixed and b is doubled the period is halved.

(c) If a and b are fixed and c is doubled the horizontal shift is doubled.

3.4 Other Trigonometric Functions

There are four additional trigonometric functions defined in terms of the sine and cosine.

$$\tan x = \frac{\sin x}{\cos x} \qquad \cot x = \frac{\cos x}{\sin x}$$

$$\sec x = \frac{1}{\cos x} \qquad \csc x = \frac{1}{\sin x}$$

The tangent and secant functions are defined whenever $x \neq \frac{\pi}{2} + n\pi$ for an integer n. The cotangent and cosecant are defined whenever $x \neq n\pi$ for an integer n.

Being familiar with the values of the trigonometric functions at the standard angles is essential to work in trigonometry. The next example reviews some of the standard material.

EXAMPLE 3.4.1   Complete the table.

SOLUTION: The table asks for the six trigonometric functions at t, where $-2\pi \leq t \leq -7\pi/6$, which are in quadrants I and II. We have added to the table the reference angles r along with their sine and cosine. To find the cosine and sine at a value t, determine the quadrant where the angle lies and adjust the signs of the sine and cosine for the corresponding reference angle. The angles $t = -7\pi/6, -5\pi/4$, and $-4\pi/3$ are in quadrant II, where $\sin t > 0$ and $\cos t < 0$. The angles $t = -5\pi/3, -7\pi/4$, and $-11\pi/6$ are in quadrant I, where $\sin t > 0$ and $\cos t > 0$. ◇

t	$-\dfrac{7\pi}{6}$	$-\dfrac{5\pi}{4}$	$-\dfrac{4\pi}{3}$	$-\dfrac{3\pi}{2}$	$-\dfrac{5\pi}{3}$	$-\dfrac{7\pi}{4}$	$-\dfrac{11\pi}{6}$	-2π
r	$\dfrac{\pi}{6}$	$\dfrac{\pi}{4}$	$\dfrac{\pi}{3}$	$\dfrac{\pi}{2}$	$\dfrac{\pi}{3}$	$\dfrac{\pi}{4}$	$\dfrac{\pi}{6}$	0
$\sin r$	$\dfrac{1}{2}$	$\dfrac{\sqrt{2}}{2}$	$\dfrac{\sqrt{3}}{2}$	1	$\dfrac{\sqrt{3}}{2}$	$\dfrac{\sqrt{2}}{2}$	$\dfrac{1}{2}$	0
$\cos r$	$\dfrac{\sqrt{3}}{2}$	$\dfrac{\sqrt{2}}{2}$	$\dfrac{1}{2}$	0	$\dfrac{1}{2}$	$\dfrac{\sqrt{2}}{2}$	$\dfrac{\sqrt{3}}{2}$	1
$\sin t$	$\dfrac{1}{2}$	$\dfrac{\sqrt{2}}{2}$	$\dfrac{\sqrt{3}}{2}$	1	$\dfrac{\sqrt{3}}{2}$	$\dfrac{\sqrt{2}}{2}$	$\dfrac{1}{2}$	0
$\cos t$	$-\dfrac{\sqrt{3}}{2}$	$-\dfrac{\sqrt{2}}{2}$	$-\dfrac{1}{2}$	0	$\dfrac{1}{2}$	$\dfrac{\sqrt{2}}{2}$	$\dfrac{\sqrt{3}}{2}$	1
$\tan t$	$-\dfrac{\sqrt{3}}{3}$	-1	$-\sqrt{3}$	$-$	$\sqrt{3}$	1	$\dfrac{\sqrt{3}}{3}$	0
$\cot t$	$-\sqrt{3}$	-1	$-\dfrac{\sqrt{3}}{3}$	0	$\dfrac{\sqrt{3}}{3}$	1	$\sqrt{3}$	$-$
$\sec t$	$-\dfrac{2\sqrt{3}}{3}$	$-\sqrt{2}$	-2	$-$	2	$\sqrt{2}$	$\dfrac{2\sqrt{3}}{3}$	1
$\csc t$	2	$\sqrt{2}$	$\dfrac{2\sqrt{3}}{3}$	1	$\dfrac{2\sqrt{3}}{3}$	$\sqrt{2}$	2	$-$

EXAMPLE 3.4.2 Find the values of all the trigonometric functions given that $\sin t = 3/5$ and $\pi/2 \le t \le \pi$.

SOLUTION: First find the $\cos t$ using the Pythagorean Identity. That is,

$$(\cos t)^2 + (\sin t)^2 = 1$$
$$(\cos t)^2 + \left(\frac{3}{5}\right)^2 = 1$$
$$(\cos t)^2 = 1 - \frac{9}{25} = \frac{16}{25}$$
$$\cos t = \pm\sqrt{\frac{16}{25}}$$
$$\cos t = \pm\frac{4}{5}.$$

Now apply the fact that $\pi/2 \le t \le \pi$, so t is in quadrant II, where $\cos t < 0$. Thus,

$$\cos t = -\frac{4}{5}, \quad \sin t = \frac{3}{5}$$

$$\tan t = \frac{\frac{3}{5}}{-\frac{4}{5}} = -\frac{3}{4}, \quad \cot t = \frac{-\frac{4}{5}}{\frac{3}{5}} = -\frac{4}{3}$$

$$\sec t = \frac{1}{-\frac{4}{5}} = -\frac{5}{4}, \quad \csc t = \frac{1}{\frac{3}{5}} = \frac{5}{3}.$$

◇

OTHER GRAPHS

	Period	x-intercepts	Vertical Asymptotes
$y = \tan x$	π	$x = \pm k\pi, k = 0, 1...$	$x = \pm(2k - 1)\frac{\pi}{2}, k = 1, 2...$
$y = \cot x$	π	$x = \pm(2k - 1)\frac{\pi}{2}, k = 1, 2...$	$x = \pm k\pi, k = 0, 1...$
$y = \sec x$	2π	none	$x = \pm(2k - 1)\frac{\pi}{2}, k = 1, 2...$
$y = \csc x$	2π	none	$x = \pm k\pi, k = 0, 1...$

EXAMPLE 3.4.3   Sketch one period of the given curve.

(a) $y = 2\tan\left(x - \frac{\pi}{2}\right)$ (b) $y = \sec(3x)$ (c) $y = \cot(2x + \pi)$

SOLUTION:

(a)

<u>Period:</u> π, the same as the period of tangent.

<u>Phase shift:</u> $\frac{\pi}{2}$ units to the right.

The strategy is to use the graph of $y = \tan x$, vertically stretch it by a factor of 2, and then shift the resulting graph to the right $\frac{\pi}{2}$ units.

<u>x-intercept:</u>

$$2\tan\left(x - \frac{\pi}{2}\right) = 0$$

$$\tan\left(x - \frac{\pi}{2}\right) = \frac{\sin\left(x - \frac{\pi}{2}\right)}{\cos\left(x - \frac{\pi}{2}\right)} = 0$$

$$\sin\left(x - \frac{\pi}{2}\right) = 0$$

$$x - \frac{\pi}{2} = 0$$

$$x = \frac{\pi}{2}$$

This is exactly what is expected since $y = \tan x$ crosses the x-axis at $(0,0)$, and the graph of $y = 2\tan\left(x - \frac{\pi}{2}\right)$ is just shifted $\frac{\pi}{2}$ units to the right. Here it crosses the x-axis at $\left(\frac{\pi}{2}, 0\right)$.

<u>Asymptotes:</u> $y = \tan x$ has asymptotes $x = -\frac{\pi}{2}$ and $x = \frac{\pi}{2}$, so $y = 2\tan\left(x - \frac{\pi}{2}\right)$ has asymptotes $x = -\frac{\pi}{2} + \frac{\pi}{2} = 0$ and $x = \frac{\pi}{2} + \frac{\pi}{2} = \pi$.

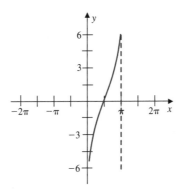

Figure 3.23

(b) One way to quickly plot $y = \sec 3x$ is to first plot $y = \cos 3x$. The graph of $y = \sec 3x$ is then a series of U-shaped curves that sit above and below the high points and low points, respectively, of the cosine wave.

<u>Period:</u> $\frac{2\pi}{3}$

<u>Phase shift:</u> none

<u>Vertical asymptotes:</u> Since $\sec 3x = \dfrac{1}{\cos 3x}$, the vertical asymptotes occur when $\cos 3x = 0$. We have

$$\cos 3x = 0$$

$$3x = \pm\frac{\pi}{2}, \pm\frac{3\pi}{2}, \pm\frac{5\pi}{2}, \ldots$$

$$x = \pm\frac{\pi}{6}, \pm\frac{\pi}{2}, \pm\frac{5\pi}{6}, \ldots$$

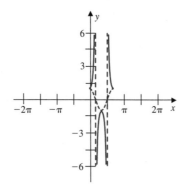

Figure 3.24

(c) Rewrite the function in form

$$y = \cot(2x + \pi) = \cot 2\left(x + \frac{\pi}{2}\right).$$

<u>Period:</u> Since the period of the cotangent function is π, the period is $\frac{\pi}{2}$.

<u>Phase shift:</u> $\frac{\pi}{2}$ units to the left.

One period of $y = \cot x$ occurs on the interval $(0, \pi)$, has vertical asymptotes at $x = 0$ and $x = \pi$, and crosses the x-axis at $x = \frac{\pi}{2}$. So, $y = \cot 2x$, has period

$\frac{\pi}{2}$, vertical asymptotes $x = 0$ and $x = \frac{\pi}{2}$, and crosses the x-axis at $x = \frac{\pi}{4}$. Now shift this graph $\frac{\pi}{2}$ units to the left.

<u>Vertical asymptotes of</u> $y = \cot 2\left(x + \frac{\pi}{2}\right) : x = -\frac{\pi}{2}$ and $x = 0$.

<u>x-intercept of</u> $y = \cot 2\left(x + \frac{\pi}{2}\right) : \left(-\frac{\pi}{4}, 0\right)$

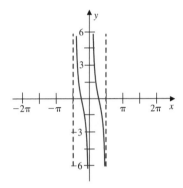

Figure 3.25

Notice one period of the graph also lies between 0 and $\frac{\pi}{2}$ with vertical asymptotes $x = 0$ and $x = \frac{\pi}{2}$, and x-intercept $\frac{\pi}{4}$. This is because the curve has period $\frac{\pi}{2}$ and is shifted to the left $\frac{\pi}{2}$ units.   ◇

EXAMPLE 3.4.4   Find all values of t in the interval $[0, 2\pi]$ that satisfy the equation $|\cot t| = 1$.

SOLUTION:

$$|\cot t| = 1$$
$$\left|\frac{\cos t}{\sin t}\right| = 1$$
$$|\sin t| = |\cos t|$$

In the interval $[0, 2\pi]$, the sine and cosine have the same magnitude, equaling $\frac{\sqrt{2}}{2}$, at $\frac{\pi}{4}, \frac{3\pi}{4}, \frac{5\pi}{4}$, and $\frac{7\pi}{4}$. So

$$|\sin t| = |\cos t| = \frac{\sqrt{2}}{2}, \quad \text{for} \quad t = \frac{\pi}{4}, \frac{3\pi}{4}, \frac{5\pi}{4}, \frac{7\pi}{4}.$$

◇

EXAMPLE 3.4.5   Find all values of t in the interval $[0, 2\pi]$ that satisfy the equation $2 \sin 2t - \sqrt{3} \tan 2t = 0$.

SOLUTION: We have the following sequence of equations:

$$2 \sin 2t - \sqrt{3} \tan 2t = 0$$

$$2 \sin 2t - \sqrt{3} \frac{\sin 2t}{\cos 2t} = 0$$

$$\frac{2 \sin 2t \cos 2t - \sqrt{3} \sin 2t}{\cos 2t} = 0$$

$$2 \sin 2t \cos 2t - \sqrt{3} \sin 2t = 0$$

$$\sin 2t (2 \cos 2t - \sqrt{3}) = 0.$$

Solutions to this equation for $0 \le t \le 2\pi \Rightarrow 0 \le 2t \le 4\pi$, occur when

$$\sin 2t = 0 \quad \text{or} \quad \cos 2t = \frac{\sqrt{3}}{2}$$

$$2t = 0, \pi, 2\pi, 3\pi, 4\pi \quad \text{or} \quad 2t = \frac{\pi}{3}, \frac{5\pi}{3}, \frac{7\pi}{3}, \frac{11\pi}{3},$$

that is,

$$t = 0, \frac{\pi}{2}, \pi, \frac{3\pi}{2}, 2\pi, \frac{\pi}{6}, \frac{5\pi}{6}, \frac{7\pi}{6}, \frac{11\pi}{6}.$$

◇

EXAMPLE 3.4.6   Determine the values of the trigonometric functions of t if $P(t)$ lies in the third quadrant and on the line $y = 3x$.

SOLUTION: To find the sine and cosine, we need to determine the x- and y-coordinates of the point $P(t)$. Since the point lies on both the line and the unit circle, it must satisfy the two equations,

$$y = 3x \quad \text{and} \quad x^2 + y^2 = 1.$$

Substituting the value of y from the first equation into the second gives

$$x^2 + (3x)^2 = 1$$

$$10x^2 = 1$$

$$x^2 = \frac{1}{10}$$

$$x = \pm\sqrt{\frac{1}{10}} = \pm\frac{\sqrt{10}}{10}.$$

Since $P(t)$ is in the third quadrant, $\cos t < 0$, and the x-coordinate of $P(t)$ is less than 0. So

$$x = -\frac{\sqrt{10}}{10} \quad \text{and} \quad y = -\frac{3\sqrt{10}}{10}.$$

The values of the trigonometric functions are

$$\sin t = -\frac{3\sqrt{10}}{10}, \quad \cos t = -\frac{\sqrt{10}}{10}$$

$$\tan t = 3, \quad \cot t = \frac{1}{3}$$

$$\sec t = -\frac{10}{\sqrt{10}} = -\sqrt{10}, \quad \csc t = -\frac{10}{3\sqrt{10}} = -\frac{\sqrt{10}}{3}.$$

◇

Solutions for Exercise Set 3.4

1. $\sin t < 0$ and $\cos t > 0 \Rightarrow P(t)$ is in quadrant IV.

3. $\sin t < 0$ and $\cot t > 0 \Rightarrow \sin t < 0$ and $\frac{\cos t}{\sin t} > 0 \Rightarrow \sin t < 0$ and $\cos t < 0 \Rightarrow P(t)$ is in quadrant III.

5. $\sin t > 0$ and $\tan t > 0 \Rightarrow \sin t > 0$ and $\frac{\sin t}{\cos t} > 0 \Rightarrow \sin t > 0$ and $\cos t > 0 \Rightarrow P(t)$ is in quadrant I.

7. A dash in the table indicates the trigonometric function is undefined at the given value.

	$\frac{7\pi}{6}$	$\frac{5\pi}{4}$	$\frac{4\pi}{3}$	$\frac{3\pi}{2}$	$\frac{5\pi}{3}$	$\frac{7\pi}{4}$	$\frac{11\pi}{6}$	2π
$\sin t$	$-\frac{1}{2}$	$-\frac{\sqrt{2}}{2}$	$-\frac{\sqrt{3}}{2}$	-1	$-\frac{\sqrt{3}}{2}$	$-\frac{\sqrt{2}}{2}$	$-\frac{1}{2}$	0
$\cos t$	$-\frac{\sqrt{3}}{2}$	$-\frac{\sqrt{2}}{2}$	$-\frac{1}{2}$	0	$\frac{1}{2}$	$\frac{\sqrt{2}}{2}$	$\frac{\sqrt{3}}{2}$	1
$\tan t$	$\frac{\sqrt{3}}{3}$	1	$\sqrt{3}$	—	$-\sqrt{3}$	-1	$-\frac{\sqrt{3}}{3}$	0
$\cot t$	$\sqrt{3}$	1	$\frac{\sqrt{3}}{3}$	0	$-\frac{\sqrt{3}}{3}$	-1	$-\sqrt{3}$	—
$\sec t$	$-\frac{2\sqrt{3}}{3}$	$-\sqrt{2}$	-2	—	2	$\sqrt{2}$	$\frac{2\sqrt{3}}{3}$	1
$\csc t$	-2	$-\sqrt{2}$	$-\frac{2\sqrt{3}}{3}$	-1	$-\frac{2\sqrt{3}}{3}$	$-\sqrt{2}$	-2	—

9. Since $\sin t = \frac{4}{5}$ and $(\cos t)^2 + (\sin t)^2 = 1$, we have

$$(\cos t)^2 = 1 - \left(\frac{4}{5}\right)^2 = \frac{9}{25} \Rightarrow \cos t = \pm\sqrt{\frac{9}{25}} = \pm\frac{3}{5},$$

and since t is in quadrant I, $\cos t > 0$, so $\cos t = \frac{3}{5}$. Then

$$\tan t = \frac{4/5}{3/5} = \frac{4}{3}, \quad \cot t = \frac{3/5}{4/5} = \frac{3}{4}, \quad \sec t = \frac{1}{3/5} = \frac{5}{3}, \quad \csc t = \frac{1}{4/5} = \frac{5}{4}.$$

11. Since $\cos t = -\frac{4}{5}$, with $\pi \le t \le \frac{3\pi}{2} \Rightarrow t$ is in quadrant III, so $\sin t < 0$. Then since $(\cos t)^2 + (\sin t)^2 = 1$, we have

$$(\sin t)^2 = 1 - \left(-\frac{4}{5}\right)^2 = 1 - \frac{16}{25} = \frac{9}{25} \Rightarrow \sin t = \pm\sqrt{\frac{9}{25}} = \pm\frac{3}{5},$$

and since $\sin t < 0$, $\sin t = -\frac{3}{5}$. Then

$$\tan t = \frac{-3/5}{-4/5} = \frac{3}{4}, \ \cot t = \frac{1}{\tan t} = \frac{4}{3}, \ \sec t = \frac{1}{-4/5} = -\frac{5}{4}, \ \csc t = \frac{1}{-3/5} = -\frac{5}{3}.$$

13. Since $\tan t = 2 \Rightarrow \dfrac{\sin t}{\cos t} = 2 \Rightarrow \sin t = 2\cos t \Rightarrow \sqrt{1 - (\cos t)^2} = 2\cos t,$

we have

$$1 - (\cos t)^2 = 4(\cos t)^2 \Rightarrow 5(\cos t)^2 = 1 \Rightarrow \cos t = \pm\sqrt{\frac{1}{5}} = \pm\frac{\sqrt{5}}{5}$$

and since $0 < t < \frac{\pi}{2}, \cos t > 0$, we have $\cos t = \frac{\sqrt{5}}{5}$ and $\sin t = 2\cos t = \frac{2\sqrt{5}}{5}$. Then

$$\cot t = \frac{1}{\tan t} = \frac{1}{2}, \ \sec t = \frac{1}{\sqrt{5}/5} = \frac{5}{\sqrt{5}} = \sqrt{5}, \ \csc t = \frac{1}{2\sqrt{5}/5} = \frac{5}{2\sqrt{5}} = \frac{\sqrt{5}}{2}.$$

15. Since $\sec t = 3 \Rightarrow \cos t = \frac{1}{3}$, and since t is in quadrant IV, $\sin t < 0$.

Since $(\cos t)^2 + (\sin t)^2 = 1$, we have

$$(\sin t)^2 = 1 - \left(\frac{1}{3}\right)^2 = 1 - \frac{1}{9} = \frac{8}{9} \Rightarrow \sin t = \pm\sqrt{\frac{8}{9}} = \pm\frac{2\sqrt{2}}{3},$$

and since $\sin t < 0$, $\sin t = -\frac{2\sqrt{2}}{3}$. Then

$$\tan t = \frac{-2\sqrt{2}/3}{1/3} = -2\sqrt{2}, \ \cot t = -\frac{1}{2\sqrt{2}} = -\frac{\sqrt{2}}{4},$$

$$\csc t = -\frac{1}{2\sqrt{2}/3} = -\frac{3\sqrt{2}}{4}.$$

17. For $y = 2\tan x$

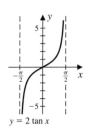

$y = 2 \tan x$

19. For $y = -2 \sec x$

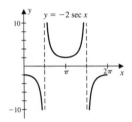

21. For $y = \tan(x + \pi/4)$

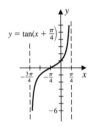

23. For $y = \frac{1}{2} \sec(x + \pi/4)$

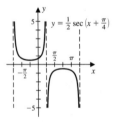

25. For $y = \tan \pi x$ the period is $\frac{\pi}{\pi} = 1$ and we have

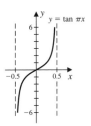

27. For $y = \tan\left(2x - \frac{\pi}{2}\right) = \tan 2\left(x - \frac{\pi}{4}\right)$ the period is $\frac{\pi}{2}$ and we have

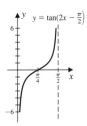

29. $\tan t + 1 = 0 \Leftrightarrow \tan t = -1 \Leftrightarrow t = \frac{3\pi}{4}, \frac{7\pi}{4}$

31. $(\tan t)^2 = \frac{1}{3} \Leftrightarrow \tan t = \pm\frac{\sqrt{3}}{3} \Leftrightarrow t = \frac{\pi}{6}, \frac{5\pi}{6}, \frac{7\pi}{6}, \frac{11\pi}{6}$

33. $|\tan t| = 1 \Leftrightarrow \tan t = 1$ or $\tan t = -1 \Leftrightarrow t = \frac{\pi}{4}, \frac{3\pi}{4}, \frac{5\pi}{4}, \frac{7\pi}{4}$

35. Since

$$2\sin 2t - \sqrt{2}\tan 2t = 0 \Leftrightarrow 2\sin 2t - \sqrt{2}\frac{\sin 2t}{\cos 2t} = 0 \Leftrightarrow \frac{2\sin 2t \cos 2t - \sqrt{2}\sin 2t}{\cos 2t} = 0$$

we have

$$2\sin 2t \cos 2t - \sqrt{2}\sin 2t = 0 \Leftrightarrow \sin 2t(2\cos 2t - \sqrt{2}) = 0 \Rightarrow \sin 2t = 0 \text{ or } \cos 2t = \frac{\sqrt{2}}{2}.$$

If $0 \leq t \leq 2\pi$, then $0 \leq 2t \leq 4\pi$, so $2t = 0, \pi, 2\pi, 3\pi, 4\pi$ or $2t = \frac{\pi}{4}, \frac{7\pi}{4}, \frac{9\pi}{4}, \frac{15\pi}{4}$, which implies that $t = 0, \frac{\pi}{2}, \pi, \frac{3\pi}{2}, 2\pi, \frac{\pi}{8}, \frac{7\pi}{8}, \frac{9\pi}{8}, \frac{15\pi}{8}$.

37. The period of $f(x) = \tan(5x)$ is $\frac{\pi}{5}$, so a reasonable viewing rectangle is $\left[-\frac{\pi}{10}, \frac{\pi}{10}\right] \times [-5, 5]$.

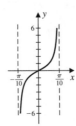

39. The period of $f(x) = \csc 100x$ is $\frac{2\pi}{100} = \frac{\pi}{50}$, so a reasonable viewing rectangle is $\left[0, \frac{\pi}{50}\right] \times [-5, 5]$.

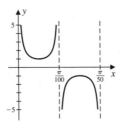

41. The period of $f(x) = \tan\left(\frac{x}{100}\right)$ is $\frac{\pi}{1/100} = 100\pi$, so a reasonable viewing rectangle is $[-50\pi, 50\pi] \times [-5, 5]$.

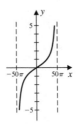

43. Let $P(t) = (x, y)$. Since $P(t)$ lies on the unit circle $x^2 + y^2 = 1$, and since the point lies on the line $y = -2x$, we have

$$x^2 + (-2x)^2 = 1 \Rightarrow 5x^2 = 1 \Rightarrow x = \pm\sqrt{\frac{1}{5}} = \pm\frac{\sqrt{5}}{5}.$$

Since $P(t)$ lies in quadrant IV, $x > 0$ and $y < 0$, so we have $x = \frac{\sqrt{5}}{5}, y = -\frac{2\sqrt{5}}{5}$. Then

$$\cos t = \frac{\sqrt{5}}{5}, \quad \sin t = -\frac{2\sqrt{5}}{5}, \quad \tan t = \frac{-2\sqrt{5}/5}{\sqrt{5}/5} = -2,$$

$$\cot t = -\frac{1}{2}, \quad \sec t = \frac{5}{\sqrt{5}} = \sqrt{5}, \quad \csc t = -\frac{5}{2\sqrt{5}} = -\frac{\sqrt{5}}{2}.$$

3.5 Trigonometric Identities

An *identity* is an equation that is true for every value of the variable. Trigonometric identities are equations that are true for all values of the variable for which both sides are defined.

FUNDAMENTAL IDENTITIES

$$\tan x = \frac{\sin x}{\cos x} \qquad\qquad \cot x = \frac{\cos x}{\sin x}$$

$$\sec x = \frac{1}{\cos x} \qquad\qquad \csc x = \frac{1}{\sin x}$$

$$(\sin x)^2 + (\cos x)^2 = 1 \qquad 1 + (\tan x)^2 = (\sec x)^2 \qquad 1 + (\cot x)^2 = (\csc x)^2$$

$$\sin(-x) = -\sin x \qquad\qquad \cos(-x) = \cos x \qquad\qquad \tan(-x) = -\tan x$$

EXAMPLE 3.5.1   Verify the identities.

(a) $(1 - (\cos x)^2)(\sec x)^2 = (\tan x)^2$ (b) $\tan x + \cot x = \sec x \csc x$

(c) $\frac{\cos x}{1 - \tan x} + \frac{\sin x}{1 - \cot x} = \sin x + \cos x$ (d) $(\cot x)^2 - (\cos x)^2 = (\cot x)^2 (\cos x)^2$

SOLUTION:

(a) One method for verifying identities is to start with one side, replace any trigonometric functions with sines and cosines, and then try to simplify the resulting expression to that on the other side of the identity. Be on the look out for the Fundamental Identities, especially the Pythagorean Identities.

In this example start with the left side. This gives

$$(1 - (\cos x)^2)(\sec x)^2 = (1 - (\cos x)^2)\frac{1}{(\cos x)^2}$$

$$= (\sin x)^2 \frac{1}{(\cos x)^2}$$

$$= \frac{(\sin x)^2}{(\cos x)^2} = (\tan x)^2.$$

We used one of the equivalent forms of the Pythagorean Identity,

$$(\sin x)^2 + (\cos x)^2 = 1,$$

in going from the first line to the second. That is,

$$(\sin x)^2 + (\cos x)^2 = 1 \Rightarrow$$

$$(\sin x)^2 = 1 - (\cos x)^2 \text{ and } (\cos x)^2 = 1 - (\sin x)^2.$$

(b) First replace the left side with sines and cosines, giving

$$\tan x + \cot x = \frac{\sin x}{\cos x} + \frac{\cos x}{\sin x}.$$

Now take the common denominator of the two fractions. Recall for any algebraic expressions

$$\frac{a}{b} + \frac{c}{d} = \frac{ad + cb}{bd}.$$

So

$$\tan x + \cot x = \frac{\sin x}{\cos x} + \frac{\cos x}{\sin x}$$

$$= \frac{\sin x \sin x + \cos x \cos x}{\cos x \sin x}$$

$$= \frac{(\sin x)^2 + (\cos x)^2}{\cos x \sin x}$$

$$= \frac{1}{\cos x \sin x} = \frac{1}{\cos x} \cdot \frac{1}{\sin x}$$

$$= \sec x \csc x.$$

(c) Changing to sines and cosines gives

$$\frac{\cos x}{1 - \tan x} + \frac{\sin x}{1 - \cot x} = \frac{\cos x}{1 - \frac{\sin x}{\cos x}} + \frac{\sin x}{1 - \frac{\cos x}{\sin x}}$$

$$= \frac{\cos x}{\frac{\cos x - \sin x}{\cos x}} + \frac{\sin x}{\frac{\sin x - \cos x}{\sin x}}$$

$$= \cos x \cdot \frac{\cos x}{\cos x - \sin x} + \sin x \cdot \frac{\sin x}{\sin x - \cos x}$$

$$= \frac{(\cos x)^2}{\cos x - \sin x} + \frac{(\sin x)^2}{\sin x - \cos x}.$$

If the last two fractions had the same denominator, they could be added. Since

$$\sin x - \cos x = -(\cos x - \sin x),$$

we have

$$\frac{\cos x}{1 - \tan x} + \frac{\sin x}{1 - \cot x} = \frac{(\cos x)^2}{\cos x - \sin x} + \frac{(\sin x)^2}{\sin x - \cos x}$$

$$= \frac{(\cos x)^2}{\cos x - \sin x} - \frac{(\sin x)^2}{\cos x - \sin x}$$

$$= \frac{(\cos x)^2 - (\sin x)^2}{\cos x - \sin x}$$

$$= \frac{(\cos x - \sin x)(\cos x + \sin x)}{\cos x - \sin x}$$

$$= \cos x + \sin x.$$

In the next to the last step, we used the factoring formula,

$$a^2 - b^2 = (a - b)(a + b),$$

with $a = \cos x$ and $y = \sin x$.

(d) We again change to sines and cosines, which gives

$$(\cot x)^2 - (\cos x)^2 = \frac{(\cos x)^2}{(\sin x)^2} - (\cos x)^2$$

$$= \frac{(\cos x)^2 - (\cos x)^2 (\sin x)^2}{(\sin x)^2}$$

$$= \frac{(\cos x)^2 (1 - (\sin x)^2)}{(\sin x)^2}$$

$$= \frac{(\cos x)^2}{(\sin x)^2} \cdot (1 - (\sin x)^2)$$

$$\left(\frac{\cos x}{\sin x}\right)^2 \cdot (1 - (\sin x)^2)$$

$$= (\cot x)^2 (\cos x)^2.$$

◇

EXAMPLE 3.5.2   Make the indicated trigonometric substitution, and simplify the expression.

(a) $\sqrt{1 - x^2}$; $x = \sin t$, for $-\frac{\pi}{2} \le t \le \frac{\pi}{2}$ (b) $\frac{\sqrt{x^2 - 1}}{x}$; $x = \sec t$, for $0 < t < \frac{\pi}{2}$

(c) $\frac{x}{(1 - x^2)^{3/2}}$; $x = \sin t$, for $-\frac{\pi}{2} < t < \frac{\pi}{2}$

SOLUTION:

(a)

$$\sqrt{1 - x^2} = \sqrt{1 - (\sin t)^2}$$

$$= \sqrt{(\cos t)^2}$$

$$= |\cos t|$$

$$= \cos t$$

The absolute value can be dropped since for $-\frac{\pi}{2} \le t \le \frac{\pi}{2}$, we have $\cos t \ge 0$.

(b)

$$\frac{\sqrt{x^2 - 1}}{x} = \frac{\sqrt{(\sec t)^2 - 1}}{\sec t}$$

$$= \frac{\sqrt{(\tan t)^2}}{\sec t}$$

$$= \frac{\tan t}{\sec t} = \frac{\frac{\sin t}{\cos t}}{\frac{1}{\cos t}}$$

$$= \frac{\sin t}{\cos t} \cdot \frac{\cos t}{1}$$

$$= \sin t$$

(c)

$$\frac{x}{(1 - x^2)^{3/2}} = \frac{\sin t}{(1 - (\sin t)^2)^{3/2}}$$

$$= \frac{\sin t}{((\cos t)^2)^{3/2}}$$

$$= \frac{\sin t}{(\cos t)^3} = \frac{\sin t}{\cos t} \cdot \frac{1}{(\cos t)^2}$$

$$= \tan t (\sec t)^2$$

◇

ADDITION AND SUBTRACTION IDENTITIES

The fundamental identities are

$$\sin(x \pm y) = \sin x \cos y \pm \cos x \sin y$$

$$\cos(x \pm y) = \cos x \cos y \mp \sin x \sin y.$$

These can be used to show that

$$\tan(x \pm y) = \frac{\tan x \pm \tan y}{1 \mp \tan x \tan y},$$

as well as identities involving other trigonometric functions.

EXAMPLE 3.5.3   Determine the exact value of the trigonometric function.

(a) $\cos\left(\frac{\pi}{6} + \frac{3\pi}{4}\right)$ (b) $\sin\left(-\frac{7\pi}{12}\right)$

SOLUTION:

(a) Applying the addition formula for cosine, we have

$$\cos\left(\frac{\pi}{6} + \frac{3\pi}{4}\right) = \cos\left(\frac{\pi}{6}\right)\cos\left(\frac{3\pi}{4}\right) - \sin\left(\frac{\pi}{6}\right)\sin\left(\frac{3\pi}{4}\right)$$

$$= \frac{\sqrt{3}}{2} \cdot \left(-\frac{\sqrt{2}}{2}\right) - \frac{1}{2} \cdot \frac{\sqrt{2}}{2}$$

$$= -\frac{\sqrt{2}}{4}\left(\sqrt{3} + 1\right).$$

(b) First express $\frac{7\pi}{12}$ as the sum of two values for which we know the exact values of the sine and cosine. This gives

$$\frac{7\pi}{12} = \frac{\pi}{3} + \frac{\pi}{4}.$$

Then,

$$\sin\left(-\frac{7\pi}{12}\right) = -\sin\left(\frac{7\pi}{12}\right) = -\sin\left(\frac{\pi}{3} + \frac{\pi}{4}\right)$$

$$= -\left(\sin\left(\frac{\pi}{3}\right)\cos\left(\frac{\pi}{4}\right) + \cos\left(\frac{\pi}{3}\right)\sin\left(\frac{\pi}{4}\right)\right)$$

$$= -\left(\frac{\sqrt{3}}{2} \cdot \frac{\sqrt{2}}{2} + \frac{1}{2} \cdot \frac{\sqrt{2}}{2}\right)$$

$$= -\frac{\sqrt{2}}{4}\left(\sqrt{3} + 1\right).$$

◇

$\sin x_1 \, \cos_{y_2} \pm \cos_{x_1} \sin x_2$

EXAMPLE 3.5.4   Express $\sin 3x$ in terms of $\sin x$ and $\cos x$.

SOLUTION: First write $3x = 2x + x$ and apply the addition formula for the sine function. Then

$$\sin 3x = \sin(2x + x)$$
$$= \sin 2x \cos x + \cos 2x \sin x.$$

We now apply the addition formula a second time to $\sin 2x = \sin(x + x)$ and $\cos 2x = \cos(x + x)$. Then

$$\sin 3x = \sin 2x \cos x + \cos 2x \sin x$$
$$= (\sin x \cos x + \cos x \sin x) \cos x + (\cos x \cos x - \sin x \sin x) \sin x$$
$$= 2 \sin x (\cos x)^2 + \sin x (\cos x)^2 - (\sin x)^3$$
$$= 3 \sin x (\cos x)^2 - (\sin x)^3.$$

◇

EXAMPLE 3.5.5   Use the addition and subtraction formulas to verify the given identity.

(a) $\sin \left(t + \frac{3\pi}{2} \right) = -\cos t$ (b) $\cos \left(t + \frac{3\pi}{2} \right) = \sin t$

SOLUTION:

(a) We have

$$\sin \left(t + \frac{3\pi}{2} \right) = \sin t \cos \left(\frac{3\pi}{2} \right) + \cos t \sin \left(\frac{3\pi}{2} \right)$$
$$= (\sin t) \cdot (0) + (\cos t) \cdot (-1)$$
$$= -\cos t.$$

(b) We have

$$\cos\left(t + \frac{3\pi}{2}\right) = \cos t \cos \frac{3\pi}{2} - \sin t \sin \frac{3\pi}{2}$$
$$= (\cos t) \cdot (0) - (\sin t) \cdot (-1)$$
$$= \sin t.$$

$\diamond$

DOUBLE ANGLE FORMULAS

The basic double-angle formulas are

$$\sin 2x = 2 \sin x \cos x$$

$$\cos 2x = (\cos x)^2 - (\sin x)^2 = 2(\cos x)^2 - 1 = 1 - 2(\sin x)^2.$$

These can be used to derive

$$\tan 2x = \frac{2 \tan x}{1 - (\tan x)^2},$$

as well as identities involving other trigonometric functions.

EXAMPLE 3.5.6   If $\cos t = \frac{3}{5}$, where $0 < t < \frac{\pi}{2}$, find $\cos 2t$, $\sin 2t$, and $\tan 2t$.

SOLUTION: First find $\sin t$, which is needed in the double-angle formulas. Since t is in the first quadrant, $\sin t > 0$. So,

$$\sin t = \sqrt{1 - (\cos t)^2}$$
$$= \sqrt{1 - \left(\frac{3}{5}\right)^2} = \sqrt{\frac{16}{25}} = \frac{\sqrt{16}}{\sqrt{25}}$$
$$= \frac{4}{5}.$$

Then,

$$\cos 2t = 2\left(\frac{3}{5}\right)^2 - 1 = \frac{18}{25} - 1 = -\frac{7}{25}$$

$$\sin 2t = 2\left(\frac{4}{5}\right)\left(\frac{3}{5}\right) = \frac{24}{25}$$

$$\tan 2t = \frac{\sin 2t}{\cos 2t} = \frac{\frac{24}{25}}{-\frac{7}{25}} = -\frac{24}{25}\cdot\frac{25}{7} = -\frac{24}{7}.$$

◇

EXAMPLE 3.5.7   Verify the identity $\dfrac{2\tan x}{1 + (\tan x)^2} = \sin 2x.$

SOLUTION: We have

$$\frac{2\tan x}{1 + (\tan x)^2} = \frac{2\frac{\sin x}{\cos x}}{1 + \left(\frac{\sin x}{\cos x}\right)^2}$$

$$= \frac{2\frac{\sin x}{\cos x}}{1 + \frac{(\sin x)^2}{(\cos x)^2}} = \frac{2\frac{\sin x}{\cos x}}{\frac{(\cos x)^2 + (\sin x)^2}{(\cos x)^2}}$$

$$= \frac{2\frac{\sin x}{\cos x}}{\frac{1}{(\cos x)^2}} = 2\frac{\sin x}{\cos x}\cdot(\cos x)^2$$

$$= 2\sin x\cos x = \sin 2x.$$

The graphs of $y = \dfrac{2\tan x}{1 + (\tan x)^2}$ and $y = \sin 2x$ are plotted together in the figure. Notice the graphs coincide showing graphically that the equation is an identity.

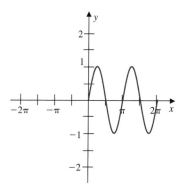

Figure 3.26

HALF ANGLE FORMULAS

The basic half-angle formulas can be written either as

$$(\sin x)^2 = \frac{1 - \cos 2x}{2} \quad \text{and} \quad (\cos x)^2 = \frac{1 + \cos 2x}{2}$$

or as

$$\sin\left(\frac{x}{2}\right) = \sqrt{\frac{1 - \cos x}{2}} \quad \text{and} \quad \cos\left(\frac{x}{2}\right) = \sqrt{\frac{1 + \cos x}{2}}.$$

From these we can derive

$$(\tan x)^2 = \frac{1 - \cos 2x}{1 + \cos 2x} \quad \text{and} \quad \tan\left(\frac{x}{2}\right) = \frac{1 - \cos x}{\sin x} = \frac{\sin x}{1 + \cos x}.$$

EXAMPLE 3.5.8   Find the exact value of $\sin\left(\frac{3\pi}{8}\right)$.

SOLUTION: If $x = \frac{3\pi}{4}$ in the formula for $\sin\left(\frac{x}{2}\right)$, then $\frac{x}{2} = \frac{3\pi}{8}$, and

$$\sin\left(\frac{3\pi}{8}\right) = \sqrt{\frac{1 - \cos\left(\frac{3\pi}{4}\right)}{2}}$$

$$= \sqrt{\frac{1 - \frac{\sqrt{2}}{2}}{2}} = \sqrt{\frac{2 - \sqrt{2}}{4}}$$

$$= \frac{\sqrt{2 - \sqrt{2}}}{2}.$$

◇

The technique used in the next example is used to reduce powers of trigonometric functions to forms that are easier to work with.

EXAMPLE 3.5.9   Rewrite the expression $(\cos x)^4$ so that it involves the sum or difference of only constants and sine and cosine functions to the first power.

SOLUTION: Use the half angle-formula for the cosine to write

$$(\cos x)^4 = ((\cos x)^2)^2$$

$$= \left(\frac{1 + \cos 2x}{2}\right)^2$$

$$= \frac{1}{4}\left(1 + 2\cos 2x + (\cos 2x)^2\right).$$

Now use the half-angle formula again on $(\cos 2x)^2$, so

$$(\cos 2x)^2 = \frac{1 + \cos 4x}{2},$$

and

$$(\cos x)^4 = \frac{1}{4}\left(1 + 2\cos 2x + (\cos 2x)^2\right)$$

$$= \frac{1}{4}\left(1 + 2\cos 2x + \frac{1 + \cos 4x}{2}\right)$$

$$= \frac{1}{4} + \frac{\cos 2x}{2} + \frac{1 + \cos 4x}{8}$$

$$= \frac{3}{8} + \frac{\cos 2x}{2} + \frac{\cos 4x}{8}.$$

◇

PRODUCT-TO-SUM AND SUM-TO-PRODUCT IDENTITIES

The basic product-to-sum identities are

(i) $\sin x \cos y = \dfrac{1}{2}[\sin(x + y) + \sin(x - y)]$

(ii) $\cos x \sin y = \dfrac{1}{2}[\sin(x + y) - \sin(x - y)]$

(iii) $\cos x \cos y = \dfrac{1}{2}[\cos(x + y) + \cos(x - y)]$

(iv) $\sin x \sin y = \dfrac{1}{2}[\cos(x - y) - \cos(x + y)].$

EXAMPLE 3.5.10   Write the product $\cos 3t \sin 5t$ as a sum or difference.

SOLUTION: We have

$$\cos 3t \sin 5t = \frac{1}{2}[\sin(3t + 5t) - \sin(3t - 5t)]$$

$$= \frac{1}{2}[\sin(8t) - \sin(-2t)]$$

$$= \frac{1}{2}\sin(8t) + \frac{1}{2}\sin(2t).$$

Recall $\sin(-x) = -\sin x.$

◇

There are four basic sum-to-product formulas that can be used to write sums of sines and cosines as products. They are

$$\text{(i)} \quad \sin x + \sin y = 2 \sin \frac{x+y}{2} \cos \frac{x-y}{2}$$

$$\text{(ii)} \quad \sin x - \sin y = 2 \cos \frac{x+y}{2} \sin \frac{x-y}{2}$$

$$\text{(iii)} \quad \cos x + \cos y = 2 \cos \frac{x+y}{2} \cos \frac{x-y}{2}$$

$$\text{(iv)} \quad \cos x - \cos y = -2 \sin \frac{x+y}{2} \sin \frac{x-y}{2}.$$

EXAMPLE 3.5.11   Rewrite $\cos 5t + \cos 2t$ as a product.

SOLUTION: We have

$$\cos 5t + \cos 2t = 2 \cos \frac{5t+2t}{2} \cos \frac{5t-2t}{2}$$

$$= 2 \cos \frac{7t}{2} \cos \frac{3t}{2}.$$

◇

SOLVING TRIGONOMETRIC EQUATIONS

EXAMPLE 3.5.12   Find all values of x in the interval $[0, 2\pi]$ that satisfy the equation $\sin 2x = \sqrt{3} \sin x$.

SOLUTION: Use the double-angle formula to write $\sin 2x = 2\sin x \cos x$. Then

$$\sin 2x = \sqrt{3}\sin x$$

$$2\sin x \cos x = \sqrt{3}\sin x$$

$$2\sin x \cos x - \sqrt{3}\sin x = 0$$

$$\sin x(2\cos x - \sqrt{3}) = 0$$

$$\sin x = 0, \quad \cos x = \frac{\sqrt{3}}{2}$$

$$x = 0, \pi, 2\pi \quad \text{or} \quad x = \frac{\pi}{6}, \frac{11\pi}{6}.$$

$\diamond$

EXAMPLE 3.5.13   Find all values of x in the interval $[0, 2\pi]$ that satisfy the equation $\cos 2x \cos 3x = \sin 2x \sin 3x$.

SOLUTION: If we bring the expression on the right to the left side of the equation, we recognize the new expression as fitting the sum formula for cosines. That is,

$$\cos 2x \cos 3x - \sin 2x \sin 3x = 0$$

$$\cos(2x + 3x) = 0$$

$$\cos(5x) = 0$$

$$5x = \frac{\pi}{2}, \frac{3\pi}{2}, \frac{5\pi}{2}, \frac{7\pi}{2}, \frac{9\pi}{2}, \frac{11\pi}{2}, \frac{13\pi}{2}, \frac{15\pi}{2}, \frac{17\pi}{2}, \frac{19\pi}{2}$$

$$x = \frac{\pi}{10}, \frac{3\pi}{10}, \frac{\pi}{2}, \frac{7\pi}{10}, \frac{9\pi}{10}, \frac{11\pi}{10}, \frac{13\pi}{10}, \frac{3\pi}{2}, \frac{17\pi}{10}, \frac{19\pi}{10}.$$

Remember, $0 \le x \le 2\pi$, if and only if $0 \le 5x \le 10\pi$. $\diamond$

Solutions for Exercise Set 3.5

1. $\sin\left(\frac{\pi}{3} - \frac{5\pi}{4}\right) = \sin\frac{\pi}{3}\cos\frac{5\pi}{4} - \cos\frac{\pi}{3}\sin\frac{5\pi}{4} = \frac{\sqrt{3}}{2}\left(-\frac{\sqrt{2}}{2}\right) - \frac{1}{2}\left(-\frac{\sqrt{2}}{2}\right) =$
$-\frac{\sqrt{2}}{4}\left(\sqrt{3} - 1\right) = \frac{\sqrt{2}}{4}\left(1 - \sqrt{3}\right)$

3. $\cos\left(\frac{7\pi}{12}\right) = \cos\left(\frac{\pi}{3} + \frac{\pi}{4}\right) = \cos\frac{\pi}{3}\cos\frac{\pi}{4} - \sin\frac{\pi}{3}\sin\frac{\pi}{4} = \frac{1}{2}\left(\frac{\sqrt{2}}{2}\right) - \frac{\sqrt{3}}{2}\left(\frac{\sqrt{2}}{2}\right) =$
$\frac{\sqrt{2}}{4}\left(1 - \sqrt{3}\right)$

5. $\tan\left(\frac{\pi}{12}\right) = \tan\left(\frac{\pi}{3} - \frac{\pi}{4}\right) = \frac{\tan\left(\frac{\pi}{3}\right) - \tan\left(\frac{\pi}{4}\right)}{1 + \tan\left(\frac{\pi}{3}\right)\tan\left(\frac{\pi}{4}\right)} = \frac{\sqrt{3}-1}{1+\sqrt{3}(1)} = \frac{\sqrt{3}-1}{\sqrt{3}+1}$

7. Since $\left(\sin\frac{7\pi}{12}\right)^2 = \frac{1}{2}\left(1 - \cos\frac{7\pi}{6}\right) = \frac{1}{2}\left(1 - \left(-\frac{\sqrt{3}}{2}\right)\right) = \frac{1}{2}\left(1 + \frac{\sqrt{3}}{2}\right)$ and

since $0 < \frac{\pi}{12} < \frac{\pi}{2}$, the sine is positive, so

$$\sin\frac{\pi}{12} = \sqrt{\frac{1}{2}\left(1 + \frac{\sqrt{3}}{2}\right)} = \sqrt{\frac{2 + \sqrt{3}}{4}} = \frac{\sqrt{2 + \sqrt{3}}}{2}.$$

9. Since $\left(\cos\frac{3\pi}{8}\right)^2 = \frac{1}{2}\left(1 + \cos\frac{3\pi}{4}\right) = \frac{1}{2}\left(1 - \frac{\sqrt{2}}{2}\right)$ and since $0 < \frac{3\pi}{8} < \frac{\pi}{2}$,

the cosine is positive, so

$$\cos\frac{3\pi}{8} = \sqrt{\frac{1}{2}\left(1 - \frac{\sqrt{2}}{2}\right)} = \sqrt{\frac{2 - \sqrt{2}}{4}} = \frac{\sqrt{2 - \sqrt{2}}}{2}.$$

11. Since $\left(\sin\frac{13\pi}{12}\right)^2 = \frac{1}{2}\left(1 - \cos\frac{13\pi}{6}\right) = \frac{1}{2}\left(1 - \frac{\sqrt{3}}{2}\right)$ and since $\pi < \frac{13\pi}{12} < \frac{3\pi}{2}$, the sine is negative, so

$$\sin\frac{13\pi}{12} = -\sqrt{\frac{1}{2}\left(1 - \frac{\sqrt{3}}{2}\right)} = -\sqrt{\frac{2 - \sqrt{3}}{4}} = -\frac{\sqrt{2 - \sqrt{3}}}{2}.$$

13. We are given $\cos t = \frac{3}{5}, 0 < t < \frac{\pi}{2}$. Since $0 < t < \frac{\pi}{2}$, $\sin t > 0$.

(a) $\cos 2t = 2(\cos t)^2 - 1 = 2\left(\frac{3}{5}\right)^2 - 1 = 2\left(\frac{9}{25}\right) - 1 = -\frac{7}{25}$

(b) To use the formula $\sin 2t = 2\sin t\cos t$, first find $\sin t$. That is,

$$(\cos t)^2 + (\sin t)^2 = 1 \Rightarrow (\sin t)^2 = 1 - \left(\frac{3}{5}\right)^2 = \frac{16}{25} \Rightarrow \sin t = \frac{4}{5}.$$

Then $\sin 2t = 2 \sin t \cos t = 2 \left(\frac{4}{5}\right)\left(\frac{3}{5}\right) = \frac{24}{25}$.

(c) $\left(\cos\frac{t}{2}\right)^2 = \frac{1}{2}(1 + \cos t) = \frac{1}{2}\left(1 + \frac{3}{5}\right) = \frac{4}{5} \Rightarrow \cos\frac{t}{2} = \pm\sqrt{\frac{4}{5}}$. Since

$0 < t < \frac{\pi}{2} \Rightarrow 0 < \frac{t}{2} < \frac{\pi}{4}$, we have $\cos\frac{t}{2} > 0 \Rightarrow \cos\frac{t}{2} = \sqrt{\frac{4}{5}} = \frac{2}{\sqrt{5}} = \frac{2\sqrt{5}}{5}$.

(d) $\left(\sin\frac{t}{2}\right)^2 = \frac{1}{2}(1 - \cos t) = \frac{1}{2}\left(1 - \frac{3}{5}\right) = \frac{1}{5} \Rightarrow \sin\frac{t}{2} = \pm\sqrt{\frac{1}{5}}$. Since

$0 < t < \frac{\pi}{2} \Rightarrow 0 < \frac{t}{2} < \frac{\pi}{4}$, we have $\sin\frac{t}{2} > 0 \Rightarrow \sin\frac{t}{2} = \sqrt{\frac{1}{5}} = \frac{1}{\sqrt{5}} = \frac{\sqrt{5}}{5}$.

15.   We are given $\tan t = \frac{5}{12}$, $\sin t < 0$. Since $\tan t > 0$ and $\sin t < 0 \Rightarrow$

$\cos t < 0$. First determine $\sin t$ and $\cos t$. From the identity

$$(\tan t)^2 + 1 = (\sec t)^2 = \frac{1}{(\cos t)^2} \Rightarrow \left(\frac{5}{12}\right)^2 + 1 = \frac{1}{(\cos t)^2}$$

and $(\cos t)^2 = \frac{144}{169} \Rightarrow \cos t = -\frac{12}{13}$. Then

$$(\sin t)^2 = 1 - (\cos t)^2 = 1 - \frac{144}{169} = \frac{25}{169} \Rightarrow \sin t = -\sqrt{\frac{25}{169}} = -\frac{5}{13}.$$

(a)
$$\cos 2t = 2(\cos t)^2 - 1 = 2\left(\frac{144}{169}\right) - 1 = \frac{119}{169}$$

(b)
$$\sin 2t = 2 \sin t \cos t = 2\left(-\frac{5}{13}\right)\left(-\frac{12}{13}\right) = \frac{120}{169}$$

(c)

$$\left(\cos\frac{t}{2}\right)^2 = \frac{1}{2}(1 + \cos t) = \frac{1}{2}\left(1 - \frac{12}{13}\right) = \frac{1}{26} \Rightarrow \cos\frac{t}{2} = \pm\sqrt{\frac{1}{26}}$$

Since $\cos t < 0$ and $\sin t < 0$, t is in quadrant III so $\pi < t < \frac{3\pi}{2} \Rightarrow \frac{\pi}{2} <$

$\frac{t}{2} < \frac{3\pi}{4}$ and we have

$$\cos\frac{t}{2} < 0 \Rightarrow \cos\frac{t}{2} = -\frac{1}{\sqrt{26}} = -\frac{\sqrt{26}}{26}.$$

(d)

$$\left(\sin\frac{t}{2}\right)^2 = \frac{1}{2}(1 - \cos t) = \frac{1}{2}\left(1 + \frac{12}{13}\right) = \frac{25}{26} \Rightarrow \sin\frac{t}{2} = \pm\sqrt{\frac{25}{26}}$$

Since $\pi < t < \frac{3\pi}{2} \Rightarrow \frac{\pi}{2} < \frac{t}{2} < \frac{3\pi}{4}$ and we have

$$\sin\frac{t}{2} > 0 \Rightarrow \sin\frac{t}{2} = \sqrt{\frac{25}{26}} = \frac{5\sqrt{26}}{26}.$$

17. $\sin(t + \pi) = \sin t \cos \pi + \cos t \sin \pi = (\sin t)(-1) + (\cos t)(0) = -\sin t$

19. $\sin\left(t + \frac{\pi}{2}\right) = \sin t \cos\frac{\pi}{2} + \cos t \sin\frac{\pi}{2} = (\sin t)(0) + (\cos t)(1) = \cos t$

21. $\sin(\pi - t) = \sin\pi\cos t - \cos\pi\sin t = (0)(\cos t) - (-1)(\sin t) = \sin t$

23. $(\cos 2x)^2 = \frac{1 + \cos 4x}{2} = \frac{1}{2} + \frac{1}{2}\cos 4x$

25.

$$(\sin x)^4 = ((\sin x)^2)^2 = \left(\frac{1 - \cos 2x}{2}\right)^2 = \frac{1}{4}\left(1 - 2\cos 2x + (\cos 2x)^2\right)$$

$$= \frac{1}{4}\left(1 - 2\cos 2x + \frac{1 + \cos 4x}{2}\right) = \frac{3}{8} - \frac{1}{2}\cos 2x + \frac{1}{8}\cos 4x$$

27. $\sin 6t \cos 5t = \frac{1}{2}(\sin(6t + 5t) + \sin(6t - 5t)) = \frac{1}{2}(\sin 11t + \sin t)$

29. $\cos 2t \cos 3t = \frac{1}{2}(\cos(2t + 3t) + \cos(2t - 3t)) = \frac{1}{2}(\cos 5t + \cos(-t)) = \frac{1}{2}(\cos 5t + \cos t)$

31. $(1 - (\cos x)^2)(\sec x)^2 = (\sin x)^2\frac{1}{(\cos x)^2} = (\tan x)^2$

33. $\cot x - \tan x = \frac{\cos x}{\sin x} - \frac{\sin x}{\cos x} = \frac{(\cos x)^2 - (\sin x)^2}{\sin x \cos x} = \frac{\cos 2x}{\sin x \cos x} = \frac{2\cos 2x}{\sin 2x} = 2\cot 2x$

35. $\sin x \sin 2x + \cos x \cos 2x = \cos(2x - x) = \cos x$

37. $\sec x - \cos x = \frac{1}{\cos x} - \cos x = \frac{1 - (\cos x)^2}{\cos x} = \frac{(\sin x)^2}{\cos x} = \sin x\frac{\sin x}{\cos x} = \sin x \tan x$

39. $\sin 2x = \sin x \Rightarrow \sin 2x - \sin x = 0 \Rightarrow 2\sin x \cos x - \sin x = 0 \Rightarrow$

$\sin x(2\cos x - 1) = 0 \Rightarrow \sin x = 0$ or $\cos x = \frac{1}{2} \Rightarrow x = 0, \pi, 2\pi, \frac{\pi}{3}, \frac{5\pi}{3}$

41. $2(\sin x)^2 + \cos x - 1 = 0 \Rightarrow 2(1 - (\cos x)^2) + \cos x - 1 = 0 \Rightarrow -2(\cos x)^2 +$

$\cos x + 1 = 0 \Rightarrow 2(\cos x)^2 - \cos x - 1 = 0 \Rightarrow (2\cos x + 1)(\cos x - 1) =$

$0 \Rightarrow \cos x = -\frac{1}{2}$ or $\cos x = 1 \Rightarrow x = \frac{2\pi}{3}, \frac{4\pi}{3}, 0, 2\pi$

43. We have $\tan x + \cot x = \dfrac{2}{\sin 2x} \Rightarrow \dfrac{\sin x}{\cos x} + \dfrac{\cos x}{\sin x} = \dfrac{2}{\sin 2x} \Rightarrow \dfrac{(\cos x)^2 + (\sin x)^2}{\cos x \sin x} =$

$\dfrac{2}{\sin 2x} \Rightarrow \sin 2x = 2\sin x \cos x.$ So the equation is an identity and hence

holds for all applicable x.

45. $2\sin \dfrac{x+y}{2} \cos \dfrac{x-y}{2} = 2 \cdot \dfrac{1}{2}\left(\sin\left(\dfrac{x+y}{2} + \dfrac{x-y}{2}\right) + \sin\left(\dfrac{x+y}{2} - \dfrac{x-y}{2}\right)\right) =$

$\sin x + \sin y$

47. $2\cos \dfrac{x+y}{2} \cos \dfrac{x-y}{2} = 2 \cdot \dfrac{1}{2}\left(\cos\left(\dfrac{x+y}{2} + \dfrac{x-y}{2}\right) + \cos\left(\dfrac{x+y}{2} - \dfrac{x-y}{2}\right)\right) =$

$\cos x + \cos y$

49. It is an identity since the graphs of the two functions coincide.

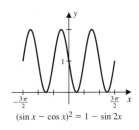

$(\sin x - \cos x)^2 = 1 - \sin 2x$

51. It is an identity since the graphs of the two functions coincide.

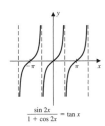

$\dfrac{\sin 2x}{1 + \cos 2x} = \tan x$

53. It is not an identity since the graphs of the two functions do not coincide.

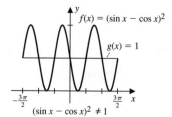

$$f(x) = (\sin x - \cos x)^2$$

$$g(x) = 1$$

$$(\sin x - \cos x)^2 \neq 1$$

3.6 Right Triangle Trigonometry

ANGLE MEASURE : RADIANS AND DEGREES

An angle formed by rotating a ray counterclockwise from the initial position one complete revolution back to the initial position has, by definition, a measure of 360 degrees, written $360°$. The radian measure of the same angle is equal to the circumference of the unit circle, which is 2π radians. Angles measured in the clockwise direction are negative.

Conversion between Degrees and Radians is

$$180° = \pi \text{ radians}, \ 1° = \frac{\pi}{180} \text{ radians}, \ 1 \text{ radian } = \frac{180°}{\pi}.$$

EXAMPLE 3.6.1   Convert from degree measure to radian measure.

(a) $45°$ (b) $300°$

SOLUTION:

(a) Since $1° = \frac{\pi}{180}$ radians,

$$45° = 45 \cdot \frac{\pi}{180}$$
$$= \frac{\pi}{4} \text{ radians.}$$

(b) Also,

$$300° = 300 \cdot \frac{\pi}{180}$$
$$= \frac{5\pi}{3} \text{ radians.}$$

EXAMPLE 3.6.2   Convert from radian measure to degree measure.

(a) $\frac{5\pi}{4}$ (b) $-\frac{7\pi}{6}$

SOLUTION:

(a) Since 1 radian $= \frac{180°}{\pi}$,

$$\frac{5\pi}{4} \text{ radians} = \frac{5\pi}{4} \cdot \frac{180}{\pi}$$
$$= 225°.$$

(b) Similarly,

$$-\frac{7\pi}{6} \text{ radians} = -\frac{7\pi}{6} \cdot \frac{180}{\pi}$$
$$= -210°.$$

TRIGONOMETRIC FUNCTIONS OF AN ANGLE IN A RIGHT TRIANGLE

For an angle θ in the right triangle shown below, we have the following values of the trigonometric functions.

$$\sin\theta = \frac{b}{c} = \frac{\text{opposite}}{\text{hypotenuse}} \qquad \cos\theta = \frac{a}{c} = \frac{\text{adjacent}}{\text{hypotenuse}}$$

$$\tan\theta = \frac{b}{a} = \frac{\text{opposite}}{\text{adjacent}} \qquad \cot\theta = \frac{a}{b} = \frac{\text{adjacent}}{\text{opposite}}$$

$$\sec\theta = \frac{c}{a} = \frac{\text{hypotenuse}}{\text{adjacent}} \qquad \csc\theta = \frac{c}{b} = \frac{\text{hypotenuse}}{\text{opposite}}$$

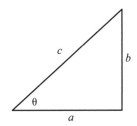

Figure 3.27

EXAMPLE 3.6.3   Find the value of the six trigonometric functions of the angle θ shown in the following right triangle.

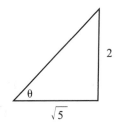

Figure 3.28

SOLUTION: The length of the hypotenuse is needed and can be found using the Pythagorean Theorem. If x denotes the length of the hypotenuse, then

$$x^2 = \left(\sqrt{5}\right)^2 + 2^2$$
$$x^2 = 9$$
$$x = 3.$$

So

$$\sin\theta = \frac{2}{3}, \quad \cos\theta = \frac{\sqrt{5}}{3}$$

$$\tan\theta = \frac{2}{\sqrt{5}} = \frac{2\sqrt{5}}{5}, \quad \cot\theta = \frac{\sqrt{5}}{2}$$

$$\sec\theta = \frac{3}{\sqrt{5}} = \frac{3\sqrt{5}}{5}, \quad \csc\theta = \frac{3}{2}.$$

◇

EXAMPLE 3.6.4   Refer to the triangle in the figure and use the information to find any missing angles or sides.

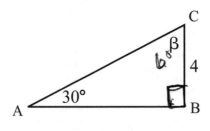

Figure 3.29

SOLUTION: Since the sum of the angles of a triangle is 180°,

$$\beta = 180° - 90° - 30° = 60°.$$

To find the side $\overline{AB}$, use the tangent of the 30° angle. So

$$\tan 30° = \frac{4}{\overline{AB}}$$

$$\overline{AB}\tan 30° = 4$$

$$\overline{AB} = \frac{4}{\tan 30°}$$

$$= \frac{4}{\sqrt{3}/3} = \frac{12}{\sqrt{3}} = 4\sqrt{3}.$$

Recall,

$$
\begin{aligned}
\tan 30^\circ &= \tan \frac{\pi}{6} \\
&= \frac{\sin \frac{\pi}{6}}{\cos \frac{\pi}{6}} = \frac{1/2}{\sqrt{3}/2} \\
&= \frac{1}{\sqrt{3}} = \frac{1}{\sqrt{3}} \cdot \frac{\sqrt{3}}{\sqrt{3}} = \frac{\sqrt{3}}{3}.
\end{aligned}
$$

The remaining side $\overline{AC}$ can be found using the Pythagorean Theorem. That is,

$$
\begin{aligned}
\overline{AC}^2 &= \left(4\sqrt{3}\right)^2 + 4^2 \\
&= 48 + 16 \\
&= 64 \\
\overline{AC} &= 8.
\end{aligned}
$$

◇

EXAMPLE 3.6.5   From the figure determine $\sin \alpha, \sin \beta, \cos \alpha,$ and $\cos \beta$.
Then find

(a) $\sin(\alpha + \beta)$ (b) $\cos(\alpha + \beta)$

(c) $\sin 2\alpha$ (d) $\cos 2\alpha$

(e) $\sin \frac{\alpha}{2}$ (f) $\cos \frac{\alpha}{2}$

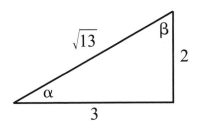

Figure 3.30

SOLUTION: From the figure we have

$$\sin \alpha = \frac{2}{\sqrt{13}}, \ \sin \beta = \frac{3}{\sqrt{13}}$$

$$\cos \alpha = \frac{3}{\sqrt{13}}, \ \cos \beta = \frac{2}{\sqrt{13}}.$$

(a)

$$\sin(\alpha + \beta) = \sin \alpha \cos \beta + \sin \beta \cos \alpha$$

$$= \frac{2}{\sqrt{13}} \cdot \frac{2}{\sqrt{13}} + \frac{3}{\sqrt{13}} \cdot \frac{3}{\sqrt{13}}$$

$$= \frac{4}{13} + \frac{9}{13} = 1$$

(b)

$$\cos(\alpha + \beta) = \cos \alpha \cos \beta - \sin \alpha \sin \beta$$

$$= \frac{3}{\sqrt{13}} \cdot \frac{2}{\sqrt{13}} - \frac{2}{\sqrt{13}} \cdot \frac{3}{\sqrt{13}}$$

$$= 0$$

(c)

$$\sin 2\alpha = 2 \sin \alpha \cos \alpha$$

$$= 2 \frac{2}{\sqrt{13}} \cdot \frac{3}{\sqrt{13}} = \frac{12}{13}$$

(d)

$$\cos 2\alpha = (\cos \alpha)^2 - (\sin \alpha)^2$$

$$= \frac{9}{13} - \frac{4}{13} = \frac{5}{13}$$

(e)

$$\sin \frac{\alpha}{2} = \sqrt{\frac{1 - \cos \alpha}{2}}$$

$$= \sqrt{\frac{1 - \frac{3}{\sqrt{13}}}{2}}$$

$$= \sqrt{\frac{\sqrt{13} - 3}{2\sqrt{13}}}$$

(f)

$$\cos \frac{\alpha}{2} = \sqrt{\frac{1 + \cos \alpha}{2}}$$

$$= \sqrt{\frac{1 + \frac{3}{\sqrt{13}}}{2}}$$

$$= \sqrt{\frac{\sqrt{13} + 3}{2\sqrt{13}}}$$

◇

APPLICATIONS

EXAMPLE 3.6.6   The angle of elevation to the top of a building is 65° from the ground when viewed 165 ft from the building. Estimate the height of the building in feet.

SOLUTION: The right triangle in the figure describes the situation.

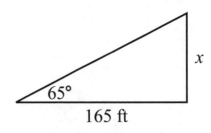

Figure 3.31

The sides of lengths x and 165 are the opposite and adjacent sides, respectively, of the 65° angle. So we can use the tangent function to get

$$\tan 65° = \frac{x}{165}$$

$$x = 165 \tan 65° \text{ feet}$$

$$\approx 354 \text{ feet}.$$

◇

EXAMPLE 3.6.7   A surveyor wants to find the distance between points A and B on opposite sides of a pond. From point A, the surveyor determines a line of sight perpendicular to AB and establishes point C along this line 150 feet from A. Angle ACB is determined to be 62°. How far is it from A to B.

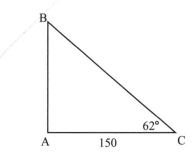

Figure 3.32

SOLUTION: The lengths $\overline{AB}$ and $\overline{AC} = 150$ are the sides opposite and adjacent, respectively, to the angle of 62° and so can be related using the tangent function. That is,

$$\tan 62° = \frac{\overline{AB}}{\overline{AC}} = \frac{\overline{AB}}{150}$$
$$\overline{AB} = 150 \tan 62° \approx 282 \text{ feet.}$$

Therefore, the distance between A and B is approximately 282 feet.   ◇

Solutions for Exercise Set 3.6

1. $60° = 60\frac{\pi}{180}$ radians $= \frac{\pi}{3}$ radians

3. $150° = 150\frac{\pi}{180}$ radians $= \frac{5\pi}{6}$ radians

5. $225° = 225\frac{\pi}{180}$ radians $= \frac{5\pi}{4}$ radians

7. $-72° = -72\frac{\pi}{180}$ radians $= -\frac{2\pi}{5}$ radians

9. $\frac{3\pi}{4}$ radians $= \frac{3\pi}{4}\frac{180}{\pi}$ degrees $= 135°$

11. $-\frac{11\pi}{6}$ radians $= -\frac{11\pi}{6}\frac{180}{\pi}$ degrees $= -330°$

13. $\frac{9\pi}{2}$ radians $= \frac{9\pi}{2}\frac{180}{\pi}$ degrees $= 810°$

15. The hypotenuse is $\sqrt{4^2 + 3^2} = \sqrt{25} = 5$, so

$$\cos\theta = \frac{4}{5}, \ \sin\theta = \frac{3}{5}, \ \tan\theta = \frac{3}{4}, \cot\theta = \frac{4}{3}, \ \sec\theta = \frac{5}{4}, \ \csc\theta = \frac{5}{3}.$$

17. The hypotenuse is $\sqrt{3+1} = 2$, so

$$\cos\theta = \frac{\sqrt{3}}{2}; \ \sin\theta = \frac{1}{2}; \ \tan\theta = \frac{\sqrt{3}}{3},$$

$$\cot\theta = \frac{3}{\sqrt{3}} = \sqrt{3}; \ \sec\theta = \frac{2}{\sqrt{3}} = \frac{2\sqrt{3}}{3}; \ \csc\theta = 2.$$

19. Let x be the missing side. Then $x^2 + 25 = 169 \Rightarrow x = \sqrt{144} = 12$, and

$$\cos\theta = \frac{12}{13}, \quad \sin\theta = \frac{5}{13}, \quad \tan\theta = \frac{5}{12},$$

$$\cot\theta = \frac{12}{5}, \quad \sec\theta = \frac{13}{12}, \quad \csc\theta = \frac{13}{5}.$$

21. Since $\cos 30° = \frac{x}{16}$ it implies that $x = 16\cos 30° = 16\frac{\sqrt{3}}{2} = 8\sqrt{3}$.

23. Since $\sin 45° = \frac{5}{x}$ it implies that $x = \frac{5}{\sin 45°} = \frac{5}{\sqrt{2}/2} = \frac{10}{\sqrt{2}} = 5\sqrt{2}$.

25. Since $\tan 60° = \frac{6}{x}$ it implies that $x = \frac{6}{\tan 60°} = \frac{6}{\sqrt{3}} = \frac{6\sqrt{3}}{3} = 2\sqrt{3}$.

27. The angle $\beta = 60°$ and

$$\tan 30° = \frac{\overline{BC}}{\overline{AB}} = \frac{4}{\overline{AB}} \Rightarrow \overline{AB} = \frac{4}{\sqrt{3}/3} = 4\sqrt{3}.$$

Also,
$$\sin 30° = \frac{\overline{BC}}{\overline{AC}} = \frac{4}{\overline{AC}} \Rightarrow \overline{AC} = \frac{4}{1/2} = 8.$$

Note that once $\overline{AB}$ is computed $\overline{AC}$ can be found using the Pythagorean Theorem as well. That is, $\overline{AC} = \sqrt{16 + 48} = \sqrt{64} = 8$.

29. The angle $\alpha = 45°$ and
$$\cos 45° = \frac{\overline{AB}}{\overline{AC}} = \frac{\overline{AB}}{5} \Rightarrow \overline{AB} = \frac{5\sqrt{2}}{2}.$$

Also,
$$\sin 45° = \frac{\overline{BC}}{5} \Rightarrow \overline{BC} = \frac{5\sqrt{2}}{2}.$$

31. The angle $\beta = 62.4°$ and
$$\tan 27.6° = \frac{15.3}{\overline{AB}} \Rightarrow \overline{AB} = \frac{15.3}{\tan 27.6°} \approx 29.3.$$

Also,
$$\sin 27.6° = \frac{15.3}{\overline{AC}} \Rightarrow \overline{AC} = \frac{15.3}{\sin 27.6°} \approx 33.$$

33. (a)

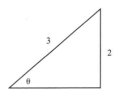

The missing side is $\sqrt{9 - 4} = \sqrt{5}$, and
$$\cos \theta = \frac{\sqrt{5}}{3}, \tan \theta = \frac{2}{\sqrt{5}} = \frac{2\sqrt{5}}{5}, \cot \theta = \frac{\sqrt{5}}{2},$$
$$\sec \theta = \frac{3}{\sqrt{5}} = \frac{3\sqrt{5}}{5}, \csc \theta = \frac{3}{2}.$$

(b)

The missing side is $\sqrt{25 - 1} = \sqrt{24} = 2\sqrt{6}$, and

$$\sin\theta = \frac{2\sqrt{6}}{5}, \tan\theta = 2\sqrt{6}, \cot\theta = \frac{1}{2\sqrt{6}} = \frac{\sqrt{6}}{12},$$

$$\sec\theta = 5, \csc\theta = \frac{5}{2\sqrt{6}} = \frac{5\sqrt{6}}{12}.$$

(c)

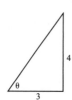

The hypotenuse is $\sqrt{16 + 9} = 5$, and

$$\sin\theta = \frac{4}{5}, \cos\theta = \frac{3}{5}, \cot\theta = \frac{3}{4}, \sec\theta = \frac{5}{3}, \csc\theta = \frac{5}{4}.$$

(d) The triangle in this part is the same as the triangle in part (a). The missing side is $\sqrt{9-4} = \sqrt{5}$, and

$$\sin\theta = \frac{2}{3}, \cos\theta = \frac{\sqrt{5}}{3}, \tan\theta = \frac{2}{\sqrt{5}} = \frac{2\sqrt{5}}{5},$$

$$\cot\theta = \frac{\sqrt{5}}{2}, \sec\theta = \frac{3}{\sqrt{5}} = \frac{3\sqrt{5}}{5}.$$

35. $\sin\alpha = \frac{3}{5}, \sin\beta = \frac{4}{5}, \cos\alpha = \frac{4}{5}, \cos\beta = \frac{3}{5}$

(a) $\sin(\alpha+\beta) = \sin\alpha\cos\beta + \sin\beta\cos\alpha = \frac{3}{5}\frac{3}{5} + \frac{4}{5}\frac{4}{5} = 1$

(b) $\cos(\alpha+\beta) = \cos\alpha\cos\beta - \sin\alpha\sin\beta = \frac{12}{25} - \frac{12}{25} = 0$

(c) $\sin 2\alpha = 2\sin\alpha\cos\alpha = 2\frac{3}{5}\frac{4}{5} = \frac{24}{25}$

(d) $\cos 2\alpha = (\cos\alpha)^2 - (\sin\alpha)^2 = \frac{16}{25} - \frac{9}{25} = \frac{7}{25}$

(e) $\sin\left(\frac{\alpha}{2}\right) = \sqrt{\frac{1-\cos\alpha}{2}} = \sqrt{\frac{1-4/5}{2}} = \sqrt{\frac{1}{10}} = \frac{\sqrt{10}}{10}$

(f) $\cos\left(\frac{\alpha}{2}\right) = \sqrt{\frac{1+\cos\alpha}{2}} = \sqrt{\frac{1+4/5}{2}} = \sqrt{\frac{9}{10}} = \frac{3}{\sqrt{10}} = \frac{3\sqrt{10}}{10}$

37. If h denotes the height of the building, then $\tan 12° = \frac{h}{1} \Rightarrow h = \tan 12° \approx 0.21$ miles. In feet, $h = 5280\tan 12° \approx 1122.3$ feet.

39. If h denotes the height of the cliff, then

$$\tan 70° = \frac{h}{40} \Rightarrow h = 40\tan 70° \approx 109.9 \text{ feet.}$$

41. From the figure we see that

$$\cot 10.5° = \frac{x}{80} \Rightarrow x = 80\cot 10.5° \approx 431.64,$$

so the campfire is approximately 431.64 feet from the base of the tower.

3.7 Inverse Trigonometric Functions

For a function to have an inverse, the function must be one-to-one. The trigonometric functions are not one-to-one, which is easily seen by the horizontal line test. For each of the trigonometric functions, there are horizontal lines that intersect the graph in many points, not just one. To define the inverse trigonometric functions, the domains have to be restricted so the functions become one-to-one.

INVERSE SINE

The domain of the sine function is restricted to the interval $\left[-\frac{\pi}{2}, \frac{\pi}{2}\right]$, making it one-to-one. The inverse sine function is then defined by

$$y = \arcsin x \Leftrightarrow \sin y = x.$$

The inverse sine undoes the process of the restricted sine function.

An important relationship to remember about all functions and their inverses, which is very helpful when trying to construct graphs, is that

the domain of the function is the range of the inverse,

and

the range of the function is the domain of the inverse.

For the sine function,

$$\text{domain of arcsin} = \text{ range of } \sin = [-1, 1]$$

$$\text{range of arcsin} = \text{ domain of } \sin = \left[-\frac{\pi}{2}, \frac{\pi}{2}\right]$$

$$\arcsin(\sin x) = x, \text{ for } x \text{ in the domain of } \sin$$

$$\sin(\arcsin x) = x, \text{ for } x \text{ in the domain of arcsin}.$$

EXAMPLE 3.7.1   Find the exact value of the quantity.

(a) $\arcsin\left(\frac{\sqrt{3}}{2}\right)$ (b) $\sin\left(\arcsin\left(\frac{1}{2}\right)\right)$

(c) $\arcsin\left(\sin\frac{5\pi}{4}\right)$ (d) $\cos\left(\arcsin\left(\frac{1}{2}\right)\right)$

SOLUTION:

(a) The problem can be read as, "find a number between $-\frac{\pi}{2}$ and $\frac{\pi}{2}$ with sine equal to $\frac{\sqrt{3}}{2}$." Since $\sin\frac{\pi}{3} = \frac{\sqrt{3}}{2}$,

$$\arcsin\left(\frac{\sqrt{3}}{2}\right) = \frac{\pi}{3}.$$

(b) The domain of arcsin is $[-1, 1]$, which contains $\frac{1}{2}$ so,

$$\sin\left(\arcsin\left(\frac{1}{2}\right)\right) = \frac{1}{2}.$$

(c) The answer to this part is *not* $\frac{5\pi}{4}$ since $\frac{5\pi}{4}$ is not in the restricted domain of the sine function. To solve this problem, we first need to find $\sin\left(\frac{5\pi}{4}\right)$, and then find a number in $\left[-\frac{\pi}{2}, \frac{\pi}{2}\right]$ whose sine is the same as that of $\sin\frac{5\pi}{4}$. So

$$\sin\frac{5\pi}{4} = -\frac{\sqrt{2}}{2}$$

$$\arcsin\left(\sin\frac{5\pi}{4}\right) = \arcsin\left(-\frac{\sqrt{2}}{2}\right) = -\frac{\pi}{4}.$$

Recall the sine is negative in quadrant IV.

(d)

$$\cos\left(\arcsin\left(\frac{1}{2}\right)\right) = \cos\left(\frac{\pi}{6}\right) = \frac{\sqrt{3}}{2}$$

◇

SUMMARY OF THE INVERSE TRIGONOMETRIC FUNCTIONS

The table summarizes the definitions of all the inverse trigonometric functions.

TABLE TO BE INSERTED HERE.

Definition	Domain	Range
$\arcsin x = y \Leftrightarrow \sin y = x$	$[-1, 1]$	$\left[-\frac{\pi}{2}, \frac{\pi}{2}\right]$
$\arccos x = y \Leftrightarrow \cos y = x$	$[-1, 1]$	$[0, \pi]$
$\arctan x = y \Leftrightarrow \tan y = x$	$(-\infty, \infty)$	$\left(-\frac{\pi}{2}, \frac{\pi}{2}\right)$
$\text{arccot} x = y \Leftrightarrow \cot y = x$	$(-\infty, \infty)$	$(0, \pi)$
$\text{arcsec} x = y \Leftrightarrow \sec y = x$	$(-\infty, -1] \bigcup [1, \infty)$	$\left[0, \frac{\pi}{2}\right) \bigcup \left(\frac{\pi}{2}, \pi\right]$
$\text{arccsc} x = y \Leftrightarrow \csc y = x$	$(-\infty, -1] \bigcup [1, \infty)$	$\left[-\frac{\pi}{2}, 0\right) \bigcup \left(0, \frac{\pi}{2}\right]$

EXAMPLE 3.7.2   Find the exact value of each expression.

(a) $\arccos\left(\cos\left(\frac{\pi}{6}\right)\right)$ (b) $\sin\left(\arctan\left(\sqrt{3}\right)\right)$ (c) $\arctan\left(\tan\left(\frac{5\pi}{6}\right)\right)$

SOLUTION:

(a) The restricted domain of the cosine function is $[0, \pi]$, which contains $\frac{\pi}{6}$ so

$$\arccos\left(\cos\left(\frac{\pi}{6}\right)\right) = \frac{\pi}{6}.$$

(b) The domain of arctan is all real numbers and since

$$\tan\left(\frac{\pi}{3}\right) = \frac{\sin\left(\frac{\pi}{3}\right)}{\cos\left(\frac{\pi}{3}\right)} = \frac{\frac{\sqrt{3}}{2}}{\frac{1}{2}} = \sqrt{3},$$

we have

$$\sin\left(\arctan\left(\sqrt{3}\right)\right) = \sin\left(\frac{\pi}{3}\right) = \frac{\sqrt{3}}{2}.$$

(c) Since the restricted domain of tan is $\left(-\frac{\pi}{2}, \frac{\pi}{2}\right)$, which does not contain $\frac{5\pi}{6}$, the answer is *not* $\frac{5\pi}{6}$. First compute the tangent and then find a value in the domain of the inverse tangent whose tangent agrees. The reference angle for $\frac{5\pi}{6}$ is $\frac{\pi}{6}$, and $\frac{5\pi}{6}$ is in the second quadrant. In this quadrant, the sine is positive and the cosine is negative, so

$$\tan\left(\frac{5\pi}{6}\right) = -\tan\left(\frac{\pi}{6}\right) = -\frac{\sin\left(\frac{\pi}{6}\right)}{\cos\left(\frac{\pi}{6}\right)} = -\frac{\frac{1}{2}}{\frac{\sqrt{3}}{2}} = -\frac{1}{\sqrt{3}} = -\frac{\sqrt{3}}{3}.$$

Since $\tan\left(-\frac{\pi}{6}\right)$ is also $-\frac{\sqrt{3}}{3}$,

$$\arctan\left(\tan\left(\frac{5\pi}{6}\right)\right) = -\frac{\pi}{6}.$$

◇

EXAMPLE 3.7.3   Find the exact value of the expression.

(a) $\sin\left(\arccos\left(\frac{4}{5}\right)\right)$ (b) $\sin\left(\arcsin\left(\frac{4}{5}\right) + \arccos\left(\frac{3}{5}\right)\right)$

SOLUTION:

(a) Let t satisfy

$$t = \arccos\left(\frac{4}{5}\right) \quad \Rightarrow \quad \cos t = \frac{4}{5}.$$

By the Pythagorean Identity,

$$(\cos t)^2 + (\sin t)^2 = 1$$

$$\left(\frac{4}{5}\right)^2 + (\sin t)^2 = 1$$

$$(\sin t)^2 = 1 - \frac{16}{25} = \frac{9}{25}$$

$$\sin t = \pm\sqrt{\frac{9}{25}} = \pm\frac{3}{5}.$$

Since $t = \arccos\left(\frac{4}{5}\right)$ and t is in the interval $[0, \pi]$, where the sine is always nonnegative, we have

$$\sin t = \sin\left(\arccos\left(\frac{4}{5}\right)\right) = \frac{3}{5}.$$

(b) To simplify the expression, first apply the sum formula for sine, $\sin(a+b) = \sin a \cos b + \cos a \sin b$. This gives

$$\sin\left(\arcsin\left(\frac{4}{5}\right) + \arccos\left(\frac{3}{5}\right)\right)$$

$$= \sin\left(\arcsin\left(\frac{4}{5}\right)\right)\cos\left(\arccos\left(\frac{3}{5}\right)\right) + \cos\left(\arcsin\left(\frac{4}{5}\right)\right)\sin\left(\arccos\left(\frac{3}{5}\right)\right)$$

$$= \left(\frac{4}{5}\right)\left(\frac{3}{5}\right) + \cos\left(\arcsin\left(\frac{4}{5}\right)\right)\sin\left(\arccos\left(\frac{3}{5}\right)\right)$$

The figure shows a triangle with angle θ satisfying $\sin\theta = \frac{4}{5}$, that is, $\theta = \arcsin\left(\frac{4}{5}\right)$.

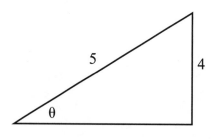

Figure 3.33

By the Pythagorean Theorem, the missing side has length

$$\sqrt{5^2 - 4^2} = \sqrt{25 - 16} = \sqrt{9} = 3,$$

so $\cos \theta = \frac{3}{5}$ or $\theta = \arccos\left(\frac{3}{5}\right)$. Then

$$\frac{3}{5} = \cos \theta = \cos\left(\arcsin\left(\frac{4}{5}\right)\right),$$

and

$$\frac{4}{5} = \sin \theta = \sin\left(\arccos\left(\frac{3}{5}\right)\right).$$

Finally,

$$\sin\left(\arcsin\left(\frac{4}{5}\right) + \arccos\left(\frac{3}{5}\right)\right) = \frac{4}{5} \cdot \frac{3}{5} + \frac{3}{5} \cdot \frac{4}{5}$$

$$= \frac{24}{25}.$$

◇

EXAMPLE 3.7.4   Solve the equation on the given interval, express the solution for x in terms of inverse functions, and use a calculator to approximate the solutions.

(a) $\sin 2x - \cos x = 0$ on $\left[0, \frac{\pi}{2}\right]$ (b) $(\tan x)^2 - 3\tan x - 4 = 0$ on $\left(-\frac{\pi}{2}, \frac{\pi}{2}\right)$

SOLUTION:

(a) Use the double-angle formula for sine, $\sin 2x = 2\sin x \cos x$. This gives

$$\sin 2x - \cos x = 0$$

$$2\sin x \cos x - \cos x = 0$$

$$\cos x (2\sin x - 1) = 0$$

$$\cos x = 0, \quad \sin x = \frac{1}{2}.$$

So

$$x = \arccos(0) = \frac{\pi}{2} \approx 1.571, \quad x = \arcsin\left(\frac{1}{2}\right) = \frac{\pi}{6} \approx 0.524.$$

(b) If we let $u = \tan x$, then

$$(\tan x)^2 - 3\tan x - 4 = u^2 - 3u - 4$$
$$= (u-4)(u+1).$$

So

$$(\tan x)^2 - 3\tan x - 4 = 0$$
$$(\tan x - 4)(\tan x + 1) = 0$$
$$\tan x = 4 \quad \text{or} \quad \tan x = -1,$$

and

$$x = \arctan(4) \approx 1.326 \quad \text{or}$$
$$x = \arctan(-1) = -\frac{\pi}{4} \approx -0.785.$$

◇

EXAMPLE 3.7.5   A lighthouse is located on an island 2 miles from the nearest point on a straight shoreline. If the light from the lighthouse is moving along the shoreline, express the angle θ formed by the beam of light and the shoreline in terms of the distance x in the figure.

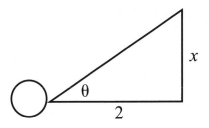

Figure 3.34

SOLUTION: The parameters θ, x, and 2 can be related by a tangent function, since the sides of length x and 2 are the opposite and adjacent sides of the angle θ. So

$$\tan \theta = \frac{x}{2},$$

and

$$\theta = \arctan \left(\frac{x}{2} \right).$$

$\diamond$

Solutions for Exercise Set 3.7

1. $y = \arccos\left(\frac{1}{2}\right) \Leftrightarrow \cos y = \frac{1}{2}$ with $0 \le y \le \pi$, so $y = \arccos\left(\frac{1}{2}\right) = \frac{\pi}{3}$.

3. $y = \arcsin(1) \Leftrightarrow \sin y = 1$ with $-\frac{\pi}{2} \le y \le \frac{\pi}{2}$, so $y = \arcsin(1) = \frac{\pi}{2}$.

5. $y = \arccos\left(-\frac{\sqrt{3}}{2}\right) \Leftrightarrow \cos y = -\frac{\sqrt{3}}{2}$ with $0 \le y \le \pi$, so $y = \arccos\left(-\frac{\sqrt{3}}{2}\right) = \frac{5\pi}{6}$.

7. $y = \arctan\left(\sqrt{3}\right) \Leftrightarrow \tan y = \sqrt{3}$ with $-\frac{\pi}{2} < y < \frac{\pi}{2}$, so $y = \arctan(\sqrt{3}) = \frac{\pi}{3}$.

9. $y = \arctan(-1) \Leftrightarrow \tan y = -1$ with $-\frac{\pi}{2} < y < \frac{\pi}{2}$, so $y = \arctan(-1) = -\frac{\pi}{4}$.

11. $y = \operatorname{arcsec}\left(\sqrt{2}\right) \Leftrightarrow \sec y = \sqrt{2} \Leftrightarrow \cos y = \frac{1}{\sqrt{2}} = \frac{\sqrt{2}}{2}$ with $0 \le y < \frac{\pi}{2}$ or $\frac{\pi}{2} < y \le \pi$, so $y = \operatorname{arcsec}(\sqrt{2}) = \frac{\pi}{4}$.

13. $y = \arccos(2) \Leftrightarrow \cos y = 2$ which can never occur since $-1 \le \cos y \le 1$.

15. Since $\cos(\arccos x) = x$ for x in $[-1, 1]$, we have $\cos\left(\arccos\left(\frac{1}{2}\right)\right) = \frac{1}{2}$.

17. The $\arccos\left(\frac{1}{2}\right)$ is the angle in $[0, \pi]$ whose cosine is $\frac{1}{2}$, so $\arccos\left(\frac{1}{2}\right) = \frac{\pi}{3}$ and $\sin\left(\arccos\left(\frac{1}{2}\right)\right) = \sin\left(\frac{\pi}{3}\right) = \frac{\sqrt{3}}{2}$.

19. The $\arcsin\left(\frac{\sqrt{3}}{2}\right)$ is the angle in $\left[-\frac{\pi}{2}, \frac{\pi}{2}\right]$ whose sine is $\frac{\sqrt{3}}{2}$, so $\arcsin\left(\frac{\sqrt{3}}{2}\right) = \frac{\pi}{3}$ and $\tan\left(\arcsin\left(\frac{\sqrt{3}}{2}\right)\right) = \tan\left(\frac{\pi}{3}\right) = \sqrt{3}$.

21. Since $\arcsin(\sin x) = x$ for x in $\left[-\frac{\pi}{2}, \frac{\pi}{2}\right]$, we have $\arcsin\left(\sin\left(-\frac{\pi}{2}\right)\right) = -\frac{\pi}{2}$.

23. The $\arccos(\cos x) = x$ for x in $[0, \pi]$, but since $-\frac{\pi}{4}$ does not lie in this interval, the identity can not be applied. However,

$$\cos\left(-\frac{\pi}{4}\right) = \cos\left(\frac{\pi}{4}\right) \Rightarrow \arccos\left(\cos\left(-\frac{\pi}{4}\right)\right) = \arccos\left(\cos\left(\frac{\pi}{4}\right)\right) = \frac{\pi}{4}.$$

25. Since $\arctan(\tan x) = x$ for x in $\left(-\frac{\pi}{2}, \frac{\pi}{2}\right)$, and $\frac{\pi}{3}$ is in this interval $\arctan\left(\tan\left(\frac{\pi}{3}\right)\right) = \frac{\pi}{3}$.

27. The arctan $(\tan x) = x$ for x in $\left(-\frac{\pi}{2}, \frac{\pi}{2}\right)$, but since $\frac{7\pi}{6}$ does not lie in this interval, the identity can not be applied. However,

$$\tan\left(\frac{7\pi}{6}\right) = \tan\left(\frac{\pi}{6}\right) \Rightarrow \arctan\left(\tan\left(\frac{7\pi}{6}\right)\right) = \arctan\left(\tan\left(\frac{\pi}{6}\right)\right) = \frac{\pi}{6}.$$

29. Let $t = \arcsin\left(\frac{3}{5}\right) \Leftrightarrow \sin t = \frac{3}{5}$. The triangle in the figure has $\sin t = \frac{3}{5}$ which implies the side adjacent t is $\sqrt{25-9} = 4$ so

$$\cos t = \frac{4}{5} \quad \text{and} \quad \cos(t) = \cos\left(\arcsin\left(\frac{3}{5}\right)\right) = \frac{4}{5}.$$

31. Let $t = \arccos\left(\frac{4}{5}\right) \Leftrightarrow \cos t = \frac{4}{5}$. The triangle in the figure has $\cos t = \frac{4}{5}$ which implies the side opposite t is $\sqrt{25-16} = 3$ so

$$\tan t = \tan\left(\arccos\left(\frac{4}{5}\right)\right) = \frac{\sin t}{\cos t} = \frac{3}{4}.$$

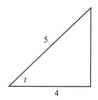

33. Let $t = \arctan 4 \Leftrightarrow \tan t = 4$. The triangle in the figure has $\tan t = 4$ which implies the hypotenuse is $\sqrt{16 + 1} = \sqrt{17}$ so

$$\cos t = \cos (\arctan 4) = \frac{1}{\sqrt{17}} = \frac{\sqrt{17}}{17}.$$

35. Apply the formula $\cos(a + b) = \cos a \cos b - \sin a \sin b$, where $a = \arcsin (3/5)$ and $b = \arccos (4/5)$. Then

$$\cos \left(\arcsin \left(\frac{3}{5} \right) + \arccos \left(\frac{4}{5} \right) \right)$$

$$= \cos \left(\arcsin \left(\frac{3}{5} \right) \right) \cos \left(\arccos \left(\frac{4}{5} \right) \right) - \sin \left(\arcsin \left(\frac{3}{5} \right) \right) \sin \left(\arccos \left(\frac{4}{5} \right) \right)$$

$$= \frac{4}{5} \cdot \frac{4}{5} - \frac{3}{5} \cdot \frac{3}{5} = \frac{7}{25}.$$

37. Let $y = \arcsin(-x)$. Then

$$-x = \sin y \Leftrightarrow x = -\sin y = \sin(-y) \Leftrightarrow -y = \arcsin x \Leftrightarrow y = -\arcsin x.$$

39. Let $y = \arccos x$. Then $\cos y = x$. A right triangle with one angle y, adjacent side x and hypotenuse 1 has opposite side $\sqrt{1 - x^2} \Rightarrow \cos y = x$, and $\sin y = \sqrt{1 - x^2}$. So $y = \arcsin \left(\sqrt{1 - x^2} \right)$.

41. Let $y = \arcsin x$. Then $\sin y = x$. A right triangle with one angle y, opposite side x and hypotenuse 1 has adjacent side $\sqrt{1 - x^2} \Rightarrow \sin y = x$, and $\tan y = \frac{x}{\sqrt{1-x^2}}$. So $\tan (\arcsin x) = \frac{x}{\sqrt{1-x^2}}$.

43. (a) We have

$$(\tan x)^2 - \tan x - 2 = 0 \Leftrightarrow (\tan x + 1)(\tan x - 2) = 0 \Leftrightarrow$$

$$\tan x = -1 \text{ or } \tan x = 2 \Leftrightarrow$$

$$x = \arctan(-1), x = \arctan(2).$$

(b) We have $x = \arctan(-1) = -\frac{\pi}{4} \approx -0.785$ or $x = \arctan(2) \approx 1.107$.

45. The illustration indicates that $\tan \theta = \frac{x}{4}$, so $\theta = \arctan\left(\frac{x}{4}\right)$.

3.8 Applications of Trigonometric Functions

LAW OF COSINES

To *solve* a triangle, that is, find the lengths of all sides and all angles, the *Law of Cosines* can be used when we know two sides and the angle between them or all three sides.

Law of Cosines:

$$a^2 = b^2 + c^2 - 2bc \cos \alpha$$

$$b^2 = a^2 + c^2 - 2ac \cos \beta$$

$$c^2 = a^2 + b^2 - 2ab \cos \gamma$$

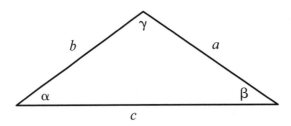

Figure 3.35

EXAMPLE 3.8.1 Let the angles of a triangle be α, β, and γ with opposite sides of lengths a, b, and c, respectively. If $\alpha = 37°, b = 16$, and $c = 24$, use the Law of Cosines to solve the triangle, rounding all answers to one decimal place.

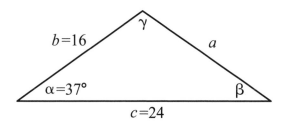

Figure 3.36

SOLUTION: We need to find the missing parts β, γ, and a. Since we are given the two sides b and c and the angle α between them, the side a can be found. That is,

$$a^2 = b^2 + c^2 - 2bc \cos \alpha$$

$$a^2 = 16^2 + 24^2 - 2(16)(24) \cos 37°$$

$$= 256 + 576 - 768 \cos 37°$$

$$a = \sqrt{832 - 768 \cos 37°} \approx 14.8.$$

Now that we also have an approximation to side a, we can find either of the angles β or γ, since we have all three sides. We elect to find β, so

$$b^2 = a^2 + c^2 - 2ac \cos \beta$$

$$16^2 = (14.8)^2 + 24^2 - 2(14.8)(24) \cos \beta$$

$$256 = 219.04 + 576 - 710.4 \cos \beta$$

$$\cos \beta = \frac{256 - 795.04}{-710.4} = \frac{539.04}{710.4}$$

$$\beta = \arccos \left(\frac{539.04}{710.4} \right) \approx 0.7094.$$

In degrees we have

$$\beta \approx 0.7094 \left(\frac{180}{\pi} \right) \approx 40.7°.$$

Since the sum of the angles of a triangle is 180°,

$$\gamma \approx 180 - 37 - 40.7 = 102.3°.$$

◇

LAW OF SINES

The *Law of Sines* states that the lengths of the sides of a triangle are proportional to the sines of the corresponding opposite angles. In a triangle with sides of lengths a, b and c, with corresponding opposite angles α, β and γ, the Law of Sines states

$$\frac{\sin \alpha}{a} = \frac{\sin \beta}{b} = \frac{\sin \gamma}{c}.$$

The Law of Sines can be used to solve a triangle when you have a side and the angle opposite it.

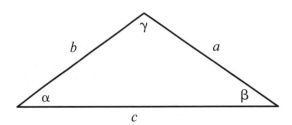

Figure 3.37

EXAMPLE 3.8.2   Let the angles of a triangle be α, β, and γ, with opposite sides of lengths a, b, and c, respectively. If $\alpha = 50°, \beta = 46°$, and $a = 25$ use the Law of Sines to solve the triangle, rounding all answers to one decimal place.

SOLUTION: We need to find the missing parts γ, b and c. Since the sum of the angles of a triangle is $180°$, and we are given two angles, α and β, we can find γ immediately by

$$\gamma = 180 - 50 - 46 = 84°.$$

To find b we have

$$\frac{\sin \beta}{b} = \frac{\sin \alpha}{a}$$
$$b = \frac{a \sin \beta}{\sin \alpha} = \frac{25 \sin 46°}{\sin 50°} \approx 23.5.$$

To find c we have

$$\frac{\sin \gamma}{c} = \frac{\sin \alpha}{a}$$
$$c = \frac{a \sin \gamma}{\sin \alpha} = \frac{25 \sin 84°}{\sin 50°} \approx 32.5.$$

$\diamond$

EXAMPLE 3.8.3   Let the angles of a triangle be α, β, and γ, with opposite sides of lengths a, b and c, respectively. Use the Law of Cosines and the Law of Sines to find the remaining parts of the triangle, if $\alpha = 150°, b = 3$ and $c = 5$.

SOLUTION: The situation is shown in the figure.

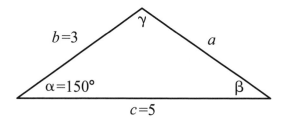

Figure 3.38

Since we have two sides and the angle α between them, the Law of Cosines can be used to find the side opposite the angle α. That is,

$$a^2 = b^2 + c^2 - 2bc\cos\alpha$$

$$a^2 = 3^2 + 5^2 - 2(3)(5)\cos 150°$$

$$a^2 = 9 + 25 - 30\left(-\frac{\sqrt{3}}{2}\right)$$

$$a^2 = 34 + 15\sqrt{3}$$

$$a = \sqrt{34 + 15\sqrt{3}} \approx 7.75.$$

Now using the Law of Sines,

$$\frac{\sin\beta}{b} = \frac{\sin\alpha}{a}$$

$$\sin\beta = \frac{b\sin\alpha}{a} = \frac{3\sin 150°}{7.75}$$

$$= \frac{3\left(\frac{1}{2}\right)}{7.75} \approx 0.4516,$$

and

$$\beta = \arcsin\left(\frac{3}{15.5}\right) \approx 11.16°.$$

Since the sum of the angles of a triangle is $180°$,

$$\gamma = 180 - \alpha - \beta \approx 180 - 150 - 11.16 = 18.84°.$$

◇

EXAMPLE 3.8.4   Show there is no triangle satisfying the conditions $a = 3$, $b = 8$, and $\alpha = 26.2°$.

SOLUTION: If there was such a triangle, then the angle β could be found by the Law of Sines from

$$\frac{\sin \beta}{b} = \frac{\sin \alpha}{a}$$

$$\sin \beta = \frac{b \sin \alpha}{a} = \frac{8 \sin 26.2°}{3} \approx 1.18.$$

Since the sine of an angle is never greater than one, we conclude no such triangle exists.   ◇

HERON'S FORMULA

Given the lengths of the three sides of a triangle, an easy formula for finding the area of the triangle is *Heron's formula*. A triangle with sides of lengths a, b and c and perimeter $P = a + b + c$ has area

$$A = \frac{1}{4}\sqrt{P(P - 2a)(P - 2b)(P - 2c)}.$$

EXAMPLE 3.8.5   Find the area of a triangle with sides of lengths 12 ft, 18 ft and 24 ft.

SOLUTION: To use Heron's formula, first compute the perimeter of the triangle. So,

$$P = a + b + c$$

$$= 12 + 18 + 24$$

$$= 54.$$

Then the area is

$$A = \frac{1}{4}\sqrt{54(54 - 24)(54 - 36)(54 - 48)}$$

$$= \frac{1}{4}\sqrt{54(30)(18)(6)}$$

$$= \frac{1}{4}\sqrt{174960} = \frac{1}{4}108\sqrt{15} = 27\sqrt{15}$$

$$\approx 104.57 \text{ ft}^2.$$

◇

EXAMPLE 3.8.6   The lengths of the sides of a triangular parcel of land are approximately 200 ft, 400 ft, and 500 ft. If the land is valued at $2,000.00 per acre, what is the value of the parcel of land?

SOLUTION: To determine how many acres in the parcel, we need to know the number of square feet of land in the parcel and also the number of square feet in one acre. The number of square feet in one acre is

$$\text{one acre} = 43,560 \text{ ft}^2.$$

The area can be found using Hero's formula. Since the perimeter is

$$P = 200 + 400 + 500 = 1100,$$

we have

$$A = \frac{1}{4}\sqrt{1100(1100 - 400)(1100 - 800)(1100 - 1000)}$$

$$= \frac{1}{4}\sqrt{1100(700)(300)(100)} = \frac{1}{4}\sqrt{23100000000}$$

$$\approx 37,997 \text{ ft}^2.$$

The number of acres in the parcel is approximately

$$\frac{37997}{43560} = 0.87 \text{ acres,}$$

and the value of the parcel of land is approximately

$$\left(\frac{37997}{43560}\right) 2000 \approx \$1,745.00.$$

◇

Solutions for Exercise Set 3.8

1. Since $\alpha = 35°, b = 15, c = 25$, we have

$$a^2 = b^2 + c^2 - 2bc \cos \alpha = 15^2 + 25^2 - 2(15)(25) \cos 35°,$$

implies

$$a = \sqrt{850 - 750 \cos 35°} \approx 15.4.$$

and

$$b^2 = a^2 + c^2 - 2ac \cos \beta$$

implies

$$\cos \beta = \frac{15^2 - (15.4)^2 - 25^2}{-2(15.4)(25)}.$$

So

$$\beta = \arccos\left(\frac{15^2 - (15.4)^2 - 25^2}{-2(15.4)(25)}\right) \approx 0.59619 = 0.59619 \left(\frac{180}{\pi}\right)^{\circ} \approx 34.2°,$$

and $\gamma \approx 180 - 35 - 34.2 = 110.8°$.

3. Since $\beta = 30°, a = 25, c = 32$ we have

$$b = \sqrt{25^2 + 32^2 - 2(25)(32) \cos 30°} \approx 16.2$$

and

$$a^2 = b^2 + c^2 - 2bc \cos \alpha$$

implies

$$\cos \alpha = \frac{25^2 - (16.2)^2 - (32)^2}{-2(16.2)(32)}.$$

Hence

$$\alpha = \arccos\left(\frac{25^2 - (16.2)^2 - (32)^2}{-2(16.2)(32)}\right) \approx 0.87895 = 0.87895 \left(\frac{180}{\pi}\right)^{\circ} \approx 50.4°,$$

and $\gamma = 180 - 30 - 50.4 = 99.6°$.

5. Since $\alpha = 50°, \beta = 76°, c = 100$ we have $\gamma = 180 - 50 - 76 = 54°$ and

$$\frac{\sin\alpha}{a} = \frac{\sin\gamma}{c} \Rightarrow a = \frac{100\sin 50°}{\sin 54°} \approx 94.7;$$

and

$$\frac{\sin\beta}{b} = \frac{\sin\gamma}{c} \Rightarrow b = \frac{100\sin 76°}{\sin 54°} \approx 119.9.$$

7. Since $\beta = 100°, \gamma = 30°, c = 20$ we have $\alpha = 180 - 100 - 30 = 50°$ and

$$\frac{\sin\alpha}{a} = \frac{\sin\gamma}{c} \Rightarrow a = \frac{20\sin 50°}{\sin 30°} \approx 30.6;$$

and

$$\frac{\sin\beta}{b} = \frac{\sin\gamma}{c} \Rightarrow b = \frac{20\sin 100°}{\sin 30°} \approx 39.4.$$

9. Since $\alpha = 120°, b = 3, c = 5$, using the Law of Cosines we have

$$a^2 = b^2 + c^2 - 2bc\cos\alpha = 3^2 + 5^2 - 2(3)(5)\cos 120° \Rightarrow$$

$$a = \sqrt{34 - 30\cos 120°} = 7.$$

Then by the Law of Sines

$$\frac{\sin\alpha}{a} = \frac{\sin\beta}{b} \Rightarrow \beta = \arcsin\left(\frac{3\sin 120°}{7}\right) \approx 21.8°$$

and $\gamma = 180 - 120 - 21.8 = 38.2°$.

11. Since $a = 12, b = 22, c = 15$ by the Law of Cosines we have

$$\cos\alpha = \frac{12^2 - (22)^2 - (15)^2}{-2(22)(15)}$$

so

$$\alpha = \arccos\left(\frac{12^2 - (22)^2 - (15)^2}{-2(22)(15)}\right) \approx 0.5432 = 0.5432\left(\frac{180}{\pi}\right)^° \approx 31.1°.$$

By the Law of Cosines

$$\gamma = \arccos\left(\frac{22^2 + 12^2 - 15^2}{2(12)(22)}\right) \approx 0.7633\left(\frac{180}{\pi}\right)^° \approx 40.2°.$$

so $\beta = 180 - 31.1 - 40.2 = 108.6°$.

13. In order for a triangle to satisfy the conditions $a = 3, b = 10, \alpha = 25.4°$, by the Law of Sines,

$$\frac{\sin \alpha}{a} = \frac{\sin \beta}{b} \Rightarrow \sin \beta = \frac{b \sin \alpha}{a} = \frac{10 \sin 25.4°}{3} \approx 1.4.$$

Since the sine of an angle is never greater than 1, no such triangle can exist.

15. Since $a = 125, b = 150, \alpha = 55°$ by the Law of Sines,

$$\frac{\sin \alpha}{a} = \frac{\sin \beta}{b} \Rightarrow \sin \beta = \frac{b \sin \alpha}{a} = \frac{150 \sin 55°}{125}.$$

There are two angles between $0°$ and $180°$ that satisfy this condition. One such angle is

$$\beta = \arcsin\left(\frac{150 \sin 55°}{125}\right) \approx 0.9829 = 0.9829\left(\frac{180}{\pi}\right)^{°} \approx 79.4°.$$

A second angle is $\beta_1 = 180 - 79.4 = 100.6°$.

Using $\beta : \gamma = 180 - 55 - 79.4 = 45.6°$ and

$$\frac{\sin \alpha}{a} = \frac{\sin \gamma}{c} \Rightarrow c = \frac{125 \sin 45.6°}{\sin 55°} \approx 109.$$

Using $\beta_1 : \gamma = 180 - 55 - 100.6 = 24.4°$ and

$$\frac{\sin \alpha}{a} = \frac{\sin \gamma}{c} \Rightarrow c = \frac{125 \sin 24.4°}{\sin 55°} \approx 63.$$

17. By Heron's Formula, the area of the triangle with sides $a = 10, b = 14, c = 18$, is

$$A = \frac{1}{4}\sqrt{P(P - 2a)(P - 2b)(P - 2c)},$$

where

$$P = a + b + c = 10 + 14 + 18 = 42.$$

So

$$A = \frac{1}{4}\sqrt{42(42-20)(42-28)(42-36)} = \frac{1}{4}\sqrt{42(22)(14)(6)}$$
$$= \frac{1}{4}84\sqrt{11} = 21\sqrt{11} \approx 69.7 \text{ cm}^2.$$

19. Since point B is directly northwest of point C the angle made by the line segment connecting C to B and the horizontal is $45°$. Then angle ACB is $180 - 45 = 135°$. By the Law of Cosines

$$\overline{AB}^2 = 2^2 + 3^2 - 2(2)(3)\cos 135° \Rightarrow \overline{AB} = \sqrt{13 - 12\cos 135°} \approx 4.6 \text{ miles.}$$

(a) The approximate cost of construction directly between points A and B is

$$125000\sqrt{13 - 12\cos 135°} \approx \$579,403.00.$$

The approximate cost of construction from A to C, then C to B is $(3+2)(100000) = \$500,000.00$. So the engineers should select the route that avoids the swamp.

(b) Let P denote the cost per mile for construction through C. If the total cost of construction from A to B to C is to equal the cost directly from A to B, then

$$5P = 579403 \Rightarrow P = \frac{579403}{5} \approx 115881.$$

So if the cost per mile through C is approximately $\$115,800.00$, then the cost of either alternative is about the same.

21. The angle at point B is $180 - 105 - 42 = 33°$, so by the Law of Sines

$$\frac{\sin 33°}{45} = \frac{\sin 42°}{\overline{AB}} \Rightarrow \overline{AB} = \frac{45\sin 42°}{\sin 33°} \approx 55.3 \text{ ft.}$$

23. At 12:00 am, two hours after the ships have left port, the first ship has traveled 40 miles and the second ship 50 miles. The angle between the two ships is $28 + 19 = 43°$. If d denotes the distance between the ships, by the Law of Cosines,

$$d^2 = 40^2 + 50^2 - 2(40)(50)\cos 43°,$$

and

$$d = \sqrt{40^2 + 50^2 - 2(40)(50)\cos 43°} \approx 83.8 \text{ miles.}$$

25. To use the Law of Sines to find the distance d from the fire to the tower B, first find the missing angle, which is $180 - 50 - 63 = 67°$. Then

$$\frac{\sin 67°}{10} = \frac{\sin 63°}{d} \Rightarrow d = \frac{10\sin 63°}{\sin 67°} \approx 9.7 \text{ miles.}$$

Solutions for Exercise Set 3 Review

1. (a) $P(t) = \left(\frac{1}{2}, \frac{\sqrt{3}}{2}\right)$

   (b) $\frac{\pi}{3}$

   (c) $\cos\frac{\pi}{3} = \frac{1}{2}$, $\sin\frac{\pi}{3} = \frac{\sqrt{3}}{2}$, $\tan\frac{\pi}{3} = \sqrt{3}$, $\cot\frac{\pi}{3} = \frac{1}{\sqrt{3}} = \frac{\sqrt{3}}{3}$, $\sec\frac{\pi}{3} = 2$, $\csc\frac{\pi}{3} = \frac{2}{\sqrt{3}} = \frac{2\sqrt{3}}{3}$

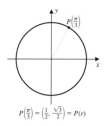

$P\left(\frac{\pi}{3}\right) = \left(\frac{1}{2}, \frac{\sqrt{3}}{2}\right) = P(r)$

3. (a) $P(t) = \left(-\frac{\sqrt{2}}{2}, -\frac{\sqrt{2}}{2}\right)$

   (b) $\frac{\pi}{4}$

   (c) $\cos\frac{5\pi}{4} = -\frac{\sqrt{2}}{2}$, $\sin\frac{5\pi}{4} = -\frac{\sqrt{2}}{2}$, $\tan\frac{5\pi}{4} = 1$, $\cot\frac{5\pi}{4} = 1$, $\sec\frac{5\pi}{4} = -\frac{2}{\sqrt{2}} = -\sqrt{2}$, $\csc\frac{5\pi}{4} = -\frac{2}{\sqrt{2}} = -\sqrt{2}$

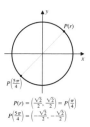

$P(r) = \left(\frac{\sqrt{2}}{2}, \frac{\sqrt{2}}{2}\right) = P\left(\frac{\pi}{4}\right)$

$P\left(\frac{5\pi}{4}\right) = \left(-\frac{\sqrt{2}}{2}, -\frac{\sqrt{2}}{2}\right)$

5. (a) $P(t) = \left(-\frac{\sqrt{2}}{2}, \frac{\sqrt{2}}{2}\right)$

   (b) $\frac{\pi}{4}$

   (c) $\cos\left(-\frac{21\pi}{4}\right) = -\frac{\sqrt{2}}{2}$, $\sin\left(-\frac{21\pi}{4}\right) = \frac{\sqrt{2}}{2}$, $\tan\left(-\frac{21\pi}{4}\right) = -1$, $\cot\left(-\frac{21\pi}{4}\right) = -1$, $\sec\left(-\frac{21\pi}{4}\right) = -\frac{2}{\sqrt{2}} = -\sqrt{2}$, $\csc\left(-\frac{21\pi}{4}\right) = \frac{2}{\sqrt{2}} = \sqrt{2}$

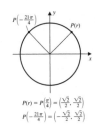

$$P(r) = P\left(\tfrac{\pi}{4}\right) = \left(\tfrac{\sqrt{2}}{2}, \tfrac{\sqrt{2}}{2}\right)$$
$$P\left(-\tfrac{21\pi}{4}\right) = \left(-\tfrac{\sqrt{2}}{2}, \tfrac{\sqrt{2}}{2}\right)$$

7. If $\cos t = \frac{3}{5}$ and $\frac{3\pi}{2} < t < 2\pi$, then $\sin t < 0$ and

$$(\cos t)^2 + (\sin t)^2 = 1 \Rightarrow (\sin t)^2 = 1 - \frac{9}{25} = \frac{16}{25} \Rightarrow \sin t = \pm\sqrt{\frac{16}{25}} = \pm\frac{4}{5}$$

so $\sin t = -\frac{4}{5}$. Then

$$\tan t = -\frac{4}{3}, \quad \cot t = -\frac{3}{4}, \quad \sec t = \frac{5}{3}, \quad \csc t = -\frac{5}{4}.$$

9. If $\tan t = \frac{1}{4}$ and $0 < t < \frac{\pi}{2}$, then $\cos t > 0$ and $\sin t > 0$ and

$$(\tan t)^2 + 1 = (\sec t)^2 \Rightarrow (\sec t)^2 = 1 + \frac{1}{16} = \frac{17}{16} \Rightarrow \sec t = \frac{1}{\cos t} = \sqrt{\frac{17}{16}} = \frac{\sqrt{17}}{4}.$$

Then

$$\cos t = \frac{4}{\sqrt{17}} = \frac{4\sqrt{17}}{17}, \quad \frac{1}{4} = \tan t = \frac{\sin t}{\cos t} \Rightarrow \sin t = \frac{1}{4}\frac{4\sqrt{17}}{17} = \frac{\sqrt{17}}{17},$$

and

$$\cot t = 4, \quad \csc t = \frac{17}{\sqrt{17}} = \sqrt{17}.$$

11. If $0 \le x \le \pi$, then $0 \le \frac{x}{3} \le \frac{\pi}{3}$ and $\cos\frac{x}{3} = \frac{1}{2} \Rightarrow \frac{x}{3} = \frac{\pi}{3} \Rightarrow x = \pi.$

13. Since

$$2(\cos x)^2 - 3\cos x + 1 = 0 \Rightarrow (2\cos x - 1)(\cos x - 1) = 0 \Rightarrow$$
$$\cos x = \frac{1}{2}, \cos x = 1 \Rightarrow x = \frac{\pi}{3}, x = 0.$$

15. Since

$$(\tan x)^3 - 4\tan x = 0 \Rightarrow \tan x[(\tan x)^2 - 4] = 0 \Rightarrow \tan x(\tan x - 2)(\tan x + 2) = 0,$$

we have

$$\tan x = 0, \tan x = 2, \tan x = -2 \Rightarrow x = 0, \pi, x = \arctan 2 \approx 1.1,$$

$$x = \arctan(-2) = \pi - \arctan(2) \approx 2.03.$$

Note $x = \arctan(-2)$ is not in $[0, \pi]$.

17. We have

$$\cot x - \csc x = 1 \Rightarrow \frac{\cos x}{\sin x} - \frac{1}{\sin x} = \frac{\cos x - 1}{\sin x} = 1 \Rightarrow \cos x - 1 = \sin x \Rightarrow \cos x - \sin x = 1$$

so $x = 0, \frac{3\pi}{2}$. But there are no solutions since $\cot x$ is not defined at $x = 0$ and $x = \frac{3\pi}{2}$ is not in the interval $[0, \pi]$.

19. The function $f(x) = (\sin x)^3$ is odd since

$$f(-x) = (\sin(-x))^3 = (-\sin x)^3 = -(\sin x)^3 = -f(x).$$

21. The function $f(x) = x(\sin x)^3$ is even since

$$f(-x) = -x(\sin(-x))^3 = -x(-\sin x)^3 = -x[-(\sin x)^3] = x(\sin x)^3 = f(x).$$

23. Since (a) and (c) have period $\frac{2\pi}{2} = \pi$ and (b) and (d) have period $\frac{2\pi}{1/2} = 4\pi$, we have (a) or (c) is either (i) or (ii) and (b) or (d) is either (iii) or (iv).

(a) Matches (ii) since $\sin 2\left(0 + \frac{\pi}{2}\right) = \sin \pi = 0$.

(b) Matches (iv) since the curve has the form of a sine wave shifted to the left $\frac{\pi}{2}$ units..

(c) Matches (i). (d) Matches (iii).

25. For $y = 5\sin\frac{x}{2}$

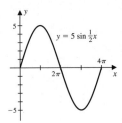

27. For $y = -3\cos 2x$

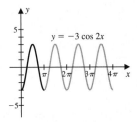

29. For $y = \cos(2x - \pi) = \cos 2\left(x - \frac{\pi}{2}\right)$

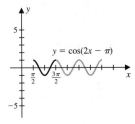

31. For $y = \cot(x + \pi/6)$

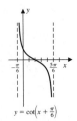

33. For $y = \sec 4\pi x$

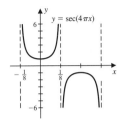

35. (a) The curve has an equation of the form $y = A\sin(Bx)$, where the amplitude is 1 and the period is $\frac{\pi}{2}$. So

$$A = 1, \frac{2\pi}{B} = \frac{\pi}{2} \Rightarrow B = 4 \quad \text{and} \quad y = \sin 4x.$$

(b) The curve has the form $y = A\cos B(x + C)$, with

$$A = 1, \frac{2\pi}{B} = \frac{\pi}{2} \Rightarrow B = 4, C = -\frac{\pi}{8} \quad \text{so} \quad y = \cos 4\left(x - \frac{\pi}{8}\right).$$

37. (a) The curve has an equation of the form $y = A\sin B(x + C)$, where the amplitude is 4, the period is 2, and the curve is shifted to the left $\frac{1}{2}$ unit. So

$$A = 4, \frac{2\pi}{B} = 2 \Rightarrow B = \pi, C = \frac{1}{2} \quad \text{and} \quad y = 4\sin \pi \left(x + \frac{1}{2}\right).$$

(b) The curve has the form $y = A\cos Bx$, with

$$A = 2, B = \pi, \quad \text{so} \quad y = 4\cos \pi x.$$

39. For $\alpha = 60°, \overline{AC} = 13$ we have

$$\sin 60° = \frac{\overline{BC}}{13} \Rightarrow \overline{BC} = 13\sin 60° = \frac{13\sqrt{3}}{2};$$

$$\cos 60° = \frac{\overline{AB}}{13} \Rightarrow \overline{AB} = 13\cos 60° = \frac{13}{2};$$

$$\beta = 90 - 60 = 30°.$$

41.

$$\cos\left(\frac{5\pi}{12}\right) = \cos\left(\frac{3\pi}{12} + \frac{2\pi}{12}\right) = \cos\left(\frac{\pi}{4} + \frac{\pi}{6}\right)$$

$$= \cos\frac{\pi}{4}\cos\frac{\pi}{6} - \sin\frac{\pi}{4}\sin\frac{\pi}{6} = \frac{\sqrt{2}}{2}\frac{\sqrt{3}}{2} - \frac{\sqrt{2}}{2}\frac{1}{2}$$

$$= \frac{\sqrt{2}}{4}\left(\sqrt{3} - 1\right)$$

43.

$$\cos\left(-\frac{13\pi}{12}\right) = \cos\left(\frac{13\pi}{12}\right) = \cos\left(\frac{9\pi}{12} + \frac{4\pi}{12}\right) = \cos\left(\frac{3\pi}{4} + \frac{\pi}{3}\right)$$

$$= \cos\frac{3\pi}{4}\cos\frac{\pi}{3} - \sin\frac{3\pi}{4}\sin\frac{\pi}{3} = -\frac{\sqrt{2}}{2}\frac{1}{2} - \frac{\sqrt{2}}{2}\frac{\sqrt{3}}{2}$$

$$= -\frac{\sqrt{2}}{4}\left(\sqrt{3} + 1\right)$$

45. $\sin\frac{\pi}{8} = \sqrt{\frac{1 - \cos\frac{\pi}{4}}{2}} = \sqrt{\frac{1 - \sqrt{2}/2}{2}} = \sqrt{\frac{2 - \sqrt{2}}{4}} = \frac{\sqrt{2 - \sqrt{2}}}{2}$

47. $\tan\frac{7\pi}{12} = \frac{\sin\frac{7\pi}{12}}{\cos\frac{7\pi}{12}} = \frac{\sqrt{\frac{1 - \cos\frac{7\pi}{6}}{2}}}{-\sqrt{\frac{1 + \cos\frac{7\pi}{6}}{2}}} = \frac{\sqrt{\frac{1 + \sqrt{3}/2}{2}}}{-\sqrt{\frac{1 - \sqrt{3}/2}{2}}} = \frac{-\sqrt{2 + \sqrt{3}}}{\sqrt{2 - \sqrt{3}}} = -\sqrt{7 + 4\sqrt{3}}$

49. $\sin(t - \pi) = \sin t \cos\pi - \cos t \sin\pi = (\sin t)(-1) - (\cos t)(0) = -\sin t$

51. $\sin\left(\frac{3\pi}{2} - t\right) = \sin\frac{3\pi}{2}\cos t - \cos\frac{3\pi}{2}\sin t = (-1)(\cos t) - (0)(\sin t) = -\cos t$

53. $(\cos 3x)^2 = \frac{1 + \cos 6x}{2} = \frac{1}{2} + \frac{1}{2}\cos 6x$

55. $(\sin x)^4 = ((\sin x)^2)^2 = \left(\frac{1 - \cos 2x}{2}\right)^2 = \frac{1}{4}\left(1 - 2\cos 2x + (\cos 2x)^2\right) =$
$\frac{1}{4}\left(1 - 2\cos 2x + \frac{1 + \cos 4x}{2}\right) = \frac{3}{8} - \frac{1}{2}\cos 2x + \frac{1}{8}\cos 4x$

57. $\sin 4t \cos 5t = \frac{1}{2}[\sin(4t + 5t) + \sin(4t - 5t)] = \frac{1}{2}[\sin 9t - \sin t]$

59. $\cos 2t \cos 4t = \frac{1}{2}[\cos(2t + 4t) + \cos(2t - 4t)] = \frac{1}{2}[\cos 6t + \cos 2t]$

61. $\sin 2t + \sin 6t = 2\sin\frac{2t + 6t}{2}\cos\frac{2t - 6t}{2} = 2\sin 4t \cos 2t$

63. $\cos 4t + \cos 2t = 2\cos\frac{4t + 2t}{2}\cos\frac{4t - 2t}{2} = 2\cos 3t \cos t$

65. $(\cos x)^4 - (\sin x)^4 = ((\cos x)^2 - (\sin x)^2)((\cos x)^2 + (\sin x)^2) = (\cos x)^2 -$

$(\sin x)^2 = \cos 2x$

67.

$$\frac{\sin x}{1 - \cos x} = \frac{\sin x}{1 - \cos x} \cdot \frac{1 + \cos x}{1 + \cos x} = \frac{\sin x(1 + \cos x)}{1 - (\cos x)^2}$$

$$= \frac{\sin x(1 + \cos x)}{(\sin x)^2} = \frac{1 + \cos x}{\sin x} = \cot x + \csc x$$

69. $\sin\left(\arctan \sqrt{3}\right) = \sin\left(\frac{\pi}{3}\right) = \frac{\sqrt{3}}{2}$

71. Since $\sin(\arcsin x) = x$ for x in $[-1, 1]$, we have

$$\sin\left(\arcsin\left(\frac{3}{5}\right) - \arcsin\left(\frac{5}{13}\right)\right) = \sin\left(\arcsin\left(\frac{3}{5}\right)\right)\cos\left(\arcsin\left(\frac{5}{13}\right)\right)$$

$$- \cos\left(\arcsin\left(\frac{3}{5}\right)\right)\sin\left(\arcsin\left(\frac{5}{13}\right)\right)$$

$$= \frac{3}{5}\cos\left(\arcsin\left(\frac{5}{13}\right)\right) - \cos\left(\arcsin\left(\frac{3}{5}\right)\right)\frac{5}{13}$$

$$= \frac{3}{5}\frac{12}{13} - \frac{4}{5}\frac{5}{13} = \frac{16}{65}.$$

73. Since $\alpha = 25°, b = 12, c = 20$ and $a^2 = b^2 + c^2 - 2bc\cos\alpha$, we have

$$a = \sqrt{12^2 + 20^2 - 2(12)(20)\cos 25°} \approx 10.4.$$

Also,

$$\frac{\sin\beta}{b} = \frac{\sin\alpha}{a} \Rightarrow \beta = \arcsin\left(\frac{12\sin 25°}{10.4}\right) \approx 0.50938 = 0.50938\left(\frac{180}{\pi}\right)^° \approx 29.2°$$

and $\gamma = 180 - 25 - 29.2 = 125.8°$.

75. Since $a = 6, b = 8, c = 10$ we have $a^2 + b^2 = c^2$ and $\gamma = 90°$. Also,

$$\cos\alpha = \frac{6^2 - 8^2 - 10^2}{-2(8)(10)} = \frac{4}{5},$$

so

$$\alpha = \arccos\left(\frac{4}{5}\right) \approx 0.6435 = 0.6435\left(\frac{180}{\pi}\right)^\circ \approx 36.9^\circ,$$

and $\beta = 90^\circ - \alpha \approx 53.1^\circ$.

77. Since $\beta = 76^\circ, \gamma = 50^\circ, b = 10.5$ we have $\alpha = 180 - 76 - 50 = 54^\circ$ and

$$\frac{\sin\alpha}{a} = \frac{\sin\beta}{b} \Rightarrow a = \frac{10.5\sin 54^\circ}{\sin 76^\circ} \approx 8.8; \quad \frac{\sin\gamma}{c} = \frac{\sin\beta}{b} \Rightarrow c = \frac{10.5\sin 50^\circ}{\sin 76^\circ} \approx 8.3.$$

79. The points of intersection of the curves $y = \sin x$ and $y = x^2$ occur at approximately $x = 0.9$, and $x = 0$.

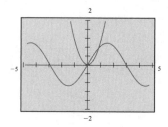

81. The period of $f(x) = 4\cos(125x)$ is $\frac{2\pi}{125} \approx 0.05$ and since the amplitude is 4, a reasonable viewing rectangle is $\left[-\frac{2\pi}{125}, \frac{2\pi}{125}\right] \times [-4, 4]$.

83. Since the area of a trapezoid is the product of its height and the average of its bases we have

$$A = \frac{1}{2}[b + (b + 2b\cos\theta)]b\sin\theta = b^2(1 + \cos\theta)\sin\theta.$$

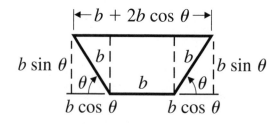

85. If the plane is traveling at 380 mi/hr, then after two and one half hours the plane is 950 mi from the airport. Its bearing from the airport is 150° measured clockwise from north, or 60° measured clockwise from east. Then the plane is

$$950 \cos 60° = 950 \left(\frac{1}{2} \right) = 475$$

miles east of the airport and

$$950 \sin 60° = 950 \frac{\sqrt{3}}{2} \approx 822.7$$

miles south of the airport.

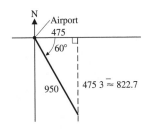

Solutions for Exercise Set 3 Calculus

1. The graph indicates that the minimum occurs at $x = -2$ with a minimum value of approximately -1.1. The maximum occurs at $x = 3$ with a maximum value of approximately 2.9.

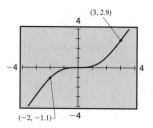

3. (a) $\sqrt{a^2 - u^2} = \sqrt{a^2 - (a\sin t)^2} = \sqrt{a^2(1 - (\sin t)^2)} = a\sqrt{(\cos t)^2} = a\cos t$

(b) $\sqrt{u^2 + a^2} = \sqrt{(a\tan t)^2 + a^2} = \sqrt{a^2((\tan t)^2 + 1)} = a\sqrt{(\sec t)^2} = a\sec t$

(c) $\sqrt{u^2 - a^2} = \sqrt{(a\sec t)^2 - a^2} = \sqrt{a^2((\sec t)^2 - 1)} = a\sqrt{(\tan t)^2} = a\tan t$

(d) $\frac{\sqrt{u^2 - a^2}}{u} = \frac{\sqrt{(a\sec t)^2 - a^2}}{a\sec t} = \frac{a\tan t}{a\sec t} = \sin t$

5. Let x be the length of the line segment connecting the two vertices that lie approximately east-west of one another. By the Law of Cosines,

$$x^2 = 10^2 + 7^2 - 2(10)(7)\cos 95° \Rightarrow x = \sqrt{10^2 + 7^2 - 2(10)(7)\cos 95°} \approx 13.$$

The perimeter of the upper triangle is $P_u = 10 + 7 + 13 = 30$ so by Heron's Formula, the area of the upper triangle is

$$A_u = \frac{1}{4}\sqrt{30(30 - 20)(30 - 14)(30 - 26)} = 80\sqrt{3} \approx 34.64.$$

For the lower triangle we have $P_l = 9 + 12 + 13 = 34$. By Heron's Formula, the area of the lower triangle is

$$A_l = \sqrt{34(34 - 18)(34 - 24)(34 - 26)} = 16\sqrt{170} \approx 52.15.$$

The area of the quadrilateral is then $A_u + A_l \approx 34.64 + 52.15 = 86.79$ ft^2.

7. If α is the angle of the line of sight from ship A to the plane, β the angle from ship B to the plane, and d is the distance from A to B, then from Exercise 6,

$$33000 = \frac{d}{\cot\alpha - \cot\beta} = \frac{d}{\cot 32° - \cot 47°} \Rightarrow d = 33000(\cot 32° - \cot 47°) \approx 22038.$$

So the two ships are approximately 22000 feet apart, rounded to the nearest 100 feet.

9. Let d be the distance between A and C and let α be the angle ACB. Then $\alpha = 180 - 80 - 59 = 41°$. By the Law of Sines

$$\frac{\sin 41°}{300} = \frac{\sin 59°}{d} \Rightarrow d = \frac{300\sin 59°}{\sin 41°} \approx 392 \text{ feet.}$$

11. (a) Let b denote the base of the inscribed triangle and h the height. Since

$$\sin\frac{\pi}{n} = \frac{b/2}{r} = \frac{b}{2r} \Rightarrow b = 2r\sin\frac{\pi}{n},$$

and since

$$\cos\frac{\pi}{n} = \frac{h}{r} \Rightarrow h = r\cos\frac{\pi}{n},$$

the area of each triangle is

$$A = \frac{1}{2}\left(2r\sin\frac{\pi}{n}\right)\left(r\cos\frac{\pi}{n}\right) = \frac{1}{2}r^2\left(2\sin\frac{\pi}{n}\cos\frac{\pi}{n}\right) = \frac{1}{2}r^2\sin\frac{2\pi}{n}.$$

(b) Since the area of a circle of radius r is known to be πr^2 we would expect $\frac{n}{2}r^2\sin\frac{2\pi}{n}$ to better approximate the area as n increases. So as n approaches ∞ we expect $\frac{n}{2}r^2\sin\frac{2\pi}{n}$ will approach πr^2.

CHAPTER 4
EXPONENTIAL AND LOGARITHM FUNCTIONS

4.1 Introduction

The exponential functions extend the notion of exponent to include all real numbers. The *natural exponential* function $f(x) = e^x$ and the *natural logarithm* function $g(x) = \ln x$ are perhaps the most important function-inverse pair in science and mathematics.

4.2 The Natural Exponential Function

For $a > 0$, the *exponential function with base a* is defined as $f(x) = a^x$, for all real numbers x. The value of the base a determines the general shape of the graph of the exponential function.

$\underline{f(x) = a^x, a > 1:}$

(i) domain: $(-\infty, \infty)$

(ii) range: $(0, \infty)$

(iii) graph increasing: for all x

(iv) graph decreasing: for no x

(v) horizontal asymptote: $y = 0$

$$f(x) \to 0 \quad \text{as} \quad x \to -\infty$$
$$f(x) \to \infty \quad \text{as} \quad x \to \infty$$

(vi) y-intercept: $(0, 1)$

(vii) x-intercept: none

$\underline{f(x) = a^x, 0 < a < 1:}$

(i) domain: $(-\infty, \infty)$

(ii) range: $(0, \infty)$

(iii) graph increasing: for no x

(iv) graph decreasing: for all x

(v) horizontal asymptote: $y = 0$

$$f(x) \to \infty \quad \text{as} \quad x \to -\infty$$

$$f(x) \to 0 \quad \text{as} \quad x \to \infty$$

(vi) y-intercept: $(0, 1)$

(vii) x-intercept: none

$\underline{f(x) = a^x = 1, \text{ for } a = 1:}$ constant function

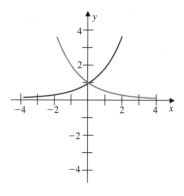

Figure 4.1

GRAPHS OF EXPONENTIAL FUNCTIONS

EXAMPLE 4.2.1   Use the graph of $y = 2^x$ shown in the figure to sketch the graph of the function.

(a) $y = 2^{(x-1)} + 1$ (b) $y = -2^{(x+1)} - 2$

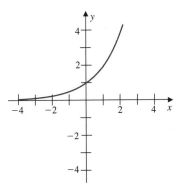

Figure 4.2

SOLUTION: The shifting and scaling properties allow us to sketch reasonable graphs quickly once we know the general shape of the basic curve.

(a) The basic function is $y = 2^x$, so if the argument is changed to $(x - 1)$, the graph of $y = 2^{(x-1)}$ is obtained by shifting the basic curve to the right 1 unit. To obtain the graph of $y = 2^{(x-1)} + 1$ from the graph of $y = 2^x$, first shift the basic curve to the right 1 unit and then shift the resulting curve upward 1 unit.

y-intercept: Setting $x = 0$, we have

$$y = 2^{-1} + 1$$
$$= \frac{1}{2} + 1 = \frac{3}{2}.$$

Horizontal asymptote: $y = 1$, since the horizontal asymptote of $y = 2^x$ is the line $y = 0$, and the graph is shifted upward 1 unit.

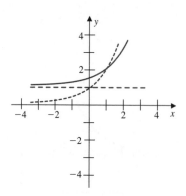

Figure 4.3

(b) The graph of $y = -2^x$ is the reflection of the graph of $y = 2^x$ about the x-axis, since introducing the minus sign reverses the signs of all y-coordinates of points on the curve. Changing the argument from x to $x + 1$ shifts the reflected graph to the left 1 unit. Finally, shift this graph downward 2 units to obtain the graph of $y = -2^{(x+1)} - 2$.

y-intercept: Setting $x = 0$, we have

$$y = -2^1 - 2$$
$$= -4.$$

Horizontal asymptote: $y = -2$, since $y = -2^{(x+1)}$ was shifted downward 2 units. Also $y = 2^x$ approaches the horizontal line $y = 0$ from above. The graph of $y = -2^{(x+1)} - 2$ was obtained from $y = 2^x$ by first reflecting it about the x-axis, so it will approach the line $y = -2$ from below.   ◇

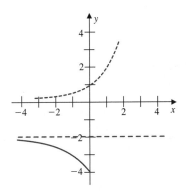

Figure 4.4

EXAMPLE 4.2.2   Use the graph of $y = \left(\frac{1}{2}\right)^x$ shown in the figure to sketch the graph of $y = 3 \cdot 2^{(1-x)} + 1$.

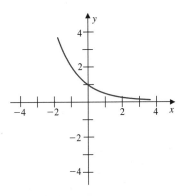

Figure 4.5

SOLUTION: First note that

$$2^{-x} = \frac{1}{2^x} = \left(\frac{1}{2}\right)^x,$$

so

$$y = 3 \cdot 2^{(1-x)} + 1$$

$$= 3 \cdot 2^{-(x-1)} + 1$$

$$= 3 \cdot \left(\frac{1}{2}\right)^{x-1} + 1.$$

Start with the graph of $y = \left(\frac{1}{2}\right)^x$, vertically stretch the graph by a factor of 3, and shift the resulting graph to the right 1 unit and upward 1 unit to obtain the graph of $y = 3 \cdot 2^{(1-x)} + 1$.

<u>y-intercept:</u> Setting $x = 0$, we have

$$y = 3 \cdot 2^1 + 1$$

$$= 7.$$

<u>Horizontal asymptote:</u> $y = 1$ ◇

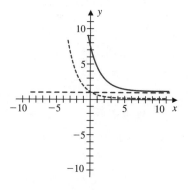

Figure 4.6

THE NATURAL EXPONENTIAL FUNCTION

As $n \to \infty$, the real numbers $\left(1 + \frac{1}{n}\right)^n$ approach an irrational number called $e \approx 2.71828$. The function

$$f(x) = e^x$$

is called the *natural exponential function*.

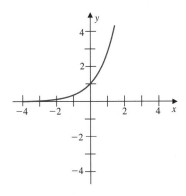

Figure 4.7

EXAMPLE 4.2.3   Sketch the graph of the function.

(a) $f(x) = 1 - e^{-(x-2)}$ (b) $f(x) = e^{-|x-1|}$

SOLUTION:

(a) For any function f, if the argument x is replaced with $-x$, the graph of $y = f(-x)$ is the reflection about the y-axis of $y = f(x)$.

Start with the graph of $y = e^x$.

1. <u>Graph of $y = e^{-x}$</u> : Reflect the graph of $y = e^x$ about the $y-$axis.

2. <u>Graph of $y = -e^{-x}$</u> : Reflect the graph of $y = e^{-x}$ about the $x-$axis.

3. <u>Graph of $y = -e^{-(x-2)}$</u> : Shift the graph of $y = -e^{-x}$ to the right 2 units.

4. <u>Graph of $y = 1 - e^{-(x-2)}$</u> : Shift the graph of $y = -e^{-(x-2)}$ upward 1 unit.

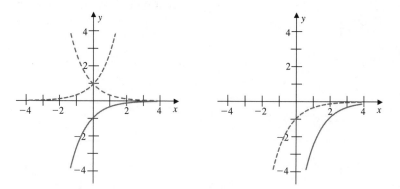

Figure 4.8   Figure 4.9

(b) Replacing x with $|x|$ changes the graph of $y = e^{-x}$ to one that is symmetric with respect to the y-axis, since

$$e^{-|x|} = e^{-|-x|}.$$

For example, if $x > 0$, the points with first coordinate x and first coordinate $-x$ have the same y-coordinates. Now shift the graph of $y = e^{-|x|}$ to the right 1 unit.

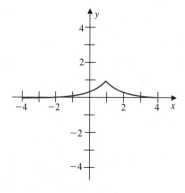

Figure 4.10

EXAMPLE 4.2.4   Match the equation with the curve in the figure.

(a) $y = e^{x-1} - 2$ (b) $y = -e^{x+1} + 1$ (c) $y = 3e^x$ (d) $y = 2e^{-x}$

(i)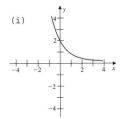

(ii)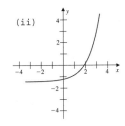

(iii)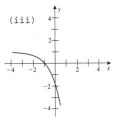

(iv)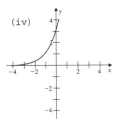

Figure 4.11   Figure 4.12   Figure 4.13   Figure 4.14

SOLUTION: The graph of $y = e^x$ is shown in the figure.

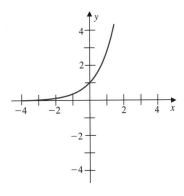

Figure 4.15

The graph of $y = e^{-x}$ is the reflection of the graph of $y = e^x$ through the y-axis, and $y = 2e^{-x}$ has the same shape so (d) matches with (i).

The graph of $y = 3e^x$ is a vertical stretching of the graph of $y = e^x$, so it has a similar shape. However, it passes through $(0, 3)$ and rises more quickly. Therefore, (c) matches with (iv).

The graph of $y = -e^x$ is the reflection of $y = e^x$ through the x-axis, and $y = -e^{x+1}$ is an additional shift to the left 1 unit. The addition of 1 shifts the graph upward 1 unit. The only curve that is a vertical flip of $y = e^x$ is (iii), so (b) matches with (iii).

(a) (ii) (b) (iii) (c) (iv) (d) (i)   ◇

EXAMPLE 4.2.5   Use a graphing device to approximate all solutions to the equation $x^2 e^x = 2x^2 + 3x - 1$.

SOLUTION: To approximate the solutions, plot $y = x^2 e^x$ and $y = 2x^2 + 3x - 1$ together on the same coordinate axes and determine the points of intersection. When using a graphing device, it is essential to have a viewing rectangle that shows the important features. A viewing rectangle of $[-5, 5] \times [-5, 10]$ shows three points of intersection. A viewing rectangle of $[-5, 5] \times [-5, 5]$ would show only two of those points. From the initial plot, the points of intersection occur at approximately $x = -1.9$, $x = 0.3$ and $x = 1.3$. Zooming in near these points we get better accuracy with the points of intersection being approximately

$$(-1.91, 0.54), (0.31, 0.14), (1.31, 6.36).$$

                                                   ◇

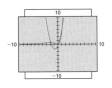 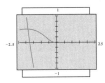 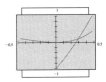 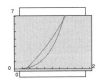

Figure 4.16   Figure 4.17   Figure 4.18   Figure 4.19

EXAMPLE 4.2.6 Use a graphing device to compare the rates of growth
of $f(x) = e^x$ and $g(x) = x^{15}$ by graphing the functions together in several
appropriate viewing rectangles. Approximate the solutions to $e^x = x^{15}$.

SOLUTION: We want to determine whether, for large x,

$$x^{15} > e^x \quad \text{or} \quad e^x > x^{15}.$$

We will sketch the graph of $y = x^{15}$ and $y = e^x$. Whichever is eventually above
the other, will be the dominant function.

Since both functions are nonnegative for all x, there is no need to show much of
the negative y-axis. The first viewing rectangle we choose is

$$[-5, 5] \times [-1, 5].$$

In this view, it appears that $g(x) = x^{15}$ eventually grows faster than $f(x) = e^x$.
We should not jump to a quick conclusion. For a second viewing rectangle, we
select

$$[0, 50] \times [0, 10^{10}].$$

It still appears that $g(x) = x^{15}$ grows faster than $f(x) = e^x$. One last viewing
rectangle reveals another story. Try

$$[0, 100] \times [0, 10^{27}].$$

This time we see that $f(x) = e^x$ eventually overtakes $g(x) = x^{15}$, and in fact, it eventually grows much faster.

The figures show points of intersection occurring when

$$x \approx 1.1, \quad \text{and} \quad x \approx 61.9,$$

which are the approximate solutions to the equation $e^x = x^{15}$. $\quad\quad\quad\quad\quad\quad\quad$ ◇

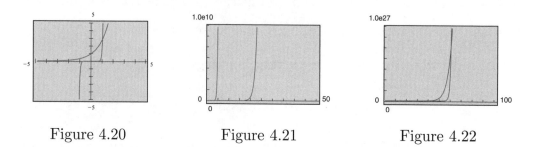

Figure 4.20 $\quad\quad\quad\quad\quad$ Figure 4.21 $\quad\quad\quad\quad\quad$ Figure 4.22

SOLVING EQUATIONS INVOLVING e^x

EXAMPLE 4.2.7   Solve the equation.

(a) $2x^3 e^x + x^4 e^x = 0$ (b) $x^2 e^x - 2x e^x = 3e^x$

SOLUTION:

(a) Since e^x is common to both terms, it can be factored from these terms. So

$$e^x(2x^3 + x^4) = 0.$$

Since $e^x > 0$ for all x, both sides of the equation can be divided by e^x, and the last equation will be 0, only when

$$2x^3 + x^4 = 0$$

$$x^3(2 + x) = 0$$

$$x = 0, \ x = -2.$$

(b) First rewrite the equation so that one side is 0. This gives

$$x^2 e^x - 2xe^x = 3e^x$$

$$x^2 e^x - 2xe^x - 3e^x = 0$$

$$e^x(x^2 - 2x - 3) = 0.$$

Since $e^x > 0$, we have

$$x^2 - 2x - 3 = 0$$

$$(x - 3)(x + 1) = 0$$

$$x = 3, \ x = -1.$$

$\diamond$

COMPOUND INTEREST

If an initial amount of A_0 dollars is invested at an interest rate i compounded n times a year, the investment after t years has a value

$$A_n(t) = A_0 \left(1 + \frac{i}{n}\right)^{nt} \quad \text{dollars.}$$

If the interest is compounded *continuously*, then the amount after t years is

$$A_c(t) = A_0 e^{it} \quad \text{dollars.}$$

EXAMPLE 4.2.8   Suppose $5,000.00$ is invested at 10% interest, and the interest rate remains fixed for 8 years. Determine the value of the investment if the interest is compounded annually, semiannually, quarterly, monthly, weekly, daily, hourly and continuously.

SOLUTION: We have the following.

Annually: $A_1(8) = 5000(1 + 0.1)^8 = \$10,717.94$

Semiannually: $A_2(8) = 5000\left(1 + \frac{0.1}{2}\right)^{16} = \$10,914.37$

Quarterly: $A_4(8) = 5000\left(1 + \frac{0.1}{4}\right)^{32} = \$11,018.79$

Monthly: $A_{12}(8) = 5000\left(1 + \frac{0.1}{12}\right)^{96} = \$11,090.88$

Weekly: $A_{52}(8) = 5000\left(1 + \frac{0.1}{52}\right)^{416} = \$11,119.16$

Daily: $A_{365}(8) = 5000\left(1 + \frac{0.1}{365}\right)^{2920} = \$11,126.49$

Hourly: There are 24 hours in a day and 365 days in the year, so the interest is computed $24 \cdot 365 = 8760$ times.

$$A_{8760}(8) = 5000\left(1 + \frac{0.1}{8760}\right)^{70080} = \$11,127.65$$

Continuously: $A_c(8) = 5000e^{0.8} = \$11,127.71$ ◇

EXAMPLE 4.2.9   What initial investment of 8% compounded semiannually for 7 years will accumulate to $25,000.00$?

SOLUTION: The value of the investment after t years is

$$A_2(t) = A_0\left(1 + \frac{0.08}{2}\right)^{2t}.$$

And after 7 years the value is

$$A_2(7) = A_0\left(1 + \frac{0.08}{2}\right)^{14}.$$

So the initial investment that will yield $25,000.00 can be found from the equation

$$25000 = A_2(7) = A_0 \left(1 + \frac{0.08}{2} \right)^{14}$$

$$25000 = A_0(1.04)^{14}$$

$$A_0 = \frac{25000}{(1.04)^{14}} \approx \$14,437.00.$$

◇

Solutions for Exercise Set 4.2

1. For $f(x) = 2^x - 3$

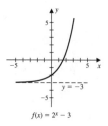

$f(x) = 2^x - 3$

3. For $f(x) = -4^x$

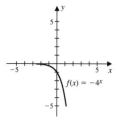

$f(x) = -4^x$

5. For $f(x) = 2 \cdot \left(\frac{1}{3}\right)^{x-1} + 1$

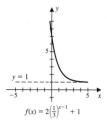

$f(x) = 2\left(\frac{1}{3}\right)^{x-1} + 1$

7. For $f(x) = -e^{x-1}$

$e^x > 1 \Leftrightarrow x > 0$

$\left(\frac{1}{2}\right)^x = \frac{1}{2^x} \geq 4 \Leftrightarrow 2^x \leq \frac{1}{4} \Leftrightarrow x \leq -2$

(a) ii (b) iii (c) iv (d) i

The points of intersection of $y = e^{x-2}$ and $y = x$ are $x \approx 0.1586$ and $x \approx 3.146$.

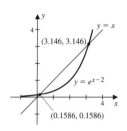

The points of intersection of $y = e^{-x}$ and $y = (x-2)^2$ are $x \approx 1.5$, $x \approx 2.3$

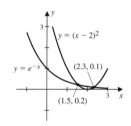

(a) $f(x) = xe^x$

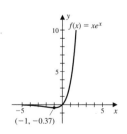

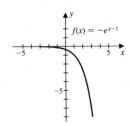

9. For $f(x) = 3 - e^{-(x-1)}$

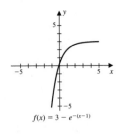

$f(x) = 3 - e^{-(x-1)}$

11. For $f(x) = e^{2x}$

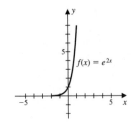

$f(x) = e^{2x}$

13. For $f(x) = e^{|x|}$

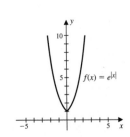

$f(x) = e^{|x|}$

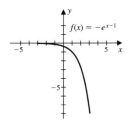

9. For $f(x) = 3 - e^{-(x-1)}$

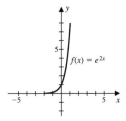

11. For $f(x) = e^{2x}$

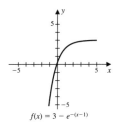

13. For $f(x) = e^{|x|}$

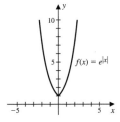

15. $e^x > 1 \Leftrightarrow x > 0$

17. $\left(\frac{1}{2}\right)^x = \frac{1}{2^x} \geq 4 \Leftrightarrow 2^x \leq \frac{1}{4} \Leftrightarrow x \leq -2$

19. (a) ii (b) iii (c) iv (d) i

21. The points of intersection of $y = e^{x-2}$ and $y = x$ are $x \approx 0.1586$ and $x \approx 3.146$.

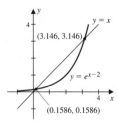

23. The points of intersection of $y = e^{-x}$ and $y = (x-2)^2$ are $x \approx 1.5, x \approx 2.3$

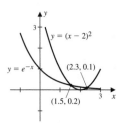

25. (a) $f(x) = xe^x$

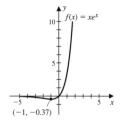

(b) Increasing: $(-1, \infty)$; decreasing: $(-\infty, -1)$.

27. (a) $f(x) = e^{-x^2 - x}$

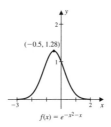

$f(x) = e^{-x^2 - x}$

(b) Increasing: $(-\infty, -0.5)$; decreasing: $(-0.5, \infty)$.

29. (a) $f(x) = e^x - e^{-x}$

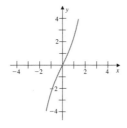

(b) Increasing for all x, $(-\infty, \infty)$.

31. Let $3^x = e^{kx}$. If $x = 1$, then $3 = e^k$. The graphs of $y = e^x$ and $y = 3$ intersect when $x \approx 1.1$, so $k \approx 1.1$.

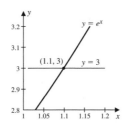

33. (a) The function $f(x) = 2^x$ eventually grows much faster than $g(x) = x^5$. In Figure (iii) we see $f(x) = 2^x$ crosses $g(x) = x^5$ and remains above $y = g(x)$ from that point on.

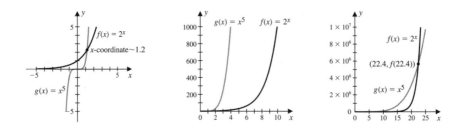

(b) The points of intersection of the two graphs occur at $x \approx 1.2$ and $x \approx 22.4$ which are approximate solutions to $2^x = x^5$.

35. The value of the CD at 6.5% interest compounded n times per year and which matures in 5 years is $A_n(5) = 5000 \left(1 + \frac{0.065}{n}\right)^{5n}$.

Interest Compounded	Value of CD
(a) annually: $n = 1$	\$6850.43
(b) monthly: $n = 12$	\$6914.09
(c) daily: $n = 365$	\$6919.95
(d) continuously:	$A_c(5) = 5000 e^{(0.065)5} = \$6920.15$

37. If \$10,000.00 is invested and the interest is compounded quarterly, after 5 years and an interest rate of

(a) 8%, the value of the investment is

$$A_4(5) = 10000 \left(1 + \frac{0.08}{4}\right)^{20} = \$14,859.47;$$

(b) 6.5%, the value of the investment is

$$A_4(5) = 10000\left(1 + \frac{0.065}{4}\right)^{20} = \$13,804.20;$$

(c) 6%, the value of the investment is

$$A_4(5) = 10000\left(1 + \frac{0.06}{4}\right)^{20} = \$13,468.55;$$

(d) 5.5%, the value of the investment is

$$A_4(5) = 10000\left(1 + \frac{0.055}{4}\right)^{20} = \$13,140.66.$$

39. Solve $10000 = A_0\left(1 + \frac{0.08}{2}\right)^{5(2)}$ for A_0 which gives $A_0 = \frac{10000}{(1+0.04)^{10}} \approx$ $6756.00.

4.3 Logarithm Functions

For $a \neq 1$, the inverse function of the exponential function to the base a, $f(x) = a^x$, is called the *logarithm function to the base a*, written $g(x) = \log_a x$. The functions are then related by the following relations.

For each x in $(0, \infty)$,

$$y = \log_a x \Leftrightarrow x = a^y.$$

As a consequence, for each x in $(0, \infty)$ and for each real number y, we have

$$\log_a a^y = y \quad \text{and} \quad a^{\log_a x} = x.$$

Recall that the domain of the exponential function is the set of all real numbers so the range of the logarithm function is also the set of all real numbers. The range of the exponential function is $(0, \infty)$ so the domain of the logarithm function is $(0, \infty)$.

The inverse to the natural exponential function $f(x) = e^x$ is the *natural logarithm function*. The natural logarithm function is the logarithm to the base e, written $y = \log_e x = \ln x$. So we have the following.

For each x in $(0, \infty)$,

$$y = \ln x \Leftrightarrow x = e^y.$$

For each x in $(0, \infty)$ and each real number y,

$$\ln e^y = y \quad \text{and} \quad e^{\ln x} = x.$$

EVALUATION OF LOGARITHMS

EXAMPLE 4.3.1   Evaluate the expression.

(a) $\log_2 64$ (b) $\log_4 16$ (c) $\log_{1/3} 3$ (d) $\log_3 \frac{1}{27}$ (e) $e^{\ln 6}$ (f) $\ln e^{\sqrt{2}}$

SOLUTION:

(a) Using the inverse relation between the logarithm function and the exponential function gives

$$x = \log_2 64 \Leftrightarrow 2^x = 64$$

$$x = 6.$$

Another way to view this is,

$$x = \log_2 64$$

$$2^x = 2^{\log_2 64}$$

$$2^x = 64$$

$$x = 6.$$

(b)

$$x = \log_4 16 \Leftrightarrow 4^x = 16$$

$$x = 2$$

(c)

$$x = \log_{1/3} 3 \Leftrightarrow x^{\frac{1}{3}} = 3$$

$$x = 3^3 = 27$$

(d)

$$x = \log_3 \frac{1}{27} \Leftrightarrow 3^x = \frac{1}{27}$$
$$x = -3$$

Note that $3^{-3} = \frac{1}{3^3} = \frac{1}{27}$.

(e) The exponential and logarithm functions are inverses of one another, and therefore, each undoes the process of the other. That is, if a value is input to ln, and the resulting number then used as input to e, the final output is the original value.

$$x \to \overset{\ln \boxed{x}}{\boxed{}} \to \overset{e^{\boxed{\ln x}}}{\boxed{}} \to x$$

So

$$e^{\ln 6} = 6.$$

(f) The inverse relation between the exponential and logarithm functions goes both ways, so

$$\ln e^{\sqrt{2}} = \sqrt{2}.$$

◇

EXAMPLE 4.3.2   Solve the equation.

(a) $\log_3(2x - 5) = 2$ (b) $\log_2(3x^2 + 10x) = 3$ (c) $e^{2x-1} = 2$

SOLUTION:

(a)

$$\log_3(2x - 5) = 2$$

$$2x - 5 = 3^2$$

$$2x = 14$$

$$x = 7$$

(b)

$$\log_2(3x^2 + 10x) = 3$$

$$3x^2 + 10x = 2^3$$

$$3x^2 + 10x = 8$$

To solve the quadratic rewrite the expression with one side 0 and factor. So

$$3x^2 + 10x = 8$$

$$3x^2 + 10x - 8 = 0$$

$$(3x - 2)(x + 4) = 0$$

$$3x - 2 = 0 \quad \text{or} \quad x + 4 = 0$$

$$x = \frac{2}{3} \quad \text{or} \quad x = -4.$$

(c) The inverse relationship between the exponential and logarithm functions is the key here. If we take the natural logarithm of the left side the result is the input to the natural exponential function, $2x - 1$. So as not to change the equation, we take the natural logarithm of both sides and simplify. So

$$e^{2x-1} = 2$$

$$\ln e^{2x-1} = \ln 2$$

$$2x - 1 = \ln 2$$

$$2x = 1 + \ln 2$$

$$x = \frac{1 + \ln 2}{2}.$$

◇

ARITHMETIC PROPERTIES OF LOGARITHMS

There are three important properties of the logarithm functions.

(i) $\log_a (x_1 x_2) = \log_a x_1 + \log_a x_2$

(ii) $\log_a \left(\dfrac{x_1}{x_2}\right) = \log_a x_1 - \log_a x_2$

(iii) $\log_a x_1^r = r \log_a x_1$

For emphasis we also list the properties for the natural logarithm function.

(i) $\ln (x_1 x_2) = \ln x_1 + \ln x_2$

(ii) $\ln \left(\dfrac{x_1}{x_2}\right) = \ln x_1 - \ln x_2$

(iii) $\ln x_1^r = r \ln x_1$

EXAMPLE 4.3.3   Use the properties of logarithms to simplify the expression so that the result does not contain logarithms of products, quotients, or powers.

(a) $\ln x(x^2 + 1)$ (b) $\ln \dfrac{1}{x^2}$ (c) $\ln \dfrac{x \sqrt[5]{x^2}}{(x + 2)^2}$ (d) $\log_5 \sqrt{\dfrac{x^3}{4x^2 - 2}}$

SOLUTION:

(a) The expression in the natural logarithm function is the product of the two terms x and $(x^2 + 1)$, so the product rule can be used to give

$$\ln x(x^2 + 1) = \ln x + \ln(x^2 + 1).$$

This is the final answer! The expression $\ln(x^2 + 1)$ cannot be further simplified.

(b) Because $e^0 = 1$, $\ln 1 = 0$, and

$$\ln \frac{1}{x^2} = \ln 1 - \ln x^2$$
$$= -2 \ln x.$$

This can also be recognized using the power rule. That is,

$$\ln \frac{1}{x^2} = \ln x^{-2} = -2 \ln x.$$

(c) This problem may look complicated, but just be careful to apply the properties in a proper sequence. The first property to apply is the quotient rule, since the entire expression is one quotient. Once the quotient rule is applied, the two resulting pieces will be treated separately and the properties applied to them individually. To simplify the steps, first rewrite

$$\sqrt[5]{x^2} = \left(x^2\right)^{1/5} = x^{\frac{2}{5}}.$$

Then we have

$$\ln \frac{x \cdot x^{2/5}}{(x + 2)^2} = \ln x^{7/5} - \ln(x + 2)^2$$
$$= \frac{7}{5} \ln x - 2 \ln(x + 2).$$

(d)

$$\log_5 \sqrt{\frac{x^3}{4x^2 - 2}} = \log_5 \left(\frac{x^3}{4x^2 - 2}\right)^{1/2}$$

$$= \frac{1}{2} \log_5 \frac{x^3}{4x^2 - 2}$$

$$= \frac{1}{2} \left[\log_5 x^3 - \log_5(4x^2 - 2)\right]$$

$$= \frac{1}{2} \log_5 x^3 - \frac{1}{2} \log_5(4x^2 - 2)$$

$$= \frac{3}{2} \log_5 x - \frac{1}{2} \log_5(4x^2 - 2)$$

It may appear that the simplification is done, but some further simplification is possible by recognizing that

$$(4x^2 - 2) = 4\left(x^2 - \frac{1}{2}\right)$$

$$= 4\left(x - \sqrt{\frac{1}{2}}\right)\left(x + \sqrt{\frac{1}{2}}\right)$$

$$= 4\left(x - \frac{\sqrt{2}}{2}\right)\left(x + \frac{\sqrt{2}}{2}\right).$$

So

$$\log_5 \sqrt{\frac{x^3}{4x^2 - 2}} = \frac{3}{2} \log_5 x - \frac{1}{2} \log_5(4x^2 - 2)$$

$$= \frac{3}{2} \log_5 x - \frac{1}{2} \log_5 4\left(x - \frac{\sqrt{2}}{2}\right)\left(x + \frac{\sqrt{2}}{2}\right)$$

$$= \frac{3}{2} \log_5 x - \frac{1}{2} \log_5 4 - \frac{1}{2} \log_5\left(x - \frac{\sqrt{2}}{2}\right)\left(x + \frac{\sqrt{2}}{2}\right)$$

$$= \frac{3}{2} \log_5 x - \frac{1}{2} \log_5 4 - \frac{1}{2} \log_5\left(x - \frac{\sqrt{2}}{2}\right) - \frac{1}{2} \log_5\left(x + \frac{\sqrt{2}}{2}\right).$$

Notice we used the product rule twice on the term

$$\frac{1}{2}\log_5 4\left(x - \frac{\sqrt{2}}{2}\right)\left(x + \frac{\sqrt{2}}{2}\right).$$

First on $4\left[\left(x - \frac{\sqrt{2}}{2}\right)\left(x + \frac{\sqrt{2}}{2}\right)\right]$, and then on $\left[\left(x - \frac{\sqrt{2}}{2}\right)\left(x + \frac{\sqrt{2}}{2}\right)\right]$.   ◇

EXAMPLE 4.3.4   Rewrite the expression as a single logarithm.

(a) $2\ln x + 3\ln(x+1)$ (b) $4\ln x - \frac{1}{2}\ln(x+1)$

SOLUTION:

(a) First use the property

$$r\ln a = \ln a^r$$

to write the expression as

$$2\ln x + 3\ln(x+1) = \ln x^2 + \ln(x+1)^3.$$

Now use the sum property

$$\ln a + \ln b = \ln ab$$

to get

$$2\ln x + 3\ln(x+1) = \ln x^2 + \ln(x+1)^3$$
$$= \ln x^2(x+1)^3.$$

(b)

$$4\ln x - \frac{1}{2}\ln(x+1) = \ln x^4 - \ln(x+1)^{1/2}$$
$$= \ln \frac{x^4}{(x+1)^{1/2}}$$

◇

EXAMPLE 4.3.5   Solve the equation for x.

(a) $\ln x + \ln(x + 1) = \ln 2$ (b) $2\ln(x + 3) - \ln x = \ln 12$ (c) $e^{3x} = 3^{2x-1}$

SOLUTION:

(a) Here, using the properties of logarithms to combine the ln expressions into one expression allows us to solve the equation. So

$$\ln x + \ln(x + 1) = \ln 2$$

$$\ln x(x + 1) = \ln 2$$

$$x(x + 1) = 2$$

$$x^2 + x - 2 = 0$$

$$(x + 2)(x - 1) = 0$$

$$x = 2, \ x = 1.$$

Since $\ln(x + 1)$ is not defined at $x = -2$, the only solution is $x = 1$, and $x = -2$ is an extraneous solution.

Note that $\ln x(x + 1) = \ln 2$ gives $x(x + 1) = 2$ for several reasons. We could use the fact that the function ln is one-to-one. Alternatively, we could note that if each side is raised to an exponent of e, then $e^{\ln x(x+1)} = e^{\ln 2}$, which by the inverse relation gives $x(x + 1) = 2$.

(b) Using the logarithm rules gives

$$2\ln(x+3) - \ln x = \ln 12$$

$$\ln(x+3)^2 - \ln x = \ln 12$$

$$\ln \frac{(x+3)^2}{x} = \ln 12$$

$$\frac{(x+3)^2}{x} = 12$$

$$x^2 + 6x + 9 = 12x$$

$$x^2 - 6x + 9 = 0$$

$$(x-3)(x-3) = 0$$

$$x = 3.$$

(c) Taking the natural logarithm of both sides gives

$$e^{3x} = 3^{2x-1}$$

$$\ln e^{3x} = \ln 3^{2x-1}$$

$$3x = (2x-1)\ln 3$$

$$3x = 2x\ln 3 - \ln 3$$

$$3x - 2x\ln 3 = -\ln 3$$

$$x(3 - 2\ln 3) = -\ln 3$$

$$x = \frac{-\ln 3}{3 - 2\ln 3} = \frac{\ln 3}{2\ln 3 - 3}$$

$$= \frac{\ln 3}{\ln 9 - 3}.$$

◇

GRAPHS OF LOGARITHM FUNCTIONS

There are several important properties to remember when sketching graphs of logarithms. The most important graphs are for $a > 1$, which have the following properties.

<u>Domain:</u> $(0, \infty)$, so the graph of $y = \log_a x$ is on the right side of the y-axis.

<u>Range:</u> $(-\infty, \infty)$, and

$$\log_a x \to \infty \quad \text{as} \quad x \to \infty.$$

Although the logarithm grows arbitrarily large as x grows large, the growth is very slow. On the other hand exponential growth is very rapid.

<u>Vertical Asymptote:</u> $x = 0$, and

$$\log_a x \to -\infty \quad \text{as} \quad x \to 0^+.$$

Notice that the logarithm function is not defined at $x = 0$, and since the domain is only positive real numbers, x can only approach 0 from the right side. This is the reason for the $+$ exponent on the 0.

<u>x-intercept:</u> $(1, 0)$, that is, the graph crosses the $x-$axis when $x = 1$, since

$$\log_a x = 0 \Leftrightarrow a^0 = x, \text{ so } x = 1.$$

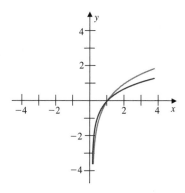

Figure 4.23

EXAMPLE 4.3.6   Sketch the graph of the function.

(a) $y = 2 - \ln(x - 1)$ (b) $y = \ln(-x + 1)$

SOLUTION:

(a) To sketch the graph we will use the following steps.

1. Use $y = \ln x$ as the basic graph.

2. Use (1) to plot $y = -\ln x$.

3. Use (2) to plot $y = -\ln(x - 1)$.

4. Use (3) to plot $y = 2 - \ln(x - 1)$.

The graph of $y = -\ln x$ is obtained by reflecting the basic graph of $y = \ln x$ about the x-axis. So

$$- \ln x \to \infty \quad \text{as} \quad x \to 0^+$$

$$- \ln x \to -\infty \quad \text{as} \quad x \to \infty.$$

Now shift $y = -\ln x$ to the right 1 unit to obtain the graph of $y = -\ln(x - 1)$. Finally, shift the graph of $y = -\ln(x - 1)$ upward 2 units to obtain the graph of $y = 2 - \ln(x - 1)$.

<u>Vertical asymptote:</u> $x = 1$

<u>x-intercept:</u> Solve

$$2 - \ln(x - 1) = 0$$
$$-\ln(x - 1) = -2$$
$$\ln(x - 1) = 2$$
$$e^{\ln(x-1)} = e^2$$
$$x - 1 = e^2$$
$$x = e^2 + 1.$$

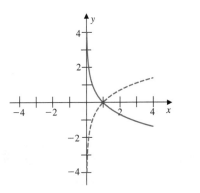

Figure 4.24

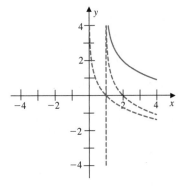

Figure 4.25

(b) The graph of $y = \ln(-x)$ is just the reflection of the graph of $y = \ln x$ about the y-axis. To obtain $y = \ln(-x + 1) = \ln(-(x - 1))$, shift the graph of $y = \ln(-x)$ to the right 1 unit.

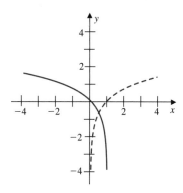

Figure 4.26

◇

EXAMPLE 4.3.7   Match the equation with the curve in the figure.

(a) $y = -\ln x + 2$ (b) $y = \ln(x - 2) - 1$

(c) $y = \ln(-x) + 1$ (d) $y = 2\ln(x - 2) - 1$

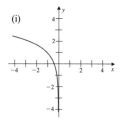

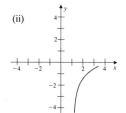

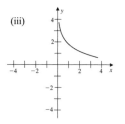

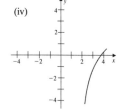

Figure 4.27   Figure 4.28   Figure 4.29   Figure 4.30

SOLUTION: The graph of $y = \ln x$ is shown in the figure.

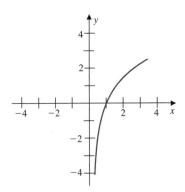

Figure 4.31

(a) A reflection of $y = \ln x$ about the x-axis and then a shift upward by 2 units, which is (iii).

(c) A reflection of $y = \ln x$ about the y-axis and a shift upward 1 unit, which is (i).

Both (b) and (d) involve a shift of $y = \ln x$ to the right 2 units and downward 1 unit. Since $y = 2\ln(x - 2) - 1$ also involves a vertical stretching, (d) matches (iv). Then (b) matches (ii).   ◇

EXAMPLE 4.3.8   Determine the length of time it takes an initial investment to triple in value if it earns 10% compounded continuously.

SOLUTION: The value of the investment after t years is given by the exponential formula

$$A_c(t) = A_0 e^{0.1t},$$

where A_0 is some initial investment and 0.1 represents the 10% interest rate. To determine the length of time it takes the initial investment to triple solve for time t in the equation

$$3A_0 = A_0 e^{0.1t}$$

$$3 = e^{0.1t}$$

$$\ln 3 = \ln e^{0.1t}$$

$$\ln 3 = 0.1t$$

$$t = \frac{\ln 3}{0.1} = \frac{\ln 3}{\frac{1}{10}} = 10 \ln 3 \text{ years} \approx 11 \text{ years.}$$

◇

Solutions for Exercise Set 4.3

1. $x = \log_3 3^5 \Leftrightarrow 3^x = 3^5 \Leftrightarrow x = 5$

3. $x = \log_4 64 \Leftrightarrow 4^x = 64 \Leftrightarrow x = 3$

5. $x = \log_4 2 \Leftrightarrow 4^x = 2 \Leftrightarrow x = \frac{1}{2}$

7. $x = \log_{10} 0.001 \Leftrightarrow 10^x = 0.001 \Leftrightarrow 10^x = \frac{1}{1000} \Leftrightarrow x = -3$

9. $x = \log_2 \frac{1}{16} \Leftrightarrow 2^x = \frac{1}{16} \Leftrightarrow x = -4$

11. Since $e^{\ln x} = x$ for $x > 0$ we have $e^{\ln 5} = 5$.

13. Since $\ln e^x = x$ for all x we have $\ln e^\pi = \pi$.

15. Since $e^{\ln x} = x$ for $x > 0$ we have $e^{2\ln \pi} = e^{\ln \pi^2} = \pi^2$.

17. $\ln x(x+2) = \ln x + \ln(x+2)$

19. $\log_3 \frac{x^4}{x+1} = \log_3 x^4 - \log_3(x+1) = 4\log_3 x - \log_3(x+1)$

21. $\ln \frac{3x^2}{(x+3)^4} = \ln 3x^2 - \ln(x+3)^4 = \ln 3 + \ln x^2 - 4\ln(x+3) = \ln 3 + 2\ln x - 4\ln(x+3)$

23. We have

$$\log_3 \frac{(3x+2)^{3/2}(x-1)^3}{x\sqrt{x+1}} = \log_3(3x+2)^{3/2}(x-1)^3 - \log_3 x\sqrt{x+1}$$

$$= \log_3(3x+2)^{3/2} + \log_3(x-1)^3 - \log_3 x - \log_3(x+1)^{1/2}$$

$$= \frac{3}{2}\log_3(3x+2) + 3\log_3(x-1) - \log_3 x - \frac{1}{2}\log_3(x+1).$$

25. We have

$$\ln \sqrt{x\sqrt{x+1}} = \ln\left(x(x+1)^{1/2}\right)^{1/2} = \frac{1}{2}\ln x(x+1)^{1/2}$$

$$= \frac{1}{2}\left[\ln x + \ln(x+1)^{1/2}\right] = \frac{1}{2}\ln x + \frac{1}{4}\ln(x+1).$$

27. $\ln x + 2\ln(x+1) = \ln x + \ln(x+1)^2 = \ln x(x+1)^2$

29. $\frac{1}{2}\ln x - 2\ln(x-1) = \ln x^{1/2} - \ln(x-1)^2 = \ln \frac{\sqrt{x}}{(x-1)^2}$

31. $\ln(x-1) + \frac{1}{2}\ln x - 2\ln x = \ln(x-1)\sqrt{x} - \ln x^2 = \ln \frac{(x-1)\sqrt{x}}{x^2} = \ln \frac{x-1}{x^{3/2}}$

33. $\log_3 x = 4 \Rightarrow 3^4 = x \Rightarrow x = 81$

35. $\log_2(3x - 4) = 3 \Rightarrow 2^3 = 3x - 4 \Rightarrow 3x = 12 \Rightarrow x = 4$

37. $\log_x 4 = 2 \Rightarrow x^2 = 4 \Rightarrow x = \pm 2 \Rightarrow x = 2$

39. $\ln(2 - x) = 4 \Rightarrow e^{\ln(2-x)} = e^4 \Rightarrow 2 - x = e^4 \Rightarrow x = 2 - e^4$

41. $\ln 2 + \ln(x + 1) = \ln(4x - 7) \Rightarrow \ln 2(x + 1) = \ln(4x - 7)$

$\Rightarrow 2x + 2 = 4x - 7 \Rightarrow 2x = 9 \Rightarrow x = \frac{9}{2}$

43. $2 \ln x = \ln(4x+6) - \ln 2 \Rightarrow \ln x^2 = \ln \frac{4x+6}{2} \Rightarrow x^2 = 2x+3 \Rightarrow x^2 - 2x - 3 = 0$

$\Rightarrow (x - 3)(x + 1) = 0 \Rightarrow x = 3, x = -1$

Hence $x = 3$, since $\ln x$ and $\ln(4x - 6)$ are not defined at $x = -1$, so $x = -1$ is not a solution.

45. $\ln(2x - 1) - \ln(x - 1) = \ln 5 \Rightarrow \ln \frac{2x-1}{x-1} = \ln 5$

$\Rightarrow \frac{2x-1}{x-1} = 5 \Rightarrow 2x - 1 = 5x - 5 \Rightarrow 3x = 4 \Rightarrow x = \frac{4}{3}$

47. $\log_3(2x^2 + 17x) = 2 \Rightarrow 2x^2 + 17x = 9 \Rightarrow 0 = 2x^2 + 17x - 9 = (2x - 1)(x + 9)$

$\Rightarrow 2x - 1 = 0, x = -9 \Rightarrow x = \frac{1}{2}, x = -9$

49. $4^x = 3 \Rightarrow \ln 4^x = \ln 3 \Rightarrow x \ln 4 = \ln 3 \Rightarrow x = \frac{\ln 3}{\ln 4} = \log_4 3$

51. $e^{2x} = 3^{x-4} \Rightarrow \ln e^{2x} = \ln 3^{x-4} \Rightarrow 2x \ln e = (x - 4) \ln 3$

$\Rightarrow 2x = (\ln 3)x - 4 \ln 3 \Rightarrow (\ln 3 - 2)x = 4 \ln 3 \Rightarrow x = \frac{4 \ln 3}{\ln 3 - 2}$

53. $2 \cdot 3^{-x} = 2^{3x} \Rightarrow \ln 2 \cdot 3^{-x} = \ln 2^{3x} \Rightarrow \ln 2 - x \ln 3 = 3x \ln 2$

$\Rightarrow \ln 2 = 3x \ln 2 + x \ln 3 \Rightarrow \ln 2 = x(3 \ln 2 + \ln 3) \Rightarrow x = \frac{\ln 2}{3 \ln 2 + \ln 3}$

55. For $y = \log_2(x - 3)$

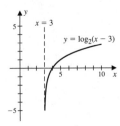

57. For $y = 2 - \log_2(x - 1)$

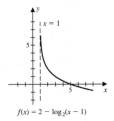

$f(x) = 2 - \log_2(x - 1)$

59. For $y = 2\ln(x + 1) - 3$

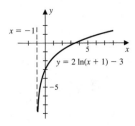

$y = 2\ln(x + 1) - 3$

61. For $y = \ln(-x)$

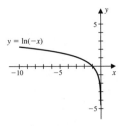

$y = \ln(-x)$

63. For $y = |\ln x|$

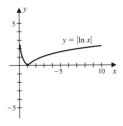

65. (a) iii (b) iv (c) i (d) ii

67. For $y = \ln(4 - x^2)$

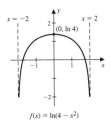

69. For $y = (\ln x)/x$

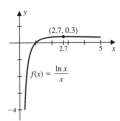

71. The function $g(x) = \sqrt[n]{x}$ grows more rapidly than $f(x) = a + \ln x$ for all $n > 0$.

4.4 Exponential Growth and Decay

If a quantity grows or decays at a rate that is directly proportional to the amount of the quantity that is present, then the quantity present at any time t can be modeled by an exponential function. If the initial amount of the quantity is Q_0, then the amount at any time t is

$$Q(t) = Q_0 e^{kt},$$

where k is the *constant of proportionality* that depends on the specific situation. If $k > 0$, we say Q *grows exponentially*, and if $k < 0$, we say Q *decays exponentially*.

EXAMPLE 4.4.1   A bacteria culture starts with 300 bacteria and 6 hours later has 3000 bacteria.

(a) Find an expression for the number of bacteria after t hours.

(b) Find the number of bacteria that will be present after 8 hours.

(c) When will the population reach 20000?

(d) How long does it take the population to double in size?

SOLUTION:

(a) Use the information given to find the specific values for Q_0 and k. Since the initial amount of bacteria is 300,

$$Q_0 = 300.$$

To find k, use the fact that after 6 hours, that is, when $t = 6$, there are 3000 bacteria present. Then $Q(6) = 3000$, so

$$3000 = Q(6) = 300e^{k(6)}$$

$$e^{6k} = \frac{3000}{300} = 10$$

$$\ln e^{6k} = \ln 10$$

$$6k = \ln 10$$

$$k = \frac{\ln 10}{6}.$$

The number of bacteria after t hours is

$$Q(t) = 300e^{\frac{\ln 10}{6}t}.$$

(b) After $t = 8$ hours the number of bacteria present is

$$Q(6) = 300e^{\frac{\ln 10}{6}(8)} \approx 6463.$$

Notice that the expression for $Q(6)$ can be rewritten in the form

$$300e^{\frac{\ln 10}{6}(8)} = 300e^{(\ln 10)\frac{8}{6}} = 300\left(e^{\ln 10}\right)^{\frac{4}{3}}$$

$$= 300(10)^{\frac{4}{3}}.$$

(c) Find the value for time t so that $Q(t) = 20000$. Solve

$$Q(t) = 300e^{\frac{\ln 10}{6}t} = 20000,$$

then

$$e^{\frac{\ln 10}{6}t} = \frac{20000}{300} = \frac{200}{3}$$

$$\ln e^{\frac{\ln 10}{6}t} = \ln \frac{200}{3}$$

$$\frac{\ln 10}{6}t = \ln \frac{200}{3},$$

and

$$t = \frac{6 \ln \frac{200}{3}}{\ln 10} \approx 11 \text{ hours.}$$

(d) Find the value for time t so that $Q(t) = 2(300) = 600$. Solve

$$Q(t) = 300e^{\frac{\ln 10}{6}t} = 600,$$

then

$$e^{\frac{\ln 10}{6}t} = 2$$

$$\frac{\ln 10}{6}t = \ln 2,$$

and

$$t = \frac{6 \ln 2}{\ln 10} \approx 1.8 \text{ hours.}$$

◇

EXAMPLE 4.4.2   A radioactive substance has a half-life of approximately 25 years.

(a) If a sample has a mass of 50 mg, find an expression for the mass after t hours.

(b) How much will remain after 100 years?

(c) When will the mass decay to 10 mg?

SOLUTION:

(a) The *half-life* of a radioactive substance is the amount of time it takes for one half of the substance to decay. To find the constant of proportionality, use the fact that the half-life is 25 years, so that at time $t = 25$ years the mass of the substance will be 25 mg. So

$$Q(t) = 50e^{kt},$$

and

$$25 = Q(25) = 50e^{25k}.$$

Therefore,

$$e^{25k} = \frac{25}{50} = \frac{1}{2}$$

$$25k = \ln\frac{1}{2} = \ln 1 - \ln 2$$

$$25k = -\ln 2$$

$$k = -\frac{\ln 2}{25},$$

and

$$Q(t) = 50e^{-\frac{\ln 2}{25}t}.$$

Note that we used the arithmetic property of logarithms,

$$\ln\left(\frac{x}{y}\right) = \ln x - \ln y$$

and the fact that $\ln 1 = 0$. Note also that $\ln 2 > 0$, so the proportionality constant is less than 0, and the exponential function represents exponential decay.

(b)

$$Q(100) = 50e^{-\frac{\ln 2}{25}(100)} = 50e^{-4\ln 2}$$

$$\approx 3.125 \text{ mg}$$

(c) We solve for t in the equation

$$10 = Q(t) = 50e^{-\frac{\ln 2}{25}t}$$

$$e^{-\frac{\ln 2}{25}t} = \frac{1}{5}$$

$$-\frac{\ln 2}{25}t = \ln\frac{1}{5} = \ln 1 - \ln 5$$

$$-\frac{\ln 2}{25}t = -\ln 5$$

$$t = \frac{25\ln 5}{\ln 2} \approx 58.1 \text{ years}$$

The computer generated graph in the figure shows the decaying mass function.

◇

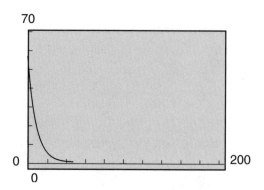

Figure 4.32

EXAMPLE 4.4.3 Find the half-life of a radioactive substance that decays 2% in 10 years.

SOLUTION: The quantity of the radioactive substance present at time t is

$$Q(t) = Q_0 e^{kt}.$$

If the substance decays 2% in 10 years, then after 10 years there still remains 98% of the original amount. This gives the equation

$$0.98Q_0 = Q(10) = Q_0 e^{10k},$$

from which the proportionality constant k can be found. We do not need to know the actual initial amount of the substance. Solving gives

$$0.98Q_0 = Q(10) = Q_0 e^{10k}$$

$$e^{10k} = 0.98$$

$$10k = \ln(0.98)$$

$$k = \frac{\ln(0.98)}{10},$$

and

$$Q(t) = Q_0 e^{\frac{\ln(0.98)}{10}t}.$$

To find the half-life, find t so that

$$\frac{1}{2}Q_0 = Q_0 e^{\frac{\ln(0.98)}{10}t}$$

$$e^{\frac{\ln(0.98)}{10}t} = \frac{1}{2}$$

$$\frac{\ln(0.98)}{10}t = -\ln 2$$

$$t = -\frac{10\ln 2}{\ln(0.98)} \approx 343.1 \text{ years.}$$

Notice the number of years is positive, since $\ln(0.98) < 0$.   ◇

EXAMPLE 4.4.4   The parents of a new born child put $20,000.00 into an account with the hope that the amount will grow to $150,000.00 when the child starts college in 18 years. What rate of continuously compounded interest is necessary for this goal to be met?

SOLUTION: If an initial investment of A_0 dollars is invested at an interest rate i compounded continuously, the investment after t years has a value

$$A_c(t) = A_0 e^{it}.$$

If the parents start with $A_0 = 20000$ and after 18 years want an accumulation of $A_c(18) = 150000$, the required interest rate can be found from the equation

$$150000 = 20000 e^{18i}$$

$$7.5 = e^{18i}$$

$$\ln 7.5 = \ln e^{18i}$$

$$\ln 7.5 = 18i$$

$$i = \frac{\ln 7.5}{18} \approx 0.11 \text{ or } 11\%.$$

◇

Solutions for Exercise Set 4.4

1. We have $Q(t) = Q_0 e^{kt}$ with $Q_0 = 1000$.

(a) Since the bacteria doubles every 4 hours,

$$2(1000) = Q(4) = 1000e^{4k} \Rightarrow e^{4k} = 2 \Rightarrow 4k = \ln 2 \Rightarrow k = \frac{\ln 2}{4}$$

and $Q(t) = 1000e^{\frac{\ln 2}{4}t}$.

(b) $Q(7) = 1000e^{\frac{\ln 2}{4}7} \approx 3364$

(c) $1000e^{\frac{\ln 2}{4}t} = 20000 \Rightarrow e^{\frac{\ln 2}{4}t} = 20 \Rightarrow \frac{\ln 2}{4}t = \ln 20 \Rightarrow t = \frac{4\ln 20}{\ln 2} \approx 17.3$ hours

3. We have $Q(t) = Q_0 e^{kt}$ with $Q_0 = 500$.

(a) Since after 5 hours there are 4000 bacteria,

$$4000 = Q(5) = 500e^{5k} \Rightarrow e^{5k} = 8 \Rightarrow \ln e^{5k} = \ln 8 \Rightarrow 5k = \ln 8 \Rightarrow k = \frac{\ln 8}{5}$$

and $Q(t) = 500e^{\frac{\ln 8}{5}t}$.

(b) $Q(6) = 500e^{\frac{\ln 8}{5}6} \approx 6063$

(c) $500e^{\frac{\ln 8}{5}t} = 15000 \Rightarrow e^{\frac{\ln 8}{5}t} = 30 \Rightarrow \frac{\ln 8}{5}t = \ln 30 \Rightarrow t = \frac{5\ln 30}{\ln 8} \approx 8.2$ hours

(d) $500e^{\frac{\ln 8}{5}t} = 1000 \Rightarrow e^{\frac{\ln 8}{5}t} = 2 \Rightarrow \frac{\ln 8}{5}t = \ln 2 \Rightarrow t = \frac{5\ln 2}{\ln 8} \approx 1.7$ hours

5. We have $Q(t) = Q_0 e^{kt}$ with $Q_0 = 50$.

(a) If the half life is 578 hours,

$$25 = Q(578) = 50e^{578k} \Rightarrow e^{578k} = \frac{1}{2} \Rightarrow 578k = \ln\frac{1}{2} = -\ln 2 \Rightarrow k = -\frac{\ln 2}{578}$$

and $Q(t) = 50e^{-\frac{\ln 2}{578}t}$.

(b) $Q(100) = 50e^{-\frac{\ln 2}{578}100} \approx 44.35$ mg

(c) $50e^{-\frac{\ln 2}{578}t} = 10 \Rightarrow e^{-\frac{\ln 2}{578}t} = \frac{1}{5} \Rightarrow -\frac{\ln 2}{578}t = \ln\frac{1}{5} = \ln 1 - \ln 5 = -\ln 5$

$\Rightarrow t = \frac{578\ln 5}{\ln 2} \approx 1342.1$ hours

7. We have $Q(t) = Q_0 e^{kt}$. If the culture doubles in size in 2 hours, then

$$2Q_0 = Q(2) = Q_0 e^{2k} \Rightarrow e^{2k} = 2 \Rightarrow 2k = \ln 2 \Rightarrow k = \frac{\ln 2}{2}$$

and $Q(t) = Q_0 e^{\frac{\ln 2}{2} t}$. The culture will triple in size when,

$$Q_0 e^{\frac{\ln 2}{2} t} = 3Q_0 \Rightarrow e^{\frac{\ln 2}{2} t} = 3 \Rightarrow \frac{\ln 2}{2} t = \ln 3 \Rightarrow t = \frac{2 \ln 3}{\ln 2} \approx 3.2 \text{ hours.}$$

9. We have $Q(t) = 200 e^{kt}$. If the mass decays to 180 g in 2 years, then

$$Q(2) = 200 e^{2k} = 180 \Rightarrow e^{2k} = \frac{180}{200} = \frac{9}{10} \Rightarrow 2k = \ln \frac{9}{10} = \ln 9 - \ln 10.$$

So $k = \frac{\ln 9 - \ln 10}{2}$ and $Q(t) = 200 e^{\frac{\ln 9 - \ln 10}{2} t}$.

The half life is the time required for half the substance to decay, so

$$100 = 200 e^{\frac{\ln 9 - \ln 10}{2} t} \Rightarrow e^{\frac{\ln 9 - \ln 10}{2} t} = \frac{1}{2} \Rightarrow \frac{\ln 9 - \ln 10}{2} t = \ln \frac{1}{2} = -\ln 2$$

So $t = -\frac{2 \ln 2}{\ln 9 - \ln 10} \approx 13.16$ years.

11. (a) Let $Q(t)$ represent the population t years after 1950. Since the initial population is the 1950 statistic, $Q(0) = 2513$. Since 1960 is ten years after the initial date of 1950,

$$3027 = Q(10) = 2513 e^{10k} \Rightarrow e^{10k} = \frac{3027}{2513} \Rightarrow 10k = \ln \frac{3027}{2513}.$$

So $k = \frac{1}{10} \ln \frac{3027}{2513}$ and $Q(t) = 2513 e^{\frac{1}{10} \ln \frac{3027}{2513} t}$. The population in 2000, which is 50 years after the initial year of 1950, is

$$Q(50) = 2513 e^{\frac{1}{10} \ln \frac{3027}{2513} 50} \approx 6372 \text{ million.}$$

The population in 2050, which is 100 years after the initial year of 1950, is

$$Q(100) = 2513 e^{\frac{1}{10} \ln \frac{3027}{2513} 100} \approx 16158 \text{ million.}$$

(b) Since

$$Q(0) = 4478 \Rightarrow 5321 = Q(10) = 4478e^{10k} \Rightarrow e^{10k} = \frac{5321}{4478} \Rightarrow k = \frac{1}{10}\ln\frac{5321}{4478},$$

we have

$$Q(t) = 4478e^{\frac{1}{10}\ln\frac{5321}{4478}t}.$$

The population in 2000, which is 20 years after the initial year of 1980, is

$$Q(20) = 4478e^{\frac{1}{10}\ln\frac{5321}{4478}20} \approx 6323 \text{ million.}$$

The population in 2050, which is 70 years after the initial year of 1980, is

$$Q(70) = 2513e^{\frac{1}{10}\ln\frac{3027}{2513}70} \approx 14978 \text{ million.}$$

13. We have $A(t) = 10000e^{0.08t}$.

(a) $20000 = 10000e^{0.08t} \Rightarrow 2 = e^{0.08t} \Rightarrow 0.08t = \ln 2 \Rightarrow$
$t = \frac{\ln 2}{0.08} = 8.7$ years

(b) $30000 = 10000e^{0.08t} \Rightarrow 3 = e^{0.08t} \Rightarrow 0.08t = \ln 3 \Rightarrow$
$t = \frac{\ln 3}{0.08} = 13.7$ years

15. We have $Q(t) = 10000e^{it}$. Then

$$25000 = Q(5) = 10000e^{5i} \Rightarrow e^{5i} = \frac{5}{2} \Rightarrow 5i = \ln\frac{5}{2} \Rightarrow i = \frac{1}{5}\ln\frac{5}{2} \approx 0.18 \text{ or } 18\%.$$

17. We have $T(t) = T_m + (T_0 - T_m)e^{kt}$, where $T_0 = -3°\text{C}$ and $T_m = 20°\text{C}$,
so $T(t) = 20 - 23e^{kt}$. Since one minute later the temperature reads $5°\text{C}$,

$$5 = T(1) = 20 - 23e^k \Rightarrow e^k = \frac{15}{23} \Rightarrow k = \ln\frac{15}{23},$$

and $T(t) = 20 - 23e^{\ln\frac{15}{23}t}$. The thermometer will read $19.5°\text{C}$, when

$$20 - 23e^{\ln\frac{15}{23}t} = 19.5 \Rightarrow e^{\ln\frac{15}{23}t} = \frac{0.5}{23} = \frac{1}{46} \Rightarrow \ln\frac{15}{23}t = \ln\frac{1}{46} = -\ln 46.$$

So

$$t = -\frac{\ln 46}{\ln \frac{15}{23}} \approx 9 \text{minutes}.$$

19. By Newton's Law of Cooling the temperature of the body at time t is

$$T(t) = T_m + (T_0 - T_m)e^{kt},$$

where $T_m = 62°\text{F}$, the constant temperature of the lake, and $T_0 = 67°\text{C}$, the temperature when the body was found at 11:00 a.m. So, $T(t) = 62 + 5e^{kt}$. At noon, or 1 hour after the body was found the temperature was $66°\text{F}$, so

$$66 = T(1) = 62 + 5e^k \Rightarrow 5e^k = 4 \Rightarrow e^k = \frac{4}{5} \Rightarrow k = \ln\frac{4}{5},$$

and $T(t) = 62 + 5e^{\ln\frac{4}{5}t}$. The victim died at the time t when the temperature of the body was $98.6°$, so

$$98.6 = 62 + 5e^{\ln\frac{4}{5}t} \Rightarrow e^{\ln\frac{4}{5}t} = \frac{36.6}{5} \Rightarrow \ln\frac{4}{5}t = \ln\frac{36.6}{5},$$

and

$$t = \frac{\ln\frac{36.6}{5}}{\ln\frac{4}{5}} \approx -8.9 \text{ hours}.$$

So the death occurred about 8.9 hours before 11:00 a.m., at about 2:06 a.m.

Solutions for Exercise Set 4 Review

1. (a) iii (b) i (c) iv (d) ii

3. For $f(x) = 2^{x-1} - 3$

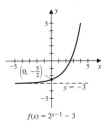

$f(x) = 2^{x-1} - 3$

5. For $f(x) = e^{-x} + 1$

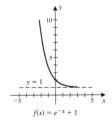

$f(x) = e^{-x} + 1$

7. For $f(x) = 3e^{1-x} = 3e^{-(x-1)}$

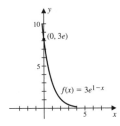

$f(x) = 3e^{1-x}$

9. For $f(x) = 2\ln x$

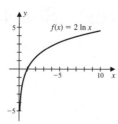

11. For $f(x) = 3 - \log_2(x+1)$

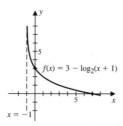

13. For $f(x) = e^{-x^2+2x-1}$

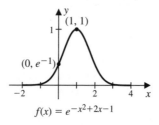

15. $x = \log_5 1 \Leftrightarrow 5^x = 1 \Leftrightarrow x = 0$

17. Since $2^{\log_2 x} = x$ for $x > 0$ we have $2^{\log_2 15} = 15$.

19. $x = \log_4 2 \Leftrightarrow 4^x = 2 \Leftrightarrow x = \frac{1}{2}$

21. Since $e^{\ln x} = x$ for $x > 0$ we have $e^{3\ln 4} = e^{\ln 4^3} = 64$.

23. $\ln \frac{3x^2}{\sqrt{x-1}} = \ln 3x^2 - \ln(x-1)^{1/2} = \ln 3 + \ln x^2 - \frac{1}{2}\ln(x-1) = \ln 3 + 2\ln x - \frac{1}{2}\ln(x-1)$

25. $\log_{10} \frac{\sqrt{x+1}\,\sqrt[3]{x-1}}{x(x+3)^{5/2}} = \log_{10} \frac{(x+1)^{1/2}(x-1)^{1/3}}{x(x+3)^{5/2}} = \log_{10}(x+1)^{1/2}(x-1)^{1/3} -$

$\log_{10} x(x+3)^{5/2} = \log_{10}(x+1)^{1/2} + \log_{10}(x-1)^{1/3} - \log_{10}x - \log_{10}(x+$

$3)^{5/2} = \frac{1}{2}\log_{10}(x+1) + \frac{1}{3}\log_{10}(x-1) - \log_{10}x - \frac{5}{2}\log_{10}(x+3)$

27. $\ln x + \frac{1}{3}\ln x(x+1) + 2\ln(x-1) = \ln x + \frac{1}{3}\ln(x^2+x) + \ln(x-1)^2 =$

$\ln x + \ln(x^2+x)^{1/3} + \ln(x-1)^2 = \ln x(x^2+x)^{1/3} + \ln(x-1)^2 = \ln x(x-$

$1)^2 \sqrt[3]{x^2+x}$

29. $3\ln(x^3+2) + \ln 5 - \frac{1}{2}\ln(x^5-1) = \ln(x^3+2)^3 + \ln 5 - \ln\sqrt{x^5-1} =$

$\ln 5(x^3+2)^3 - \ln\sqrt{x^5-1} = \ln\frac{5(x^3+2)^3}{\sqrt{x^5-1}}$

31. $\ln(2x-3) = 4 \Rightarrow e^{\ln(2x-3)} = e^4 \Rightarrow 2x-3 = e^4 \Rightarrow x = \frac{e^4+3}{2}$

33. We have $\ln(2x-1) + \ln(3x-2) = \ln 7 \Rightarrow \ln(2x-1)(3x-2) = \ln 7 \Rightarrow$

$(2x-1)(3x-2) = 7 \Rightarrow 6x^2 - 7x - 5 = 0 \Rightarrow (2x+1)(3x-5) = 0$

$\Rightarrow x = -\frac{1}{2}$ or $x = \frac{5}{3}$. Hence $x = \frac{5}{3}$, since the natural logarithms are not

defined at $x = -\frac{1}{2}$.

35. $3^x \cdot 5^{x-2} = 3^{4x} \Rightarrow \ln 3^x 5^{x-2} = \ln 3^{4x} \Rightarrow \ln 3^x + \ln 5^{x-2} = 4x\ln 3 \Rightarrow$

$x\ln 3 + (x-2)\ln 5 = 4x\ln 3 \Rightarrow x(\ln 3 + \ln 5 - 4\ln 3) = 2\ln 5 \Rightarrow$

$x = \frac{2\ln 5}{\ln 3 + \ln 5 - 4\ln 3} = \frac{2\ln 5}{\ln 5 - 3\ln 3}$

37. We have $2e^x x^2 - e^x x = e^x \Rightarrow 2e^x x^2 - e^x x - e^x = 0 \Rightarrow e^x(2x^2 - x - 1) =$

$0 \Rightarrow$

$2x^2 - x - 1 = 0$, since $e^x > 0$ for all x. So $(2x+1)(x-1) = 0 \Rightarrow x = -\frac{1}{2}$

or $x = 1$.

39. For $e^{x^2} = x - 2$ there are no solutions since the graphs never intersect.

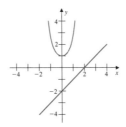

41. We have $e^x > x^4$ on $(-0.8, 1.4) \cup (8.6, \infty)$.

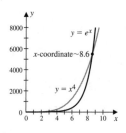 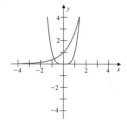

43. We have $e^{x-1} - 3 < x^5$ on $(-1.3, 14.3)$.

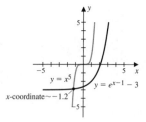 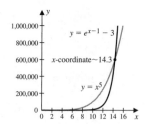

45. For $f(x) = x^2 e^{1-x^2}$ we have the following. Increasing: $(-\infty, -1) \cup (0, 1)$; decreasing: $(-1, 0) \cup (1, \infty)$; local maximums: $(-1, 1)$ and $(1, 1)$; local minimum: $(0, 0)$.

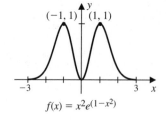

$$f(x) = x^2 e^{(1-x^2)}$$

47. The value of an initial investment A_0 deposited at 6% compounded continuously is $A(t) = A_0 e^{0.06t}$. If the investment is to double,

$$2A_0 = A_0 e^{0.06t} \Rightarrow e^{0.06t} = 2 \Rightarrow 0.06t = \ln 2 \Rightarrow t = \frac{\ln 2}{0.06} \approx 11.6 \text{ years.}$$

49. We have $Q(t) = Q_0 e^{kt}$.

(a) To find k use the initial information given. That is,

$$1000 = Q(1) = Q_0 e^{k(1)} \Rightarrow e^k = \frac{1000}{Q_0} \Rightarrow k = \ln \frac{1000}{Q_0},$$

and

$$3000 = Q(4) = Q_0 e^{k(4)} \Rightarrow e^{4k} = \frac{3000}{Q_0} \Rightarrow 4k = \ln \frac{3000}{Q_0}$$

so $4 \ln \frac{1000}{Q_0} = \ln \frac{3000}{Q_0}$. Hence

$$\left(\frac{1000}{Q_0}\right)^4 = \frac{3000}{Q_0} \Rightarrow Q_0^3 = \frac{10^{12}}{3000} = \frac{10^9}{3} \Rightarrow Q_0 = \frac{10^3}{\sqrt[3]{3}}.$$

Then $k = \ln \sqrt[3]{3} = \frac{1}{3} \ln 3$, and $Q(t) = \frac{10^3}{\sqrt[3]{3}} e^{\frac{\ln 3}{3} t}$.

(b) $Q(5) = \frac{10^3}{\sqrt[3]{3}} e^{\frac{\ln 3}{3} 5} \approx 4327$.

(c) Since

$$20000 = \frac{10^3}{\sqrt[3]{3}} e^{\frac{\ln 3}{3} t} \Rightarrow e^{\frac{\ln 3}{3} t} = 20\sqrt[3]{3} \Rightarrow \frac{\ln 3}{3} t = \ln 20\sqrt[3]{3}$$

we have

$$t = \frac{3 \ln 20\sqrt[3]{3}}{\ln 3} \approx 9.2 \text{ hours.}$$

(d) $3Q_0 = Q(t) = Q_0 e^{\frac{\ln 3}{3} t} \Rightarrow 3 = e^{\frac{\ln 3}{3} t} \Rightarrow \ln 3 = \frac{\ln 3}{3} t \Rightarrow t = 3$ hours

51. We have $A_4(t) = A_0 \left(1 + \frac{0.1}{4}\right)^{4t}$ and the time it takes for the investment to double can be found from

$$2A_0 = A_0 \left(1 + \frac{0.1}{4}\right)^{4t} \Rightarrow 2 = \left(1 + \frac{0.1}{4}\right)^{4t} \Rightarrow$$

$$\ln 2 = 4t \ln\left(1 + \frac{0.1}{4}\right) \Rightarrow t = \frac{\ln 2}{4 \ln\left(1 + \frac{0.1}{4}\right)} \approx 7 \text{ years.}$$

53.   We have $A_c(t) = A_0 e^{0.09t}$ and the time it takes for the investment to double can be found from

$$2A_0 = A_0 e^{0.09t} \Rightarrow 2 = e^{0.09t} \Rightarrow \ln 2 = 0.09t \Rightarrow t = \frac{\ln 2}{0.09} \approx 7 \text{ years.}$$

Solutions for Exercise Set 4 Calculus

1. (a) The graphs of $f(x) = 2\ln x$ and $g(x) = e^{\frac{x}{2}}$ are reflections of one another through $y = x$, so $f = g^{-1}$.

    (b) The graphs of $f(x) = \ln\frac{x}{2}$ and $g(x) = e^{2x}$ are not reflections of one another through $y = x$, so $f \neq g^{-1}$.

    (c) Neither $f(x) = \ln|x|$ nor $g(x) = e^{|x|}$ are $1-1$ functions, and hence can not have inverses.

    (d) The graphs of $f(x) = -\ln x$ and $g(x) = e^{-x}$ are reflections of one another through $y = x$, so $f = g^{-1}$.

    (e) The graphs of $f(x) = 1 + \ln x$ and $g(x) = e^{x-1}$ are reflections of one another through $y = x$, so $f = g^{-1}$.

    (f) The graphs of $f(x) = 2\ln x$ and $g(x) = \frac{1}{2}e^x$ are not reflections of one another through $y = x$, so $f \neq g^{-1}$.

3. To order the functions according to how fast they grow as $x \to \infty$, plot pairs of functions together for large x to place them in order. For example, to order $\ln x, x^x$ and e^{3x}, first plot $y = \ln x$ and $y = x^x$ to see that $y = x^x$ grows faster. Then compare $y = e^{3x}$ and $y = x^x$ to see that $y = x^x$ grows faster. Finally compare $y = e^{3x}$ and $y = \ln x$ to see that $y = e^{3x}$ grows faster, so the ordering of these three functions is $\ln x, e^{3x}, x^x$. The complete ordering from smallest to largest as $x \to \infty$ is

$$\frac{x^{10}}{e^x}, \quad \frac{1}{x^4}, \quad \ln x, \quad x^{1/20}, \quad x^{20}, \quad e^{3x}, \quad \frac{e^{6x}}{x^8}, \quad x^x.$$

5. We have $y = 3e^{x-2} = 3e^x e^{-2} = \frac{3}{e^2}e^x$, so the graph of $y = 3e^{x-2}$ is just a vertical scaling of the graph of $y = e^x$. Since $0 < \frac{3}{e^2} < 1$, it is a vertical compression.

7. We have $y = 3 + \ln 2x = 3 + \ln 2 + \ln x = (3 + \ln 2) + \ln x$, so the graph of $y = 3 + \ln 2x$ is just a vertical translation of the graph of $y = \ln x$. Since $3 + \ln 2 > 0$, the shift is upward.

9. If the interest on an initial investment of A_0 dollars is compounded continuously at a fixed rate of $r\%$, then the value after t years is $A(t) = A_0 e^{\frac{r}{100}t}$. Then the time at which the investment doubles is given by,

$$2A_0 = A_0 e^{\frac{r}{100}t} \Rightarrow 2 = e^{\frac{r}{100}t} \Rightarrow \frac{r}{100}t = \ln 2 \Rightarrow t = \frac{100 \ln 2}{r} \approx \frac{70}{r}.$$

So $\frac{70}{r}$ is a reasonable estimate for the time it takes for the investment to double in value. For example, if the interest rate is 8.75%, then $\frac{70}{8.75} = 8.0$.

11. (a) The concentration of the drug in the bloodstream can be modeled using exponential decay. So the concentration at time t has the form

$$C(t) = C_0 e^{kt}.$$

Since the initial concentration of the drug in the bloodstream is 20, we have $C_0 = 20$. If 3 hours later the concentration is 12,

$$12 = C(3) = 20e^{3k} \Rightarrow$$
$$e^{3k} = \frac{12}{20} = \frac{3}{5} \Rightarrow 3k = \ln\left(\frac{3}{5}\right) \Rightarrow$$
$$k = \frac{1}{3}\ln\left(\frac{3}{5}\right).$$

Then the concentration at time t is

$$C(t) = 20e^{\frac{1}{3}\ln\left(\frac{3}{5}\right)t}.$$

(b) To find the half-life set $C(t) = 10$ and solve for t. We have

$$10 = 20e^{\frac{1}{3}\ln\left(\frac{3}{5}\right)t} \Rightarrow$$

$$e^{\frac{1}{3}\ln\left(\frac{3}{5}\right)t} = \frac{1}{2} \Rightarrow \frac{1}{3}\ln\left(\frac{3}{5}\right)t = \ln\frac{1}{2} = -\ln 2 \Rightarrow$$

$$t = -\frac{3\ln 2}{\ln\left(\frac{3}{5}\right)} \approx 4.07 \text{ hours.}$$

(c) Since the half-life is 5 hours, we have $C(t) = C_0 e^{kt}$ and

$$\frac{1}{2}C_0 = C(5) = C_0 e^{5k} \Rightarrow$$

$$e^{5k} = \frac{1}{2} \Rightarrow 5k = \ln\left(\frac{1}{2}\right) = -\ln 2 \Rightarrow$$

$$k = -\frac{\ln 2}{5}.$$

So

$$C(t) = C_0 e^{-\frac{\ln 2}{5}t}.$$

For a 25 kilogram dog, the amount $Q(t)$ of phenobarbital in the blood at time t is $25C(t)$. That is,

$$Q(t) = 25C_0 e^{-\frac{\ln 2}{5}t}.$$

When $t = 1$ hour, we want $Q(t) = (30)(25) = 750$. So

$$750 = Q(1) = 25C_0 e^{-\frac{\ln 2}{5}} \Rightarrow$$

$$C_0 = 30e^{\frac{\ln 2}{5}} \approx 34.46 \text{ mg/kg.}$$

Since the dog's weight is equivalent to 25 kg, the initial dose should be $(34.46)(25) = 861.5$ mg.

CHAPTER 5
CONIC SECTIONS, POLAR COORDINATES, AND PARAMETRIC EQUATIONS

5.1 Introduction

The graphs of the general *quadratic equation* in x and y,

$$Ax^2 + Bxy + Cy^2 + Dx + Ey + F = 0,$$

are called conic sections. The three basic figures are the parabola, ellipse and hyperbola, though certain special, degenerate curves can also occur. When $B = 0$ and $AC = 0$, the curve is a parabola, when $B = 0$ and $AC > 0$, the curve is an ellipse, and when $B = 0$ and $AC < 0$, the curve is a hyperbola. When $B \neq 0$, the curve is a rotated conic in the plane.

Polar coordinates and parametric equations provide two additional methods for describing curves in the plane. They allow for the visualization of a greater variety of curves.

5.2 Parabolas

The graph of the familiar equation of the form $y = ax^2 + bx + c$ is a *parabola*, with axis parallel to the y-axis. A more general geometric definition of a parabola is the set of points equidistant from a given point, called the *focus*, and a given line, called the *directrix*. When the directrix is one of the coordinate axes, the parabola is said to be in *standard form*. The *axis* of the parabola is the line through the focal point that is perpendicular to the directrix. The point of intersection of the axis and the parabola is the *vertex*. A useful tip to remember is that a parabola *never* crosses its directrix.

STANDARD POSITION PARABOLAS

Equation	Vertex	Focus	Directrix
$y = \frac{1}{4c}x^2$	$(0,0)$	$(0,c)$	Horizontal: $y = -c$
$x = \frac{1}{4c}y^2$	$(0,0)$	$(c,0)$	Vertical: $x = -c$

Using the simple shifting properties, if the vertex is shifted to the point (h, k), that is, h units in the horizontal direction and k units in the vertical direction, then the parabola has an equation of the form

$$(y - k) = \frac{1}{4c}(x - h)^2 \text{ or } (x - h) = \frac{1}{4c}(y - k)^2.$$

EXAMPLE 5.2.1   Find the vertex, directrix and focus, and sketch the graph of the parabola.

(a) $y = -\frac{1}{8}x^2$ (b) $4y^2 = 9x$

SOLUTION:

(a) The equation of the parabola is already in standard form, so we only need to determine the value of c. So we have

$$y = -\frac{1}{8}x^2 = \frac{1}{4c}x^2$$
$$\frac{1}{4c} = -\frac{1}{8}$$
$$4c = -8$$
$$c = -2.$$

Vertex: $(0,0)$, since the parabola is in standard form with vertex at the origin.

Focus: $(0, c) = (0, -2)$, since the axis of the parabola is along the y-axis .

Directrix: $y = -c = 2$

Since the parabola never crosses its directrix, the parabola opens downward as shown in the figure.

Maximum or Minimum Value: Since the curve is opening downward, the vertex $(0,0)$ is a *maximum* point on the curve.

Increasing: The curve increases on the interval $(-\infty, 0)$.

Decreasing: The curve is decreasing on the interval $(0, \infty)$.

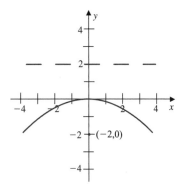

Figure 5.1

(b) A slight rewriting of the equation will put it in standard form. We have

$$4y^2 = 9x$$
$$x = \frac{4}{9}y^2$$
$$x = \frac{1}{9/4}y^2$$
$$4c = \frac{9}{4}$$
$$c = \frac{9}{16}.$$

The parabola is in standard form with axis along the x-axis.

<u>Vertex:</u> $(0,0)$

<u>Focus:</u> $(c,0) = \left(\frac{9}{16},0\right)$

<u>Directrix:</u> $x = -c = -\frac{9}{16}$

Since the parabola never crosses its directrix, the parabola opens to the right as shown in the figure. Notice also the equation does not define a function. This is seen from the figure, which shows that the graph does not satisfy the vertical line test. That is, for each $a > 0$, the vertical line $x = a$ crosses the curve in two points.   ◇

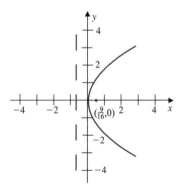

Figure 5.2

EXAMPLE 5.2.2   Find the vertex, directrix and focus, and sketch the graph of the parabola.

(a) $y^2 + 4y + 6 - 2x = 0$ (b) $2x^2 + 8x - 3y + 11 = 0$

SOLUTION:

(a) The first step is to group the x and y terms together in order to rewrite the equation in standard form, with perhaps the vertex shifted. Since there are both y and y^2 terms present, completing the square on these terms is necessary. Completing the square on the y terms gives

$$y^2 + 4y = y^2 + 4y + \left(\frac{4}{2}\right)^2 - \left(\frac{4}{2}\right)^2$$
$$= y^2 + 4y + 4 - 4$$
$$= (y + 2)^2 - 4.$$

Then

$$y^2 + 4y + 6 - 2x = 0$$
$$y^2 + 4y = 2x - 6$$
$$(y + 2)^2 - 4 = 2x - 6$$
$$(y + 2)^2 = 2x - 2$$
$$(y + 2)^2 = 2(x - 1)$$
$$x - 1 = \frac{1}{2}(y + 2)^2.$$

The vertex of the parabola is $(1, -2)$, and the graph of the parabola can be obtained from the graph of

$$x = \frac{1}{2}y^2,$$

which is a parabola in standard form with vertex at the origin and axis along the x-axis. Since

$$4c = 2, \text{ we have } c = \frac{1}{2}.$$

The focus, vertex and directrix of the parabola $x - 1 = \frac{1}{2}(y + 2)^2$ are obtained from shifting the focus, vertex and directrix of the parabola $x = \frac{1}{2}y^2$. For example, the focus of $x = \frac{1}{2}y^2$ is $(c, 0) = \left(\frac{1}{2}, 0\right)$, so the focus of $x - 1 = \frac{1}{2}(y + 2)^2$ is

$$\left(\frac{1}{2} + 1, 0 - 2\right) = \left(\frac{3}{2}, -2\right),$$

that is, shift the point $\left(\frac{1}{2}, 0\right)$ right 1 unit and downward 2 units.

Parabola	$x = \frac{1}{2}y^2$	$x - 1 = \frac{1}{2}(y + 2)^2$
c	$\frac{1}{2}$	$\frac{1}{2}$
Vertex	$(0, 0)$	$(1, -2)$
Focus	$(c, 0) = \left(\frac{1}{2}, 0\right)$	$\left(\frac{1}{2} + 1, 0 - 2\right) = \left(\frac{3}{2}, -2\right)$
Directrix	$x = -c = -\frac{1}{2}$	$x = -c + 1 = -\frac{1}{2} + 1 = \frac{1}{2}$

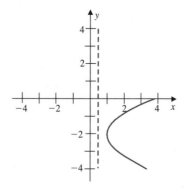

Figure 5.3

(b)

<u>Rewrite the equation:</u> Completing the square on the x terms gives

$$2x^2 + 8x - 3y + 11 = 0$$

$$2(x^2 + 4x) = 3y - 11$$

$$2(x^2 + 4x + 4 - 4) = 3y - 11$$

$$2(x + 2)^2 - 8 = 3y - 11$$

$$2(x + 2)^2 = 3y - 3$$

$$2(x + 2)^2 = 3(y - 1)$$

$$y - 1 = \frac{2}{3}(x + 2)^2.$$

The parabola is obtained by shifting, to the left 2 units and upward 1 unit, the parabola $y = \frac{2}{3}x^2$ that is in standard form with axis along the y-axis and

$$\frac{1}{4c} = \frac{2}{3}$$

$$4c = \frac{3}{2}$$

$$c = \frac{3}{8}.$$

◇

Parabola	$y = \frac{2}{3}x^2$	$y - 1 = \frac{2}{3}(x + 2)^2$
c	$\frac{3}{8}$	$\frac{3}{8}$
Vertex	$(0, 0)$	$(-2, 1)$
Focus	$(0, c) = \left(0, \frac{3}{8}\right)$	$\left(-2, 1 + \frac{3}{8}\right) = \left(-2, \frac{11}{8}\right)$
Directrix	$y = -c = -\frac{3}{8}$	$y = -c + 1 = -\frac{3}{8} + 1 = \frac{5}{8}$

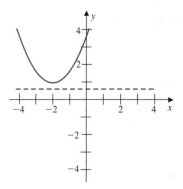

Figure 5.4

EXAMPLE 5.2.3   Determine the equation of the parabola that satisfies the given conditions.

(a) Focus at $(-1, 1)$, directrix $x = 3$.   (b) Vertex at $(3, 4)$, focus at $(3, 6)$.

SOLUTION:

(a) The equation of a parabola requires the vertex and the value of c. The vertex of a parabola lies *midway* between the focus and the directrix on the line through the focus and perpendicular to the directrix. The horizontal distance between $(-1, 1)$ and the vertical line $x = 3$ is 4, so the vertex is 2 units to the right of the focus $(-1, 1)$, and hence is the point $(1, 1)$.

If the focus had been $(-2, 0)$ rather than $(-1, 1)$ and the directrix $x = 2$, then the conditions would describe a parabola in standard position with vertex at the origin and equation

$$x = \frac{1}{4(-2)}y^2 = -\frac{1}{8}y^2.$$

The parabola described by the conditions is obtained by a vertical shift upward 1 unit and a horizontal shift to the right one unit, of the standard position parabola. So

$$x - 1 = -\frac{1}{8}(y - 1)^2.$$

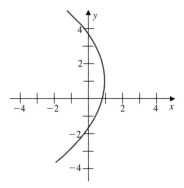

Figure 5.5

(b) Since the axis of the parabola is parallel to one of the coordinate axes and passes through the vertex $(3, 4)$ and the focus $(3, 6)$, the axis is vertical. The distance between the vertex and the focus is $6 - 4 = 2$ (both points are on the vertical line $x = 3$), and since the focus is above the vertex, the parabola opens upward, with $c = 2$. The directrix is 2 units below the vertex and has equation $y = 2$. The equation of the parabola is

$$y - 4 = \frac{1}{4(2)}(x - 3)^2$$
$$y - 4 = \frac{1}{8}(x - 3)^2.$$

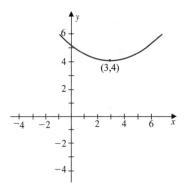

Figure 5.6

◇

EXAMPLE 5.2.4   Determine the equation of the parabola that satisfies the given conditions.

(a) Vertex: $(-1, 0)$; axis parallel to y-axis; passing through the point $(4, 10)$.

(b) Vertex: $(-1, 0)$; axis parallel to x-axis; passing through the point $(4, 10)$.

SOLUTION:

(a) The parabola in standard position with axis the y-axis has equation

$$y = \frac{1}{4c}x^2,$$

and the parabola with vertex at $(-1, 0)$ is given by

$$y = \frac{1}{4c}(x + 1)^2.$$

This uses the first two pieces of information. To use the third, observe that if the curve passes through $(4, 10)$, then this point must satisfy the equation for the curve, so

$$10 = \frac{1}{4c}(4+1)^2$$

$$10 = \frac{1}{4c} \cdot 25$$

$$\frac{2}{5} = \frac{1}{4c}$$

$$c = \frac{5}{8}.$$

The equation of the parabola is then

$$y = \frac{1}{4\left(\frac{5}{8}\right)}(x+1)^2$$

$$y = \frac{2}{5}(x+1)^2.$$

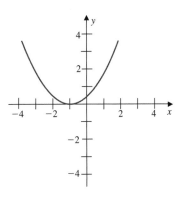

Figure 5.7

(b) The parabola in standard position with axis the x-axis has equation

$$x = \frac{1}{4c}y^2,$$

and the parabola with vertex at $(-1, 0)$ is given by

$$x + 1 = \frac{1}{4c}y^2.$$

The parabola passes through $(4, 10)$, so

$$4 + 1 = \frac{1}{4c}(10)^2$$
$$5 = \frac{1}{4c} \cdot 100 = \frac{25}{c}$$
$$c = 5.$$

The equation of the parabola is then

$$x + 1 = \frac{1}{4(5)}y^2$$
$$x + 1 = \frac{1}{20}y^2.$$

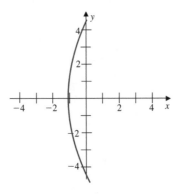

Figure 5.8

◇

EXAMPLE 5.2.5 A flash light has a parabolic cross section with a depth of 1 in. and a cross section width of 2.5 in. Where should the light source be placed to produce a parallel beam of light?

SOLUTION: Light rays emitted from the focus of a parabolic surface are reflected off the surface in parallel rays creating a concentrated beam of light. So the light source should be placed at the focus of the parabolic reflector. The information given does not allow us to find the location of the focus directly, but we can find the general form for the equation of the parabola. This will allow us to determine the value of c, which will tell us where to place the light.

The parabolic cross section of the light is shown in the figure. Since it is a parabola in standard position with axis along the x-axis, the equation has the form

$$x = \frac{1}{4c}y^2.$$

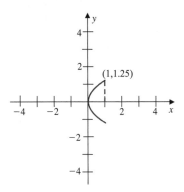

Figure 5.9

The parabola passes through the point $\left(1, \frac{1}{2}(2.5)\right) = (1, 1.25)$, so

$$1 = \frac{1}{4c}(1.25)^2 = \frac{1}{4c}\left(\frac{25}{16}\right)$$
$$\frac{64}{25} = \frac{1}{c}$$
$$c = \frac{25}{64}.$$

Therefore the focus of the parabola is $(25/64, 0)$. To produce a parallel beam of light, the light source is placed at the focus, that is,

$$\frac{25}{64} \approx 0.39 \text{ inches}$$

from the vertex of the light.   ◇

EXAMPLE 5.2.6   A rock thrown horizontally from a bridge into a river follows a parabolic curve with vertex at the bridge and axis along a perpendicular from the river to the bridge. The rock passes through a point 50 ft from the perpendicular from the river to the bridge when it is a vertical distance of 20 ft from the top. How far is the bridge above the river if the rock lands a horizontal distance of 100 ft from the perpendicular from the river to the bridge?

SOLUTION: If we knew the equation of the parabolic path of the rock, then the solution is the value of x when $y = 0$. The information gives one point on the parabola and the standard form for the parabola, which is enough to find the equation.

Set up a xy-coordinate system so that the x-axis runs out from the bridge above the path of the rock and the positive y-axis coincides with the perpendicular from the river to the bridge, as shown in the figure.

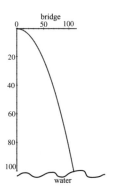

Figure 5.10

The path of the rock is a parabola in standard position with axis along the y-axis. The equation of the parabola is then

$$y = \frac{1}{4c}x^2.$$

Since the rock is 20 ft below the bridge when it is horizontally 50 ft away from the edge, the parabola passes through the point $(50, 20)$. So

$$20 = \frac{1}{4c}(50)^2$$
$$4c = \frac{2500}{20} = 125,$$

and the equation of the parabola is

$$y = \frac{1}{125}x^2.$$

The rock hits the river at the point $(100, h)$, where h is the distance from the water to the bridge. Hence

$$h = \frac{1}{125}(100)^2 = \frac{10000}{125} = 80 \text{ ft.}$$

◇

Solutions for Exercise Set 5.2

1. $y = 2x^2 = \frac{1}{4c}x^2 \Rightarrow \frac{1}{4c} = 2 \Rightarrow c = \frac{1}{8}$

    $V(0,0); F(0, 1/8)$; directrix: $y = -1/8$

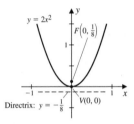

3. $9y = -16x^2 \Rightarrow y = -\frac{16}{9}x^2 \Rightarrow \frac{1}{4c} = -\frac{16}{9} \Rightarrow c = -\frac{9}{64}$

    $V(0,0); F(0, -9/64)$; directrix: $y = 9/64$

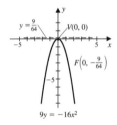

5. $y^2 = 2x \Rightarrow x = \frac{1}{2}y^2 \Rightarrow \frac{1}{4c} = \frac{1}{2} \Rightarrow c = \frac{1}{2}$

    $V(0,0); F(1/2, 0)$; directrix: $x = -1/2$

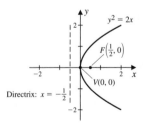

7. $9y^2 = -16x \Rightarrow x = -\frac{9}{16}y^2 \Rightarrow \frac{1}{4c} = -\frac{9}{16} \Rightarrow c = -\frac{4}{9}$

$V(0,0); F(-4/9, 0)$; directrix: $x = 4/9$

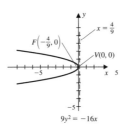

9. $x^2 - 6x + 9 = 2y \Leftrightarrow (x-3)^2 = 2y \Leftrightarrow y = \frac{1}{2}(x-3)^2 \Rightarrow \frac{1}{4c} = \frac{1}{2} \Rightarrow c = \frac{1}{2}$

$V(3,0); F(3, 1/2)$; directrix: $y = -1/2$

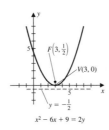

11. $x^2 - 4x - 2y + 2 = 0 \Leftrightarrow x^2 - 4x + 4 - 4 - 2y + 2 = 0$

$\Leftrightarrow (x-2)^2 = 2y + 2 \Leftrightarrow y + 1 = \frac{1}{2}(x-2)^2 \Rightarrow \frac{1}{4c} = \frac{1}{2} \Rightarrow c = \frac{1}{2}$

$V(2, -1); F(2, -1/2)$; directrix: $y = -3/2$

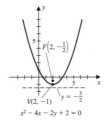

13. $y^2 - 8y + 12 = 2x \Leftrightarrow y^2 - 8y + 16 - 16 = 2x - 12 \Leftrightarrow (y-4)^2 = 2x + 4$

$\Leftrightarrow x + 2 = \frac{1}{2}(y-4)^2 \Rightarrow \frac{1}{4c} = \frac{1}{2} \Rightarrow c = \frac{1}{2}$

$V(-2,4); F(-3/2,4);$directrix: $x = -5/2$

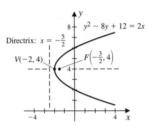

15. $2x^2 + 4x - 9y + 20 = 0 \Leftrightarrow 2(x^2 + 2x + 1 - 1) = 9y - 20$

$\Leftrightarrow 2(x+1)^2 = 9y - 18 \Leftrightarrow y - 2 = \frac{2}{9}(x+1)^2 \Rightarrow \frac{1}{4c} = \frac{2}{9} \Rightarrow c = \frac{9}{8}$

$V(-1,2); F(-1,25/8);$directrix: $y = 7/8$

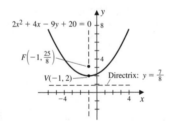

17. $3x^2 - 12x + 4y + 8 = 0 \Leftrightarrow 3(x^2 - 4x + 4 - 4) = -4y - 8$

$\Leftrightarrow 3(x-2)^2 = -4y + 4 \Leftrightarrow y - 1 = -\frac{3}{4}(x-2)^2 \Rightarrow \frac{1}{4c} = -\frac{3}{4} \Rightarrow c = -\frac{1}{3}$

$V(2,1); F(2,2/3);$directrix: $y = 4/3$

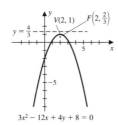

$3x^2 - 12x + 4y + 8 = 0$

19. The parabola is in standard position with axis along the y-axis, so has equation in the form $y = \frac{1}{4c}x^2$ and focus $(0, c), c > 0$. Since the point $(2, c)$ lies on the curve, $c = \frac{1}{4c}(2)^2 \Rightarrow c^2 = 1 \Rightarrow c = 1$. The equation is $y = \frac{1}{4}x^2$.

21. The parabola is in standard position with axis along the x-axis, so has equation in the form $x = \frac{1}{4c}y^2$, and since the directrix is $x = -1$, the focus is $(1, 0)$. The equation is $x = \frac{1}{4(1)}y^2 \Rightarrow x = \frac{1}{4}y^2$.

23. Focus: $(-2, 2)$; Directrix: $y = -2$

Since the vertex lies midway between the focus and the directrix, and the distance from the focus to the directrix is 4, the vertex is $(-2, 0)$. The axis of the parabola is vertical with $c = 2$, so $y = \frac{1}{8}(x + 2)^2$.

25. Vertex: $(-2, 2)$; Directrix: $x = 2$

Since the vertex lies midway between the focus and the directrix, and the distance from the directrix to the vertex is 4, the focus is $(-6, 2)$. The axis of the parabola is horizontal with $c = -4$, so $x + 2 = \frac{1}{4(-4)}(y-2)^2 \Rightarrow x + 2 = -\frac{1}{16}(y - 2)^2$.

27. Vertex: $(-2, 2)$; Focus: $(-2, 0)$

The distance between the vertex and the focus is 2, so the directrix is $y = 4$. The axis of the parabola is vertical and $c = -2$. The equation is $y - 2 = -\frac{1}{8}(x + 2)^2$.

29. Vertex: $(-2, 2)$; Focus: $(-4, 2)$

The distance between the vertex and the focus is 2, so the directrix is $x = 0$. The axis of the parabola is horizontal and $c = -2$. The equation is

$$x + 2 = -\tfrac{1}{8}(y - 2)^2.$$

31. (a) Vertex: $(0, 0)$; Point on parabola: $(4, 6)$

The axis is parallel to the y-axis, so since the vertex is at the origin, the axis is the y-axis, and the parabola has the form $y = \tfrac{1}{4c}x^2$. Since the parabola passes through the point $(4, 6)$, $6 = \tfrac{1}{4c}(4)^2 \Rightarrow 6 = \tfrac{4}{c} \Rightarrow c = \tfrac{2}{3}$, and the equation of the parabola is $y = \tfrac{1}{4(2/3)}x^2 = \tfrac{3}{8}x^2$.

(b) Vertex: $(1, 0)$; Point on parabola: $(5, 6)$

The axis is parallel to the y-axis, so since the vertex is at the point $(1, 0)$, the axis is the line $x = 1$, and the parabola has the form $y = \tfrac{1}{4c}(x - 1)^2$. Since the parabola passes through the point $(5, 6)$, $6 = \tfrac{1}{4c}(5 - 1)^2 \Rightarrow 6 = \tfrac{4}{c} \Rightarrow c = \tfrac{2}{3}$, and the equation of the parabola is $y = \tfrac{1}{4(2/3)}(x - 1)^2 = \tfrac{3}{8}(x - 1)^2$.

(c) Vertex: $(1, 2)$; Point on parabola: $(5, 8)$

The axis is parallel to the y-axis, so since the vertex is at the point $(1, 2)$, the axis is the line $x = 1$, and the parabola has the form $y - 2 = \tfrac{1}{4c}(x - 1)^2$. Since the parabola passes through the point $(5, 8)$, $8 - 2 = \tfrac{1}{4c}(5 - 1)^2 \Rightarrow 6 = \tfrac{4}{c} \Rightarrow c = \tfrac{2}{3}$, and the equation of the parabola is $y - 2 = \tfrac{3}{8}(x - 1)^2$.

(d) Vertex: $(0, 2)$; Point on parabola: $(4, 8)$

The axis is parallel to the y-axis, so since the vertex is at the point $(0, 2)$, the axis is the y-axis, and the parabola has the form $y - 2 = \tfrac{1}{4c}x^2$. Since the parabola passes through the point $(4,8), 8 - 2 = \frac{1}{4c}(4)^2 \Rightarrow 6 = \frac{4}{c} \Rightarrow$ $c = \frac{2}{3}$, and the equation of the parabola is $y - 2 = \frac{3}{8}x^2$.

33. Since the axis of the parabola is the y-axis, the vertex lies on the y-axis and hence is of the form $(0, b)$. The equation of the parabola is $y - b = \frac{1}{4c}x^2$. Since the parabola passes through $(1, 2), 2 - b = \frac{1}{4c}(1)^2 \Rightarrow$ $c = \frac{1}{4(2-b)}$, and the equation is $y - b = \frac{1}{4(1/(4(2-b)))}x^2 \Rightarrow y - b = (2 - b)x^2, b \neq 2$.

35. To describe the light, use a parabola in standard position and axis along the x-axis, so the equation has the form $x = \frac{1}{4c}y^2, c > 0$. The information implies the point $(2, 2)$ lies on the parabola, so $2 = \frac{1}{4c}(2)^2 \Rightarrow c = \frac{1}{2}$. So the focal point of the parabola is $\left(\frac{1}{2}, 0\right)$,, and to produce a parallel beam of light, the light source should be placed $\frac{1}{2}$ inch from the vertex of the light.

37. (a) The parabolic path of the ball is a parabola with axis along the y-axis and vertex at $(0, 64)$. The equation of the path has the form $y - 64 = \frac{1}{4c}x^2$. The information implies the parabola passes through the point $(100, 64 - 16) = (100, 48)$. So

$$48 - 64 = \frac{1}{4c}(100)^2 \Rightarrow 4c = -\frac{10000}{16} = -625$$

and

$$y - 64 = -\frac{1}{625}x^2.$$

To find where the ball hits the ground, find x when $y = 0$. That is,

$$-64 = -\frac{1}{625}x^2 \Rightarrow x^2 = 64(625) = 40000 \Rightarrow x = \sqrt{40000} = 200.$$

The ball hits the ground 200 feet from the building.

(b) Using the same analysis as in part (a), the equation of the parabolic path is

$$y - 1450 = -\frac{1}{625}x^2,$$

and if the distance above the ground is to be 0,

$$-1450 = -\frac{1}{625}x^2 \Rightarrow x = \sqrt{(1450)(625)} \approx 952.$$

The ball hits the ground about 952 feet from the building.

5.3 Ellipses

An *ellipse* is a set of points in a plane for which the sum of the distances from two fixed points is a given constant. The two fixed points are called the *focal points*, the line passing through the focal points is called the *axis*, and the points of intersection of the axis and the ellipse are called the *vertices*. An ellipse centered at the origin and with axis along one of the coordinate axes is said to be in *standard position*.

STANDARD POSITION ELLIPSES

Equation: $a > b$	$\frac{x^2}{a^2} + \frac{y^2}{b^2} = 1$	$\frac{y^2}{a^2} + \frac{x^2}{b^2} = 1$
Axis	x-axis	y-axis
Vertices	$(-a, 0), (a, 0)$	$(0, -a), (0, a)$
Focal Points: $c^2 = a^2 - b^2$	$(-c, 0), (c, 0)$	$(0, -c), (0, c)$
Other Intercepts	$y : (0, -b), (0, b)$	$x : (-b, 0), (b, 0)$
Center	$(0, 0)$	$(0, 0)$
Eccentricity: $c = \sqrt{a^2 - b^2}$	$e = \frac{c}{a}$	$e = \frac{c}{a}$

The *center* of an ellipse in standard position is the origin, which is the midpoint of the line segment connecting the vertices. The axis of an ellipse is also called the *major axis*, and the line segment connecting the other intercepts and perpendicular to the axis is called the *minor axis*.

If the center is shifted to the point (h, k), then the ellipse will have an equation of the form

$$\frac{(x - h)^2}{a^2} + \frac{(y - k)^2}{b^2} = 1 \quad \text{or} \quad \frac{(y - k)^2}{a^2} + \frac{(x - h)^2}{b^2} = 1.$$

EXAMPLE 5.3.1   Find the vertices and focal points and sketch the graph of the ellipse.

(a) $\frac{x^2}{16} + \frac{y^2}{9} = 1$ (b) $16x^2 + 9y^2 = 144$

SOLUTION:

(a) The equation is in standard form, and since the denominator of the x^2 term is larger than the denominator of the y^2 term, the axis of the ellipse is on the x-axis , with center at the origin. So

$$a^2 = 16, \ a = 4$$

$$b^2 = 9, \ b = 3$$

$$c = \sqrt{a^2 - b^2} = \sqrt{16 - 9} = \sqrt{7}.$$

<u>Vertices:</u> $(-a, 0) = (-4, 0), (a, 0) = (4, 0)$

<u>Focal Points:</u> $(-c, 0) = (-\sqrt{7}, 0), (c, 0) = (\sqrt{7}, 0)$

<u>y-intercepts:</u> $(0, -b) = (0, -3), (0, b) = (0, 3)$

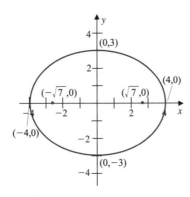

Figure 5.11

(b) First get the equation in standard form by dividing both sides of the equation by 144. This gives

$$16x^2 + 9y^2 = 144$$
$$\frac{x^2}{9} + \frac{y^2}{16} = 1.$$

Notice that the equation is similar to the equation in part (a), with the roles of a and b reversed. That is, the axis of the ellipse is the y-axis, although we still have $a = 4, b = 3$, and $c = \sqrt{7}$.

<u>Vertices:</u> $(0, -a) = (0, -4), (0, a) = (0, 4)$

<u>Focal Points:</u> $(0, -c) = (0, -\sqrt{7}), (0, c) = (0, \sqrt{7})$

<u>x-intercepts:</u> $(-b, 0) = (-3, 0), (b, 0) = (3, 0)$

Notice in both parts (a) and (b) the length of the major axis is 8, and the length of the minor axis is 6.

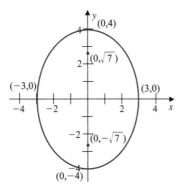

Figure 5.12

EXAMPLE 5.3.2   Find the vertices and focal points and sketch the graph of $4x^2 + 9y^2 + 8x - 36y + 4 = 0$.

SOLUTION: The key to recognizing the ellipse is to group the x terms and group the y terms and complete the square on both. So

$$4x^2 + 8x + 9y^2 - 36y = -4$$

$$4(x^2 + 2x) + 9(y^2 - 4y) = -4$$

$$4(x^2 + 2x + 1 - 1) + 9(y^2 - 4y + 4 - 4) = -4$$

$$4(x + 1)^2 + 9(y - 2)^2 = -4 + 4 + 36$$

$$4(x + 1)^2 + 9(y - 2)^2 = 36$$

$$\frac{(x + 1)^2}{9} + \frac{(y - 2)^2}{4} = 1.$$

The equation describes an ellipse that is obtained from an ellipse in standard form with the center shifted to the point $(-1, 2)$. The ellipse in standard form is

$$\frac{x^2}{9} + \frac{y^2}{4} = 1,$$

which has major axis on the x-axis, with $a = 3, b = 2$, center $(0, 0)$, vertices $(-3, 0)$ and $(3, 0)$. The focal points are $(-\sqrt{5}, 0)$ and $(\sqrt{5}, 0)$, since $c = \sqrt{a^2 - b^2} = \sqrt{9 - 4} = \sqrt{5}$.

The shifted ellipse then satisfies:

Major axis: On the horizontal line $y = 2$.

Center: The point $(-1, 2)$ obtained by shifting the origin $(0, 0)$ to the left 1 unit and upward 2 units from the origin.

Vertices: $(-a - 1, 2) = (-4, 2), (a - 1, 2) = (2, 2)$

Focal points: $(-c - 1, 2) = (-\sqrt{5} - 1, 2), (c - 1, 2) = (\sqrt{5} - 1, 2)$

<u>Minor axis:</u> On the line vertical $x = -1$.

<u>Minor axis intercepts:</u> $(-1, -b+2) = (-1, 0), (-1, b+2) = (-1, 4)$

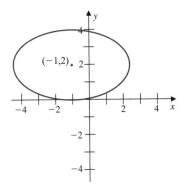

Figure 5.13

◇

EXAMPLE 5.3.3   Find an equation of the ellipse that has foci at $(3, 1)$ and $(5, 1)$ and a vertex at $(2, 1)$.

SOLUTION: Since the foci are on the line $y = 1$, the major axis of the ellipse is also on $y = 1$. We first find c, which is the distance from the midpoint of the line segment connecting the foci to a focus. This midpoint, which is also the center of the ellipse, occurs at

$$\left(\frac{5+3}{2}, \frac{1+1}{2} \right) = (4, 1),$$

and

$$c = 5 - 4 = 1.$$

Since one vertex is $(2, 1)$, which is 1 unit horizontally to the left of the focus $(3, 1)$, the other vertex is one unit to the right of the focus $(5, 1)$ and is $(6, 1)$.

The length of the major axis, which is the distance between the vertices, is 4. So a, which is half the length of the major axis, is

$$a = 2.$$

Having both a and c, we can now find b^2 and write the equation of the ellipse. So

$$c^2 = a^2 - b^2$$

$$b^2 = a^2 - c^2$$

$$b^2 = 4 - 1 = 3,$$

and the equation of the ellipse is

$$\frac{(x-4)^2}{4} + \frac{(y-1)^2}{3} = 1.$$

◇

Solutions for Exercise Set 5.3

1. $\dfrac{x^2}{4} + y^2 = 1 \Rightarrow a = 2, b = 1, c = \sqrt{4-1} = \sqrt{3}$

Focal points: $\left(\sqrt{3}, 0\right), \left(-\sqrt{3}, 0\right)$

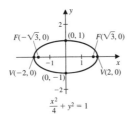

3. $x^2 + \dfrac{y^2}{9} = 1 \Rightarrow a = 3, b = 1, c = \sqrt{9-1} = \sqrt{8} = 2\sqrt{2}$

Focal points: $\left(0, 2\sqrt{2}\right), \left(0, -2\sqrt{2}\right)$

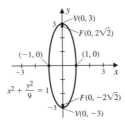

5. $16x^2 + 25y^2 = 400 \Leftrightarrow \dfrac{x^2}{25} + \dfrac{y^2}{16} = 1 \Rightarrow a = 5, b = 4, c = \sqrt{25-16} = \sqrt{9} = 3$

Focal points: $(3, 0), (-3, 0)$

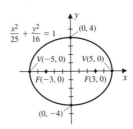

7. $3x^2 + 2y^2 = 6 \Leftrightarrow \dfrac{x^2}{2} + \dfrac{y^2}{3} = 1 \Rightarrow a = \sqrt{3}, b = \sqrt{2}, c = 1$

Focal points: $(0, 1), (0, -1)$; Vertices: $(0, \sqrt{3}), (0, -\sqrt{3})$

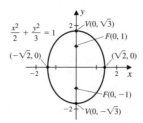

9. $4x^2 + y^2 + 16x + 12 = 0 \Leftrightarrow 4(x^2 + 4x + 4 - 4) + y^2 = -12 \Leftrightarrow$

$4(x + 2)^2 + y^2 = 4 \Leftrightarrow (x + 2)^2 + \dfrac{y^2}{4} = 1 \Rightarrow a = 2, b = 1, c = \sqrt{8 - 4} = 2$

Focal points: $\left(-2, \sqrt{3}\right), \left(-2, -\sqrt{3}\right)$; Center: $(-2, 0)$

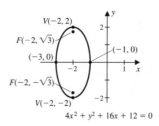

11. $x^2 + 4y^2 - 2x - 16y + 13 = 0 \Leftrightarrow x^2 - 2x + 1 - 1 + 4(y^2 - 4y + 4 - 4) = -13$

$\Leftrightarrow (x - 1)^2 + 4(y - 2)^2 = 4 \Leftrightarrow \dfrac{(x - 1)^2}{4} + (y - 2)^2 = 1$

$\Rightarrow a = 2, b = 1, c = \sqrt{4 - 1} = \sqrt{3}$

Focal points: $(1 - \sqrt{3}, 2), (1 + \sqrt{3}, 2)$; Center: $(1, 2)$

Vertices: $(-1, 2), (3, 2)$

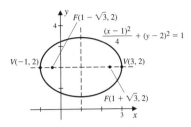

13. $2x^2 + 4y^2 + 4x - 16y + 2 = 0$

$\Leftrightarrow 2(x^2 + 2x + 1 - 1) + 4(y^2 - 4y + 4 - 4) = -2$

$\Leftrightarrow 2(x+1)^2 + 4(y-2)^2 = 16 \Leftrightarrow \dfrac{(x+1)^2}{8} + \dfrac{(y-2)^2}{4} = 1$

$\Rightarrow a = \sqrt{8} = 2\sqrt{2}, b = 2, c = \sqrt{8-4} = 2$

Focal points: $(1, 2), (-3, 2)$; Center: $(-1, 2)$

Vertices: $\left(-1 + 2\sqrt{2}, 2\right), \left(-1 - 2\sqrt{2}, 2\right)$

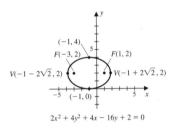

15. x-intercepts: $(\pm 4, 0)$; y-intercepts: $(0, \pm 3)$

The intercepts give the vertices of the ellipse, so the ellipse is in standard position, center at the origin, major axis along the x-axis, and $a = 4, b = 3$. The equation is $\dfrac{x^2}{16} + \dfrac{y^2}{9} = 1$.

17. Foci: $(\pm 2, 0)$; Vertices: $(\pm 3, 0)$

Since the foci are centered about the origin and on the x-axis, the ellipse is in standard position with center at the origin and major axis on the

x-axis. So

$$a = 3, c = 2, \quad \text{and} \quad c^2 = a^2 - b^2 \Rightarrow 4 = 9 - b^2 \Rightarrow b^2 = 5.$$

The equation is $\dfrac{x^2}{9} + \dfrac{y^2}{5} = 1.$

19. Foci: $(0, \pm 1)$; x-intercepts: $(\pm 2, 0)$

Since the foci are centered about the origin and on the y-axis, the ellipse is in standard position with center at the origin and major axis on the y-axis. So

$$b = 2, c = 1, \quad \text{and} \quad c^2 = a^2 - b^2 \Rightarrow 1 = a^2 - 4 \Rightarrow a^2 = 5.$$

The equation is $\dfrac{y^2}{5} + \dfrac{x^2}{4} = 1.$

21. Length of major axis 5; Length of minor axis 3; Foci on the x-axis;

Then $a = \dfrac{5}{2}, b = \dfrac{3}{2}$ and equation is $\dfrac{x^2}{25/4} + \dfrac{y^2}{9/4} = 1 \Leftrightarrow \dfrac{4x^2}{25} + \dfrac{4y^2}{9} = 1.$

23. Foci: $(3, 0), (1, 0)$; Vertex: $(0, 0)$

Since the foci are on the x-axis, the major axis of the ellipse is on the x-axis. The foci are centered about $(2, 0)$, so $c = 1$. Since a vertex is $(0, 0)$, the other vertex is $(4, 0)$, and the length of the major axis is 4. This gives $a = 2$. Then

$$c^2 = a^2 - b^2 \Rightarrow 1 = 4 - b^2 \Rightarrow b^2 = 3.$$

The equation is $\dfrac{(x - 2)^2}{4} + \dfrac{y^2}{3} = 1.$

25. Focus: $(-4, 0)$; Vertices: $(-4, -2), (-4, 8)$

Since the foci are on the vertical line $x = -4$, the major axis of the ellipse is on the line $x = -4$. The vertices are 10 units apart which is the length of the major axis, so $a = 5$. The center is $(-4, 3)$, so $c = 3$. Then

$$c^2 = a^2 - b^2 \Rightarrow 9 = 25 - b^2 \Rightarrow b^2 = 16.$$

The equation is $\dfrac{(x+4)^2}{16} + \dfrac{(y-3)^2}{25} = 1.$

27. Vertices: $(3,3), (3,-1)$; Passing through: $(2,1)$

The center is midway between the two vertices, so is $(3,1)$, and $a = 4$.

The ellipse has the form

$$\frac{(x-3)^2}{b^2} + \frac{(y-1)^2}{16} = 1.$$

If the ellipse passes through $(2,1)$, then

$$\frac{(2-3)^2}{b^2} + \frac{(1-1)^2}{16} = 1 \Rightarrow b^2 = 1.$$

The equation is $(x-3)^2 + \dfrac{(y-1)^2}{16} = 1.$

29. Let P be the point of intersection of the latus rectum and the upper half of the ellipse. The length of the latus rectum is then $2y$, where the coordinates of $P = (c, y)$. Since the equation of the ellipse is

$$\frac{x^2}{a^2} + \frac{y^2}{b^2} = 1, \Rightarrow \frac{y^2}{b^2} = 1 - \frac{x^2}{a^2} \Rightarrow y = \pm\sqrt{b^2\left(1 - \frac{x^2}{a^2}\right)}.$$

Letting $x = c$, multiplying by 2, and noting that $b^2 = a^2 - c^2$, the length of the latus rectum is

$$2\sqrt{b^2\left(1 - \frac{c^2}{a^2}\right)} = 2\sqrt{b^2\left(\frac{a^2 - c^2}{a^2}\right)} = 2\sqrt{\frac{b^4}{a^2}} = \frac{2b^2}{a}.$$

31. Since the major axis has length 480,

$$2a = 480 \Rightarrow a = 240 \Rightarrow a^2 = 57600,$$

and since the minor axis has length 280,

$$2b = 280 \Rightarrow b = 140 \Rightarrow b^2 = 19600.$$

The equation is $\dfrac{x^2}{57600} + \dfrac{y^2}{19600} = 1.$

33.   The length of the major axis of the satellites orbit is the diameter of the earth, 12760, plus 160 plus 16000, which equals 28920. Then $a = \frac{28920}{2} = 14460 \Rightarrow c = 14460 - 6540 = 7920 \Rightarrow e = \frac{c}{a} = \frac{7920}{14460} \approx 0.6$. Then $b^2 = a^2 - c^2 \approx 146365200$. Since $a^2 = 14460^2 = 209091600$ the equation of the orbit is

$$\frac{x^2}{209091600} + \frac{y^2}{14365200} = 1.$$

5.4 Hyperbolas

A *hyperbola* is a set of points in the plane for which the magnitude of the difference between the distances from two fixed points is a given constant. The two fixed points are called the *foci*, the line passing through the focal points is called the *axis*, and the points of intersection of the axis with the hyperbola are called the *vertices*. A hyperbola centered at the origin and with axis along one of the coordinate axes is in *standard position*.

STANDARD POSITION HYPERBOLAS

Equation	$\frac{x^2}{a^2} - \frac{y^2}{b^2} = 1$	$\frac{y^2}{a^2} - \frac{x^2}{b^2} = 1$
Axis	x-axis	y-axis
Vertices	$(-a, 0), (a, 0)$	$(0, -a), (0, a)$
Focal points: $c^2 = a^2 + b^2$	$(-c, 0), (c, 0)$	$(0, -c), (0, c)$
Center	$(0, 0)$	$(0, 0)$
Asymptotes	$y = \pm \frac{b}{a} x$	$y = \pm \frac{a}{b} x$
Eccentricity: $c = \sqrt{a^2 + b^2}$	$e = \frac{c}{a}$	$e = \frac{c}{a}$

If the center of the hyperbola in standard position is shifted to the point (h, k), then the hyperbola will have an equation of the form

$$\frac{(x-h)^2}{a^2} - \frac{(y-k)^2}{b^2} = 1 \text{ or } \frac{(y-k)^2}{a^2} - \frac{(x-h)^2}{b^2} = 1,$$

and the asymptotes become

$$y - k = \pm \frac{b}{a}(x - h) \text{ or } y - k = \pm \frac{a}{b}(x - h).$$

EXAMPLE 5.4.1   Find the vertices, focal points, eccentricity, and equations of the asymptotes, and sketch the graph of the hyperbola.

(a) $\frac{x^2}{9} - \frac{y^2}{16} = 1$ (b) $4y^2 - 25x^2 = 100$

SOLUTION:

(a) The equation is in standard form. Since the y^2 term is negative, the axis of the hyperbola is along the x-axis with center at the origin. So

$$a^2 = 9, \ a = 3$$

$$b^2 = 16, \ b = 4$$

$$c = \sqrt{a^2 + b^2} = \sqrt{9 + 16} = 5.$$

Vertices: $(-a, 0) = (-3, 0), (a, 0) = (3, 0)$

Focal points: $(-c, 0) = (-5, 0), (c, 0) = (5, 0)$

Eccentricity: $e = \frac{c}{a} = \frac{5}{3}$

Asymptotes: $y = \pm\frac{b}{a}x = \pm\frac{4}{3}x$

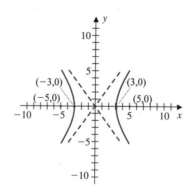

Figure 5.14

(b) First place the equation in standard form by dividing both sides of the equation by 100. This gives

$$4y^2 - 25x^2 = 100$$
$$\frac{y^2}{25} - \frac{x^2}{4} = 1.$$

Since the x^2 term is negative, the axis is along the y-axis with center at the origin. So

$$a^2 = 25, \ a = 5$$
$$b^2 = 4, \ b = 2$$
$$c = \sqrt{a^2 + b^2} = \sqrt{29}.$$

<u>Vertices:</u> $(0, -a) = (0, -5), (0, a) = (0, 5)$

<u>Focal points:</u> $(0, -c) = (0, -\sqrt{29}), (0, c) = (0, \sqrt{29})$

<u>Eccentricity:</u> $e = \frac{c}{a} = \frac{\sqrt{29}}{5}$

<u>Asymptotes:</u> $y = \pm\frac{a}{b}x = \pm\frac{5}{2}x$

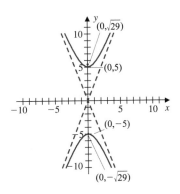

Figure 5.15

EXAMPLE 5.4.2   Find the vertices, focal points, eccentricity, and equations of the asymptotes, and sketch the graph of the hyperbola $9x^2 - 4y^2 + 18x + 8y = 31$.

SOLUTION: To compare the equation with the equation of a hyperbola in standard form, complete the square on both the x and the y terms. So

$$9x^2 - 4y^2 + 18x + 8y = 31$$

$$9x^2 + 18x - 4y^2 + 8y = 31$$

$$9(x^2 + 2x) - 4(y^2 - 2y) = 31$$

$$9(x^2 + 2x + 1 - 1) - 4(y^2 - 2y + 1 - 1) = 31$$

$$9(x + 1)^2 - 9 - 4(y - 1)^2 + 4 = 31$$

$$9(x + 1)^2 - 4(y - 1)^2 = 36$$

$$\frac{(x + 1)^2}{4} - \frac{(y - 1)^2}{9} = 1.$$

The graph of the hyperbola is obtained from shifting the graph of the hyperbola in standard position, $\frac{x^2}{4} - \frac{y^2}{9} = 1$, to the left 1 unit and upward 1 unit. The standard position hyperbola has $a = 2, b = 3, c = \sqrt{13}$, vertices $(-2, 0), (2, 0)$, focal points $(-\sqrt{13}, 0), (\sqrt{13}, 0)$, and asymptotes $y = \pm\frac{b}{a}x = \pm\frac{3}{2}x$. The hyperbola in this problem has the following properties.

Vertices: $(-a - 1, 1) = (-3, 1), (a - 1, 1) = (1, 1)$

Center: $(-1, 1)$

Focal points: $(-c - 1, 1) = (-\sqrt{13} - 1, 1), (c - 1, 1) = (\sqrt{13} - 1, 1)$

Eccentricity: $e = \frac{c}{a} = \frac{\sqrt{13}}{2}$

Asymptotes:

$$y - 1 = \pm \frac{b}{a}(x+1)$$

$$y - 1 = \pm \frac{3}{2}(x+1)$$

$$y = \frac{3}{2}x + \frac{5}{2}, y = -\frac{3}{2}x - \frac{1}{2}$$

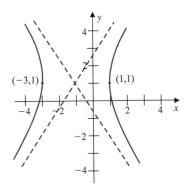

Figure 5.16

◇

EXAMPLE 5.4.3   Find an equation of the hyperbola that satisfies the stated conditions.

(a) Foci at $(\pm 7, 0)$, vertices at $(\pm 3, 0)$.

(b) Foci at $(6, 1)$ and $(-2, 1)$, equations of asymptotes $y = \frac{3}{4}x - \frac{1}{2}$ and $y = -\frac{3}{4}x + \frac{5}{2}$.

SOLUTION:

(a) Since the foci and vertices are on the x-axis and centered about the origin, the hyperbola is in standard form with axis on the x-axis and equation of the form

$$\frac{x^2}{a^2} - \frac{y^2}{b^2} = 1.$$

The vertices are $(\pm a, 0) = (\pm 3, 0)$, so $a = 3$. The foci are $(\pm c, 0) = (\pm 7, 0)$, so $c = 7$. To find the equation we need the value b^2, which we obtain from the equation

$$c^2 = a^2 + b^2$$

$$b^2 = c^2 - a^2$$

$$= 49 - 9$$

$$= 40.$$

The equation of the hyperbola is

$$\frac{x^2}{9} - \frac{y^2}{40} = 1.$$

(b) Since the foci both lie on the horizontal line $y = 1$, the axis of the hyperbola is parallel to the x-axis and the equation has the form

$$\frac{(x - h)^2}{a^2} - \frac{(y - k)^2}{b^2} = 1,$$

where (h, k) is the center. The center is midway between the foci, so the center is the point

$$\left(\frac{6 - 2}{2}, \frac{1 + 1}{2} \right) = (2, 1),$$

and the equation can be written as

$$\frac{(x - 2)^2}{a^2} - \frac{(y - 1)^2}{b^2} = 1.$$

To find a and b use the information from the asymptotes. That is, the asymptotes are of the form

$$y - 1 = \pm \frac{b}{a}(x - 2),$$

so

$$\frac{b}{a} = \frac{3}{4}$$
$$b = \frac{3a}{4}.$$

Another equation is needed to find the two unknowns a and b. But the value c is half the distance between the foci, so

$$c = \frac{1}{2} \cdot 8 = 4,$$

and

$$c^2 = a^2 + b^2$$
$$16 = a^2 + \left(\frac{3a}{4}\right)^2$$
$$16 = a^2 + \frac{9a^2}{16}$$
$$16 = \frac{25a^2}{16}$$
$$a^2 = \frac{256}{25}$$
$$b^2 = \frac{9a^2}{16} = \frac{9}{16} \cdot \frac{256}{25} = \frac{144}{25}.$$

The equation of the hyperbola is

$$\frac{25(x - 2)^2}{256} - \frac{25(y - 1)^2}{144} = 1.$$

◇

Solutions for Exercise Set 5.4

1. $\dfrac{x^2}{4} - \dfrac{y^2}{9} = 1 \Rightarrow a = 2, b = 3 \Rightarrow c^2 = a^2 + b^2 = 13 \Rightarrow c = \sqrt{13};$

Vertices: $(2,0), (-2,0);$ Foci: $\left(\sqrt{13},0\right), \left(-\sqrt{13},0\right);$

Asymptotes: $y = \pm\dfrac{3}{2}x$

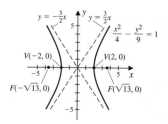

3. $\dfrac{y^2}{4} - \dfrac{x^2}{9} = 1 \Rightarrow a = 2, b = 3 \Rightarrow c^2 = 13 \Rightarrow c = \sqrt{13};$

Vertices: $(0,2), (0,-2);$ Foci: $\left(0,\sqrt{13}\right), \left(0,-\sqrt{13}\right);$

Asymptotes: $y = \pm\dfrac{2}{3}x$

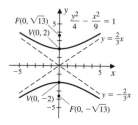

5. $x^2 - y^2 = 1 \Rightarrow a = 1, b = 1 \Rightarrow c^2 = 2 \Rightarrow c = \sqrt{2};$

Vertices: $(1,0), (-1,0);$ Foci: $\left(\sqrt{2},0\right), \left(-\sqrt{2},0\right);$

Asymptotes: $y = \pm x$

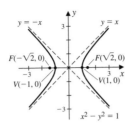

7. We have $9y^2 - 18y - 4x^2 = 27 \Leftrightarrow 9(y^2 - 2y + 1 - 1) - 4x^2 = 27$
$\Leftrightarrow 9(y-1)^2 - 4x^2 = 36 \Leftrightarrow \dfrac{(y-1)^2}{4} - \dfrac{x^2}{9} = 1.$
So $a = 2, b = 3 \Rightarrow c^2 = 13 \Rightarrow c = \sqrt{13}.$ Vertices: $(0,3), (0,-1);$
Foci: $\left(0, 1 + \sqrt{13}\right), \left(0, 1 - \sqrt{13}\right);$
Asymptotes: $y - 1 = \pm\dfrac{2}{3}x \Leftrightarrow y = \pm\dfrac{2}{3}x + 1.$

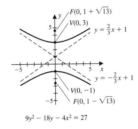

9. We have $3x^2 - y^2 = 6x \Leftrightarrow 3(x^2 - 2x + 1 - 1) - y^2 = 0 \Leftrightarrow 3(x-1)^2 - y^2 = 3$
$\Leftrightarrow (x-1)^2 - \dfrac{y^2}{3} = 1.$ So $a = 1, b = \sqrt{3} \Rightarrow c^2 = 4 \Rightarrow c = 2.$
Vertices: $(0,0), (2,0);$ Foci: $(3,0), (-1,0);$
Asymptotes: $y = \pm\sqrt{3}(x-1)$

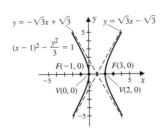

11. $9x^2 - 4y^2 - 18x - 8y = 31 \Leftrightarrow 9(x^2 - 2x + 1 - 1) - 4(y^2 + 2y + 1 - 1) = 31$
$\Leftrightarrow 9(x - 1)^2 - 4(y + 1)^2 = 36 \Leftrightarrow \dfrac{(x - 1)^2}{4} - \dfrac{(y + 1)^2}{9} = 1.$
So $a = 2, b = 3 \Rightarrow c^2 = 13 \Rightarrow c = \sqrt{13}.$
Vertices: $(3, -1), (-1, -1)$; Foci: $\left(1 - \sqrt{13}, -1\right), \left(1 + \sqrt{13}, -1\right)$;
Asymptotes: $y + 1 = \pm\dfrac{3}{2}(x - 1)$

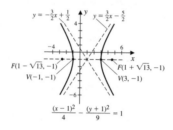

13. We have $9y^2 - 4x^2 - 36y + 16x - 16 = 0 \Leftrightarrow 9(y^2 - 4y + 4 - 4) - 4(x^2 - 4x + 4 - 4) = 16$
$\Leftrightarrow 9(y - 2)^2 - 4(x - 2)^2 = 36 \Leftrightarrow \dfrac{(y - 2)^2}{4} - \dfrac{(x - 2)^2}{9} = 1.$
So $a = 2, b = 3 \Rightarrow c = \sqrt{13}.$
Vertices: $(2, 4), (2, 0)$; Foci: $\left(2, 2 + \sqrt{13}\right), \left(2, 2 - \sqrt{13}\right)$
Asymptotes: $y - 2 = \pm\dfrac{2}{3}(x - 2)$

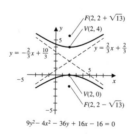

15. Foci: $(\pm 5, 0)$; Vertices: $(\pm 3, 0)$

Since the foci are centered about the origin and on the x-axis, the axis is on the x-axis, $a = 3, c = 5$, and

$$c^2 = a^2 + b^2 \Rightarrow 25 = 9 + b^2 \Rightarrow b^2 = 16.$$

The equation is $\dfrac{x^2}{9} - \dfrac{y^2}{16} = 1$.

17. Foci: $(0, \pm 5)$; Vertices: $(0, \pm 4)$

Since the foci are centered about the origin and on the y-axis, the axis is on the y-axis, $a = 4, c = 5$, and

$$c^2 = a^2 + b^2 \Rightarrow 25 = 16 + b^2 \Rightarrow b^2 = 9.$$

The equation is $\dfrac{y^2}{16} - \dfrac{x^2}{9} = 1$.

19. Focus: $(2, 2)$; Vertices: $(2, 1), (2, -3)$

Since the focus and the vertices are on the line $x = 2$, the axis is parallel to the y-axis. Since the vertices are centered about $(2, -1)$, $a = 2$ and since the focus $(2, 2)$ is 3 units above the center $(2, -1)$, $c = 3$. Then

$$c^2 = a^2 + b^2 \Rightarrow 9 = 4 + b^2 \Rightarrow b^2 = 5.$$

The equation is $\dfrac{(y + 1)^2}{4} - \dfrac{(x - 2)^2}{5} = 1$.

21. Foci: $(-1, 4), (5, 4)$; Vertex: $(0, 4)$

Since the foci and the one vertex are on the line $y = 4$, the axis is parallel to the x-axis. Since the foci are centered about the point $(2, 4)$, $c = 3$ and since the vertex $(0, 4)$ is 2 units to the left of the center $(2, 4)$, $a = 2$. Then

$$c^2 = a^2 + b^2 \Rightarrow 9 = 4 + b^2 \Rightarrow b^2 = 5.$$

The equation is $\dfrac{(x - 2)^2}{4} - \dfrac{(y - 4)^2}{5} = 1$.

23. Vertices: $(0, \pm 2)$; Passing through: $(3, 4)$

Since the vertices are centered about the origin and on the y-axis, the hyperbola is in standard position with axis on the y-axis and has the form $\dfrac{y^2}{a^2} - \dfrac{x^2}{b^2} = 1$. Then $a = 2$, and since the hyperbola passes through $(3, 4)$, we have

$$\frac{(4)^2}{4} - \frac{(3)^2}{b^2} = 1 \Rightarrow \frac{9}{b^2} = 3 \Rightarrow b = \sqrt{3}.$$

The equation is $\dfrac{y^2}{4} - \dfrac{x^2}{3} = 1$.

25. Vertices: $(\pm 3, 0)$; Asymptotes: $y = \pm\frac{4}{3}x$.

Since the vertices are centered about the origin on the x-axis, the hyperbola is in standard position with axis on the x-axis. The vertices imply $a = 3$, and the asymptotes imply,

$$\frac{b}{a} = \frac{4}{3} \Rightarrow \frac{b}{3} = \frac{4}{3} \Rightarrow b = 4.$$

The equation is $\dfrac{x^2}{9} - \dfrac{y^2}{16} = 1$.

27. Let $P(x, y)$ be the point of intersection of the latus rectum and the hyperbola as shown in the figure. The length is then $2y$, where the coordinates of $P = (c, y)$. Since the equation of the hyperbola is

$$\frac{x^2}{a^2} - \frac{y^2}{b^2} = 1, \Rightarrow \frac{y^2}{b^2} = \frac{x^2}{a^2} - 1 \Rightarrow y = \pm\sqrt{b^2\left(\frac{x^2}{a^2} - 1\right)}.$$

Letting $x = c$, multiplying by 2, and noting that $b^2 = c^2 - a^2$, the length of the latus rectum is

$$2\sqrt{b^2\left(\frac{c^2}{a^2} - 1\right)} = 2\sqrt{b^2\left(\frac{c^2 - a^2}{a^2}\right)} = 2\sqrt{\frac{b^4}{a^2}} = \frac{2b^2}{a}.$$

29. The axes of the hyperbolas are perpendicular to each other.

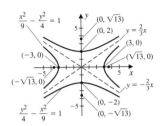

31. If the total cost of production and delivery from plants A and B to a destination C are denoted T_A and T_B, then

$$T_A - T_B = 130 + d(A, C) - d(B, C).$$

If $d(B, C) = d(A, C) + 130$, then $T_A - T_B = 0$ and it makes no difference from which plant the car is shipped. If $d(B, C) < d(A, C) + 130$, then

$$T_A - T_B > 0 \Rightarrow T_A > T_B,$$

and the car should be shipped from plant B. If $d(B, C) > d(A, C) + 130$, then

$$T_A - T_B < 0 \Rightarrow T_A < T_B,$$

and the car should be shipped from plant A.

5.5 Polar Coordinates

In the rectangular coordinate system, a point in the plane is specified by an ordered pair (x, y) that describes the location of the point using a rectangular grid. The first coordinate specifies the vertical distance to the x-axis along the line perpendicular to the x-axis, and the second coordinate specifies the horizontal distance to the y-axis along the line perpendicular to the y-axis.

The *polar coordinate* system represents a point in the plane as an ordered pair (r, θ). The location of the point uses a distance, r, from the point to a fixed point called the *pole*, and the angle, θ, made by the ray from the pole to the point and a fixed half ray extending from the pole. This fixed ray is called the *polar axis*.

It is also convenient to allow negative entries for the first coordinate of a point given in polar coordinates. If $r > 0$, then

$$(r, \theta) \quad \text{and} \quad (-r, \theta + \pi) \quad \text{represent the same point.}$$

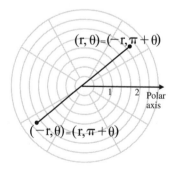

Figure 5.17

That is, the point $(-r, \theta)$, is obtained by reflecting the point (r, θ) through the origin.

EXAMPLE 5.5.1   Plot the point with the given polar coordinates. Then give two other pairs of polar coordinates that represent the point, one with $r > 0$ and one with $r < 0$.

(a) $\left(2, \frac{\pi}{4}\right)$ (b) $\left(-3, \frac{\pi}{3}\right)$ (c) $\left(1, -\frac{\pi}{6}\right)$

SOLUTION:

(a) Rotate the polar axis $\frac{\pi}{4}$ radians, or $45°$, in the counterclockwise direction, and place the point on the rotated ray 2 units from the pole.

The polar representation of a point in the plane is *not* unique, since every additional rotation of 2π radians, clockwise or counterclockwise, ends up at the same point. So, another representation of the point is given by

$$\left(2, \frac{\pi}{4} + 2\pi\right) = \left(2, \frac{9\pi}{4}\right).$$

To find a representation of the point with $r < 0$, first find the point with $r > 0$, that is, the reflection through the origin of $\left(2, \frac{\pi}{4}\right)$. Then make the first coordinate negative. So,

$$\left(2, \frac{\pi}{4} + \pi\right) = \left(2, \frac{5\pi}{4}\right) \text{ is the point opposite } \left(2, \frac{\pi}{4}\right),$$

and

$$\left(-2, \frac{5\pi}{4}\right) \text{ is opposite } \left(2, \frac{5\pi}{4}\right) \text{ and so represents } \left(2, \frac{\pi}{4}\right).$$

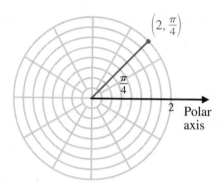

Figure 5.18

(b) Possibilities are:

<u>Second Representation, $r < 0$</u> : $\left(-3, \frac{\pi}{3} + 2\pi\right) = \left(-3, \frac{7\pi}{3}\right)$

<u>Third Representation, $r > 0$</u> : $\left(3, \frac{\pi}{3} + \pi\right) = \left(3, \frac{4\pi}{3}\right)$

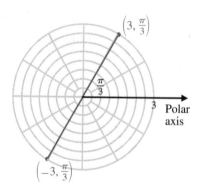

Figure 5.19

(c) The negative angle simply means rotate the polar axis in the clockwise direction to find the point. Subtracting 2π makes an additional rotation in the clockwise direction.

<u>Second Representation, $r > 0$</u> : $\left(1, -\frac{\pi}{6} - 2\pi\right) = \left(1, -\frac{13\pi}{6}\right)$

<u>Third Representation, $r < 0$:</u> $\left(-1, -\frac{\pi}{6} + \pi\right) = \left(-1, \frac{5\pi}{6}\right)$

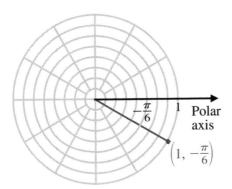

Figure 5.20

POLAR COORDINATES TO RECTANGULAR COORDINATES

If the polar coordinates (r, θ) of a point in the plane are given, then the rectangular coordinates of the same point are

$$x = r \cos \theta \quad \text{and} \quad y = r \sin \theta.$$

These relationships are immediate consequences of the definitions of the cosine and sine of the angle θ.

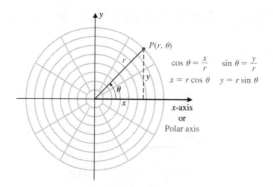

Figure 5.21

EXAMPLE 5.5.2   Convert the polar coordinates to rectangular coordinates.

(a) $\left(1, \frac{\pi}{3}\right)$ (b) $\left(-3, -\frac{7\pi}{6}\right)$

SOLUTION:

(a) Since $r = 1$ and $\theta = \frac{\pi}{3}$,

$$x = r \cos \theta = 1 \cos \frac{\pi}{3} = \frac{1}{2}$$

$$y = r \sin \theta = 1 \sin \frac{\pi}{3} = \frac{\sqrt{3}}{2}.$$

(b) Since the reference angle for $-\frac{7\pi}{6}$ is $\frac{\pi}{6}$ and $-\frac{7\pi}{6}$ is in quadrant II,

$$\cos\left(-\frac{7\pi}{6}\right) = -\cos\frac{\pi}{6} \quad \text{and} \quad \sin\left(-\frac{7\pi}{6}\right) = \sin\frac{\pi}{6}.$$

So

$$x = -3\cos\left(-\frac{7\pi}{6}\right) = -3\left(-\cos\frac{\pi}{6}\right) = -3\left(-\frac{\sqrt{3}}{2}\right) = \frac{3\sqrt{3}}{2}$$

$$y = -3\sin\left(-\frac{7\pi}{6}\right) = -3\sin\frac{\pi}{6} = -3 \cdot \frac{1}{2} = -\frac{3}{2}.$$

◇

EXAMPLE 5.5.3   Convert the polar equation to a rectangular equation.

(a) $r = 4\sin\theta$ (b) $r^2 = \cos 2\theta$

SOLUTION: The terms to look for are $x = r\cos\theta$, $y = r\sin\theta$ and $r^2 = x^2 + y^2$. This last relationship is just the Pythagorean Theorem. If these terms are not present

multiply both sides of the equation by r,

then make the replacements. So,

$$r = 4\sin\theta$$

$$r^2 = 4r\sin\theta$$

$$x^2 + y^2 = 4y$$

$$x^2 + y^2 - 4y = 0.$$

We recognize this as the equation of a circle, but it is not in standard form. Complete the square on the x term. Then,

$$x^2 + y^2 - 4y = 0$$

$$x^2 + y^2 - 4y + 4 = 4$$

$$x^2 + (y - 2)^2 = 4,$$

and this is the equation of the circle with radius 2 and center at the point $(0, 2)$.

(b) We can immediately make the substitution $r^2 = x^2 + y^2$ and get

$$r^2 = \cos 2\theta$$

$$x^2 + y^2 = \cos 2\theta.$$

In the current form there is no way to replace the $\cos 2\theta$ term, but there is a trigonometric identity that is useful. If we rewrite the $\cos 2\theta$ in terms of just cosines and sines, then we can use the substitutions $x = r\cos\theta$ and $y = r\sin\theta$. The identity we need is

$$\cos 2\theta = (\cos\theta)^2 - (\sin\theta)^2.$$

Then

$$r^2 = \cos 2\theta$$

$$x^2 + y^2 = \cos 2\theta$$

$$x^2 + y^2 = (\cos\theta)^2 - (\sin\theta)^2$$

$$= \left(\frac{x}{r}\right)^2 - \left(\frac{y}{r}\right)^2$$

$$x^2 + y^2 = \frac{x^2}{r^2} - \frac{y^2}{r^2}$$

$$= \frac{x^2}{x^2 + y^2} - \frac{y^2}{x^2 + y^2}$$

$$\left(x^2 + y^2\right)^2 = x^2 - y^2.$$

◇

RECTANGULAR COORDINATES TO POLAR COORDINATES

If the rectangular coordinates (x, y) of a point in the plane are given, then one set polar coordinates of the same point is obtained from

$$\tan\theta = \frac{y}{x}, \ x \neq 0, \ r^2 = x^2 + y^2,$$

where the sign of r is chosen to ensure that the point is in the correct quadrant. When $x = 0$, we have $\theta = \frac{\pi}{2}$ and $r = y$.

EXAMPLE 5.5.4   Convert the rectangular coordinates to polar coordinates.

(a) $(-2, 2)$ (b) $(3\sqrt{3}, 3)$

SOLUTION:

(a) Since $x = -2$ and $y = 2$,

$$\tan \theta = \frac{2}{-2} = -1,$$

and so

$$\theta = \frac{3\pi}{4} \quad \text{or} \quad \theta = -\frac{\pi}{4}.$$

To find r

$$r^2 = x^2 + y^2$$
$$= (-2)^2 + (2)^2 = 8$$
$$r = \pm\sqrt{8} = \pm 2\sqrt{2}.$$

The point $(-2, 2)$ lies in quadrant II, and so two different representations can be given as

$$\left(2\sqrt{2}, \frac{3\pi}{4}\right) \quad \text{and} \quad \left(-2\sqrt{2}, -\frac{\pi}{4}\right).$$

(b) Since

$$\tan \theta = \frac{3}{3\sqrt{3}} = \frac{1}{\sqrt{3}}$$
$$= \frac{\sqrt{3}}{3},$$

we have

$$\theta = \frac{\pi}{6} \text{ or } \frac{7\pi}{6},$$

and

$$r^2 = \left(3\sqrt{3}\right)^2 + 3^2$$
$$= 36.$$

So

$$r = \pm 6.$$

Since the point $(3\sqrt{3}, 3)$ lies in quadrant I, two possible polar representations are

$$\left(6, \frac{\pi}{6}\right), \ \left(-6, \frac{7\pi}{6}\right).$$

◇

EXAMPLE 5.5.5   Convert the rectangular equation to polar coordinates.

(a) $x^2 + y^2 = 2x$ (b) $y = x^2$

SOLUTION:

(a) Using the relations $r^2 = x^2 + y^2$ and $x = r\cos\theta$,

$$x^2 + y^2 = 2x$$
$$r^2 = 2r\cos\theta$$
$$r = 2\cos\theta.$$

(b) Using the relations $x = r\cos\theta$ and $y = r\sin\theta$,

$$y = x^2$$
$$r\sin\theta = (r\cos\theta)^2$$
$$= r^2(\cos\theta)^2,$$

and

$$r = \frac{\sin\theta}{(\cos\theta)^2}$$
$$= \frac{\sin\theta}{\cos\theta} \cdot \frac{1}{\cos\theta}.$$

So

$$r = \tan \theta \sec \theta.$$

◇

GRAPHS OF POLAR EQUATIONS

The graph of a polar equation $r = f(\theta)$ is the collection of all points that have at least one polar representation that satisfies the equation.

Circles	$r = a, r = a\cos\theta, r = a\sin\theta$
Lines	$\theta = a$
Cardioid	$r = a + a\cos\theta, r = a + a\sin\theta$
Limacon without a loop: $a > b$	$r = a + b\cos\theta, r = a + b\sin\theta$
Limacon with a loop: $a < b$	$r = a + b\cos\theta, r = a + b\sin\theta$
Lemniscates	$r^2 = a^2\cos 2\theta, r^2 = a^2\sin 2\theta$
Roses	$r = a\cos n\theta, r = a\sin n\theta,$ leafs$=\begin{cases} 2n, & \text{if } n \text{ is even} \\ n, & \text{if } n \text{ is odd} \end{cases}$

EXAMPLE 5.5.6   Sketch the graph of the polar equation.

(a) $r = 3\cos\theta$ (b) $r = 1 + \sin\theta$

SOLUTION:

(a) Since the period of the cosine function is 2π, the entire curve will be traced if θ varies between 0 and 2π.

θ	0	$\frac{\pi}{6}$	$\frac{\pi}{4}$	$\frac{\pi}{3}$	$\frac{\pi}{2}$	$\frac{2\pi}{3}$	$\frac{3\pi}{4}$	$\frac{5\pi}{6}$	π
$\cos\theta$	1	$\frac{\sqrt{3}}{2}$	$\frac{\sqrt{2}}{2}$	$\frac{1}{2}$	0	$-\frac{1}{2}$	$-\frac{\sqrt{2}}{2}$	$-\frac{\sqrt{3}}{2}$	-1
$r = 3\cos\theta$	3	$\frac{3\sqrt{3}}{2}$	$\frac{3\sqrt{2}}{2}$	$\frac{3}{2}$	0	$-\frac{3}{2}$	$-\frac{3\sqrt{2}}{2}$	$-\frac{3\sqrt{3}}{2}$	-3
θ	$\frac{7\pi}{6}$	$\frac{5\pi}{4}$	$\frac{4\pi}{3}$	$\frac{3\pi}{2}$	$\frac{5\pi}{3}$	$\frac{7\pi}{4}$	$\frac{11\pi}{6}$	2π	
$\cos\theta$	$-\frac{\sqrt{3}}{2}$	$-\frac{\sqrt{2}}{2}$	$-\frac{1}{2}$	0	$\frac{1}{2}$	$\frac{\sqrt{2}}{2}$	$\frac{\sqrt{3}}{2}$	1	
$r = 3\cos\theta$	$-\frac{3\sqrt{3}}{2}$	$-\frac{3\sqrt{2}}{2}$	$-\frac{3}{2}$	0	$\frac{3}{2}$	$\frac{3\sqrt{2}}{2}$	$\frac{3\sqrt{3}}{2}$	3	

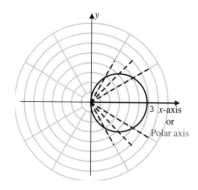

Figure 5.22

Notice how the points with $r < 0$ are reflected through the origin before they are plotted. Also, the graph is symmetric with respect to the polar axis, and in this example it would have been sufficient to only plot θ ranging from 0 to π. The curve appears to be the circle with center $\left(\frac{3}{2}, 0\right)$ and radius $\frac{3}{2}$. To see this algebraically, we can convert the polar equation to a rectangular equation.

That is,

$$r = 3\cos\theta$$

$$r^2 = 3r\cos\theta$$

$$x^2 + y^2 = 3x$$

$$x^2 - 3x + y^2 = 0$$

$$x^2 - 3x + \frac{9}{4} + y^2 = \frac{9}{4}$$

$$\left(x - \frac{3}{2}\right)^2 + y^2 = \left(\frac{3}{2}\right)^2.$$

(b) We could plot a number of points as in part (a), but another way to construct the graph is to use the values on the sine curve for $0 \le \theta \le 2\pi$.   ◇

θ	$\sin\theta$	$1 + \sin\theta$
Increases from 0 to $\frac{\pi}{2}$	Increases from 0 to 1	Increases from 1 to 2
Increases from 0 to π	Decreases from 1 to 0	Decreases from 2 to 1
Increases from π to $\frac{3\pi}{2}$	Decreases from 0 to -1	Decreases from 1 to 0
Increases from $\frac{3\pi}{2}$ to 2π	Increases from -1 to 0	Increases from 0 to 1

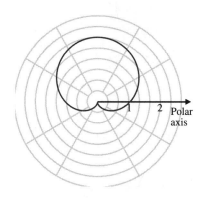

Figure 5.23

INTERSECTION OF POLAR CURVES

EXAMPLE 5.5.7   Find the polar and rectangular coordinates of all points of intersection of the curves $r = 1 + \sin\theta$ and $r = -\sin\theta$.

SOLUTION: We first take an algebraic approach and find all values of θ for which the two equations yield the same values of r. So we equate the two expressions for r and solve for θ to give

$$1 + \sin\theta = -\sin\theta$$

$$2\sin\theta = -1$$

$$\sin\theta = -\frac{1}{2}$$

$$\theta = \frac{7\pi}{6} \quad \text{or} \quad \theta = \frac{11\pi}{6}.$$

Both of these values give $r = \frac{1}{2}$. Do we have all the values of θ? The answer is we may or we may not. The figure shows that the curves also intersect at the pole! The algebra did not yield this point.

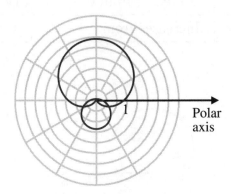

Figure 5.24

The reason is that the expressions $1 + \sin\theta$ and $-\sin\theta$ are 0 for different values of θ, so these values will not be found by equating the expressions. In this case, when

$$r = 1 + \sin\theta = 0,$$

$$\sin\theta = -1$$

$$\theta = \frac{3\pi}{2},$$

and when

$$r = -\sin\theta = 0,$$

$$\sin\theta = 0$$

$$\theta = 0.$$

The polar coordinates of the three points of intersection are

$$(0,0), \quad \left(\frac{1}{2}, \frac{7\pi}{6}\right) \quad \text{and} \quad \left(\frac{1}{2}, \frac{11\pi}{6}\right).$$

Rectangular coordinates:

$(0,0)$: The rectangular coordinates are also $(0,0)$.

$\left(\frac{1}{2}, \frac{7\pi}{6}\right) : x = \frac{1}{2}\cos(7\pi/6) = -\sqrt{3}/4,\ y = \frac{1}{2}\sin(7\pi/6) = -1/4$

$\left(\frac{1}{2}, \frac{11\pi}{6}\right) : x = \frac{1}{2}\cos(11\pi/6) = \sqrt{3}/4,\ y = \frac{1}{2}\sin(11\pi/6) = -1/4$ ◇

Solutions for Exercise Set 5.5

1. (a)

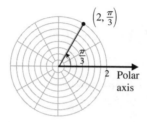

(b) $x = r \cos \theta = 2 \cos \frac{\pi}{3} = 1$; $y = r \sin \theta = 2 \sin \frac{\pi}{3} = \sqrt{3}$

(c) $\left(2, \frac{7\pi}{3}\right), \left(-2, \frac{4\pi}{3}\right)$

3. (a)

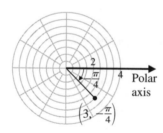

(b) $x = r \cos \theta = 3 \cos \left(-\frac{\pi}{4}\right) = \frac{3\sqrt{2}}{2}$; $y = r \sin \theta = 3 \sin \left(-\frac{\pi}{4}\right) = -\frac{3\sqrt{2}}{2}$

(c) $\left(3, \frac{7\pi}{4}\right), \left(-3, \frac{3\pi}{4}\right)$

5. (a)

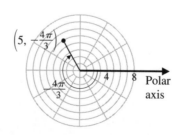

(b) $x = r\cos\theta = 5\cos\left(-\frac{4\pi}{3}\right) = -\frac{5}{2}$; $y = r\sin\theta = 5\sin\left(-\frac{4\pi}{3}\right) = \frac{5\sqrt{3}}{2}$

(c) $\left(5, \frac{2\pi}{3}\right), \left(-5, \frac{5\pi}{3}\right)$

7. (a)

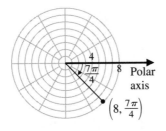

(b) $x = r\cos\theta = 8\cos\left(\frac{7\pi}{4}\right) = 4\sqrt{2}$; $y = r\sin\theta = 8\sin\left(\frac{7\pi}{4}\right) = -4\sqrt{2}$

(c) $\left(8, -\frac{\pi}{4}\right), \left(-8, \frac{3\pi}{4}\right)$

9. (a)

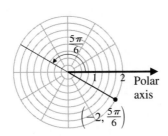

(b) $x = r\cos\theta = -2\cos\left(\frac{5\pi}{6}\right) = \sqrt{3}$; $y = r\sin\theta = -2\sin\left(\frac{5\pi}{6}\right) = -1$

(c) $\left(-2, -\frac{7\pi}{6}\right), \left(2, \frac{11\pi}{6}\right)$

11. (a)

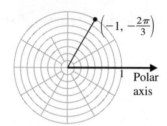

(b) $x = r\cos\theta = -\cos\left(-\frac{2\pi}{3}\right) = \frac{1}{2}$; $y = r\sin\theta = -\sin\left(-\frac{2\pi}{3}\right) = \frac{\sqrt{3}}{2}$

(c) $\left(1, \frac{2\pi}{3}\right), \left(-1, \frac{5\pi}{3}\right)$

13. The point $(2, 0)$ lies on the positive x-axis which coincides with the ray $\theta = 0$, so the point is also $(2, 0)$ in polar coordinates.

15. The point $(0, -4)$ lies on the negative y-axis which coincides with the ray $\theta = \frac{3\pi}{2}$, so the point is $\left(4, \frac{3\pi}{2}\right)$ in polar coordinates.

17. Since

$$x = 1, y = -\sqrt{3} \Rightarrow r = \sqrt{x^2 + y^2} = \sqrt{4} = 2,$$

and since the point lies in quadrant IV,

$$\tan\theta = \frac{y}{x} = -\sqrt{3} \Rightarrow \theta = \frac{5\pi}{3}.$$

So the point can be represented as $\left(2, \frac{5\pi}{3}\right)$ in polar coordinates.

19. Since

$$x = -4, y = 4 \Rightarrow r = \sqrt{x^2 + y^2} = \sqrt{32} = 4\sqrt{2},$$

and since the point lies in quadrant II,

$$\tan\theta = \frac{y}{x} = -1 \Rightarrow \theta = \frac{3\pi}{4}.$$

So the point can be represented as $\left(4\sqrt{2}, \frac{3\pi}{4}\right)$ in polar coordinates.

21. $r = 4 \Rightarrow r^2 = 16 \Rightarrow x^2 + y^2 = 16$

23. $\theta = \frac{3\pi}{4} \Rightarrow \frac{y}{x} = \tan\frac{3\pi}{4} = -1 \Rightarrow y = -x$

25. Since $r^2 = x^2 + y^2$ and $x = r \cos \theta$, we have

$$r = 2 \cos \theta \Rightarrow r^2 = 2r \cos \theta \Rightarrow x^2 + y^2 = 2x \Rightarrow x^2 - 2x + y^2 = 0,$$

so $x^2 - 2x + 1 + y^2 = 1$ and $(x-1)^2 + y^2 = 1$.

27. Since $x = r \cos \theta$ and $y = r \sin \theta$, we have

$$y = x \Rightarrow r \cos \theta = r \sin \theta \Rightarrow \tan \theta = 1 \Rightarrow \theta = \frac{\pi}{4}.$$

29. Since $r^2 = x^2 + y^2$, we have

$$x^2 + y^2 = 9 \Rightarrow r^2 = 9 \Rightarrow r = 3.$$

31. Since $r^2 = x^2 + y^2$ and $y = r \sin \theta$, we have

$$x^2 + y^2 = 2y \Rightarrow r^2 = 2r \sin \theta \Rightarrow r = 2 \sin \theta.$$

33. For $r = 3$

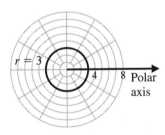

35. For $\theta = 5\pi/3$

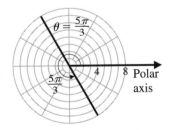

37.   For $r = 3\cos\theta$

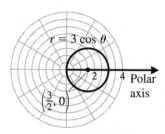

39.   For $r = -4\sin\theta$

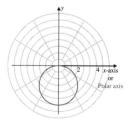

41.   For $r = 2 + 2\cos\theta$

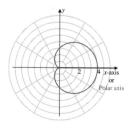

43.   For $r = 2 + \sin\theta$

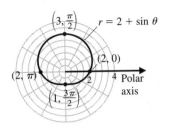

45. For $r = 1 - 2\cos\theta$

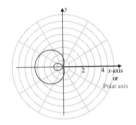

47. For $r = 3\sin 3\theta$

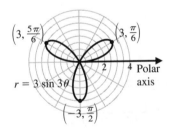

49. For $r = 3\cos 2\theta$

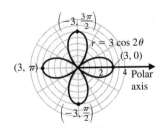

51. For $r^2 = 16\cos 2\theta$

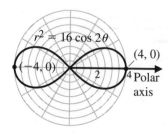

53. For $r = \theta$

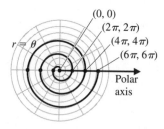

55. For $r = e^\theta$

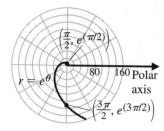

57. The graphs are all circles. The graph in (a) and (c) are symmetric about the y-axis and are reflections of each other about the x-axis. The graph in (b) and (d) are symmetric about the x-axis and are reflections of each other about the y-axis.

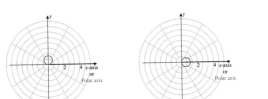 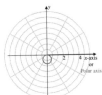 

59. Graph (a) is a circle with center $\left(0, \frac{1}{2}\right)$ and radius $\frac{1}{2}$, and graphs (b)-(d) are cardioids with axis along the y-axis which approach a circle.

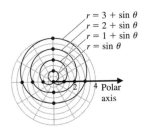

61. The graphs are all cardioids with axis along the y-axis. The graphs in (a) and (c) coincide, as do the graphs in (b) and (d). The graph in (b) is the reflection of graph (a) about the x-axis.

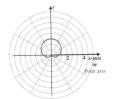 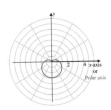  

63. (a) $(0,0), \left(90, \frac{\pi}{4}\right), \left(90\sqrt{2}, \frac{\pi}{2}\right), \left(90, \frac{3\pi}{4}\right)$
(b) $(0,0), (90,0), \left(90\sqrt{2}, \frac{\pi}{4}\right), \left(90, \frac{\pi}{2}\right)$

65. Setting the equations for r equal, we have $1 + 2\cos\theta = 1 \Rightarrow 2\cos\theta = 0 \Rightarrow \theta = \frac{\pi}{2}$ or $\theta = \frac{3\pi}{2}$. The figure indicates that the curves also intersect at the point $(1, 0)$. The polar coordinates of the points of intersection are $\left(1, \frac{\pi}{2}\right), \left(1, \frac{3\pi}{2}\right), (1, 0)$. The rectangular coordinates of the points of intersection are $(0, 1), (0, -1), (1, 0)$.

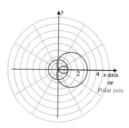

67. Setting the equations for r equal, we have $2 - 2\sin\theta = 2\sin\theta \Rightarrow 4\sin\theta = 2 \Rightarrow \sin\theta = \frac{1}{2} \Rightarrow \theta = \frac{\pi}{6}$ or $\theta = \frac{5\pi}{6}$. The figure indicates the curves also intersect at the pole. The polar coordinates of the points of intersection are $\left(1, \frac{\pi}{6}\right), \left(1, \frac{5\pi}{6}\right), (0, 0)$. The rectangular coordinates of the points of intersection are $\left(\frac{\sqrt{3}}{2}, \frac{1}{2}\right), \left(-\frac{\sqrt{3}}{2}, \frac{1}{2}\right), (0, 0)$.

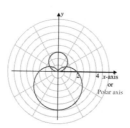

69. The graphs of $r = 1 + \sin(n\theta) + (\cos(2n\theta))^2$ for $n = 1, 2, 3, 4$ are shown below.

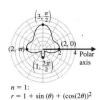

$n = 1$:
$r = 1 + \sin(\theta) + (\cos(2\theta))^2$

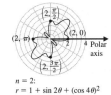

$n = 2$:
$r = 1 + \sin 2\theta + (\cos 4\theta)^2$

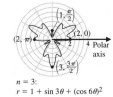

$n = 3$:
$r = 1 + \sin 3\theta + (\cos 6\theta)^2$

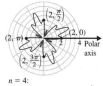

$n = 4$:
$r = 1 + \sin 4\theta + (\cos 8\theta)^2$

71. The graphs of $r = \sin m\theta$ and $r = |\sin m\theta|$ are shown below.

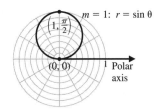

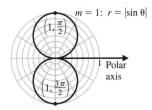

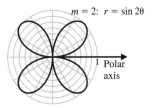

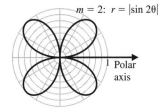

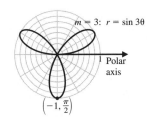

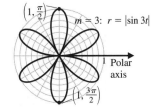

5.6 Conic Sections in Polar Coordinates

The graph of a polar equation of the form

$$r = \frac{ed}{1 \pm e \cos \theta} \quad \text{or} \quad r = \frac{ed}{1 \pm e \sin \theta}$$

is a conic section with eccentricity e. If

$e = 1$, the graph is a parabola;

$e < 1$, the graph is an ellipse;

$e > 1$, the graph is a hyperbola.

The focus is at the pole and, the directrix is d units from the pole.

Equation	Directrix
$r = \frac{ed}{1+e\cos\theta}$	Vertical, right of pole
$r = \frac{ed}{1-e\cos\theta}$	Vertical, left of pole
$r = \frac{ed}{1+e\sin\theta}$	Horizontal, above the pole
$r = \frac{ed}{1-e\sin\theta}$	Horizontal, below the pole

EXAMPLE 5.6.1   Sketch the graph of the conic section and find a corresponding rectangular equation.

(a) $r = \frac{3}{1-\cos\theta}$ (b) $r = \frac{3}{3-\sin\theta}$

SOLUTION:

(a) First determine e and d. The polar equation is in the form

$$r = \frac{ed}{1 - e\cos\theta},$$

with

$$e = 1$$

$$ed = 3 \Rightarrow d = 3.$$

Since $e = 1$, the conic is a parabola, with a vertical directrix, 3 units to the left of the pole. The directrix has rectangular equation $x = -3$. Since the vertex is midway between the directrix and the pole, the vertex is at the point $(-3/2, 0)$ (in rectangular coordinates). The parabola can not cross the directrix and hence opens to the right. When

$$\theta = \frac{\pi}{2}, \quad \cos\theta = 0 \text{ and } r = 3;$$

$$\theta = \frac{3\pi}{2}, \quad \cos\theta = 0 \text{ and } r = 3.$$

The graph is shown in the figure.

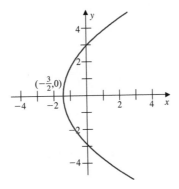

Figure 5.25

The rectangular equation of the parabola is of the form

$$x - h = \frac{1}{4c}(y - k)^2.$$

To determine the rectangular equation, convert the polar equation to the corresponding rectangular form using the relations $x = r\cos\theta$, $y = r\sin\theta$, and $r^2 = x^2 + y^2$. Then

$$r = \frac{3}{1 - \cos\theta}$$

$$r - r\cos\theta = 3$$

$$r = 3 + r\cos\theta = 3 + x$$

$$r^2 = (x + 3)^2$$

$$x^2 + y^2 = x^2 + 6x + 9$$

$$y^2 = 6x + 9$$

$$y^2 = 6\left(x + \frac{3}{2}\right)$$

$$x + \frac{3}{2} = \frac{1}{6}y^2.$$

In rectangular form, the vertex of the parabola is $(-3/2, 0)$ and $4c = 6$, so $c = 3/2$. The parabola is obtained from the parabola in standard form

$$x = \frac{1}{6}y^2,$$

shifted to the left $3/2$ units.

(b) First write the polar equation in the form

$$r = \frac{ed}{1 - e\sin\theta}.$$

Then

$$r = \frac{3}{3 - \sin\theta}$$

$$= \frac{3}{3(1 - \frac{1}{3}\sin\theta)},$$

so

$$r = \frac{1}{1 - \frac{1}{3}\sin\theta}.$$

Since $e = \frac{1}{3}$, the conic is an ellipse and

$$ed = 1 \Rightarrow d = 3.$$

The conic is an ellipse with directrix horizontal and 3 units below the pole. The graph passes through the points

$$(1, 0), \ (1, \pi), \ \left(\frac{3}{2}, \frac{\pi}{2}\right), \ \text{and} \ \left(\frac{3}{4}, \frac{3\pi}{2}\right).$$

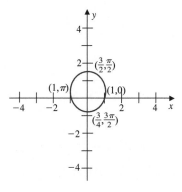

Figure 5.26

Rectangular equation:

$$r = \frac{3}{3 - \sin\theta}$$

$$3r - r\sin\theta = 3$$

$$3r = 3 + r\sin\theta$$

$$3r = 3 + y$$

$$9r^2 = (3 + y)^2$$

$$9x^2 + 9y^2 = 9 + 6y + y^2$$

$$9x^2 + 8y^2 - 6y = 9$$

$$9x^2 + 8\left(y^2 - \frac{3}{4}y\right) = 9$$

$$9x^2 + 8\left(y^2 - \frac{3}{4}y + \frac{9}{64} - \frac{9}{64}\right) = 9$$

$$9x^2 + 8\left(y - \frac{3}{8}\right)^2 = 9 + \frac{9}{8}$$

$$9x^2 + 8\left(y - \frac{3}{8}\right)^2 = \frac{81}{8}$$

◇

APPLICATION

EXAMPLE 5.6.2   The earth moves in an elliptical orbit with the sun at one focal point and an eccentricity $e = 0.0167$. The major axis of the elliptical orbit is approximately 2.99×10^8 kilometers. Write a polar equation for this ellipse, assuming that the pole is at the sun.

SOLUTION: The elliptic orbit of the earth about the sun has an equation of the form

$$r = \frac{ed}{1 + e\cos\theta},$$

where the earth is closest the sun when $\theta = 0$ and furthest from the sun when $\theta = \pi$. The length of the major axis is then

$$\frac{ed}{1 + e\cos\pi} - \frac{ed}{1 + e\cos 0} = \frac{ed}{1 - e} - \frac{ed}{1 + e}$$

$$= \frac{ed(1 + e) - ed(1 - e)}{(1 - e)(1 + e)}$$

$$= \frac{ed + e^2 d - ed + e^2 d}{1 - e^2}$$

$$= \frac{2e^2 d}{1 - e^2}.$$

If the length of the major axis is approximately 2.99×10^8 and $e = 0.0167$, then

$$2.99 \times 10^8 = \frac{2e^2 d}{1 - e^2}$$

$$d = \frac{\left(2.99 \times 10^8\right)\left(1 - e^2\right)}{2e^2}$$

$$= \frac{\left(2.99 \times 10^8\right)\left(1 - 0.0167^2\right)}{2(0.0167)^2}$$

$$\approx 5.4 \times 10^{11}.$$

The polar equation of the earths orbit about the sun is

$$r = \frac{0.0167\left(5.4 \times 10^{11}\right)}{1 + 0.0167\cos\theta} = \frac{9 \times 10^9}{1 + 0.0167\cos\theta}.$$

◇

Solutions for Exercise Set 5.6

1. (a) The equation $r = \dfrac{2}{1 + \cos\theta}$ is in the standard form $r = \dfrac{ed}{1 + e\cos\theta}$, where $e = 1 \Rightarrow d = 2$, so the equation describes a parabola opening to the left with vertex $(1, 0)$, focus $(0, 0)$ and directrix along the line $x = 2$.

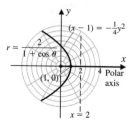

(b) Since $r^2 = x^2 + y^2$ and $x = r\cos\theta$, we have

$$r = \frac{2}{1 + \cos\theta} \Rightarrow r + r\cos\theta = 2 \Rightarrow$$
$$r = 2 - r\cos\theta = 2 - x \Rightarrow r^2 = 4 - 4x + x^2.$$

So

$$x^2 + y^2 = 4 - 4x + x^2 \Rightarrow x - 1 = -\frac{1}{4}y^2.$$

3. (a) The equation

$$r = \frac{2}{2 - \sin\theta} = \frac{2}{2\left(1 - \frac{1}{2}\sin\theta\right)} = \frac{1}{1 - \frac{1}{2}\sin\theta}$$

is in the standard form $r = \frac{ed}{1 - e\sin\theta}$, where $e = \frac{1}{2} \Rightarrow d = 2$. Since $e < 1$, the equation describes an ellipse with major axis along the y-axis and with horizontal directrix $y = -2$.

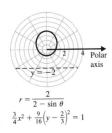

$$r = \frac{2}{2 - \sin\theta}$$

$$\tfrac{3}{4}x^2 + \tfrac{9}{16}\left(y - \tfrac{2}{3}\right)^2 = 1$$

(b) Since $r^2 = x^2 + y^2$ and $y = r\sin\theta$, we have

$$r = \frac{2}{2 - \sin\theta} \Rightarrow 2r - r\sin\theta = 2 \Rightarrow$$

$$2r = 2 + r\sin\theta = 2 + y \Rightarrow 4r^2 = 4 + 4y + y^2.$$

So

$$4x^2 + 4y^2 = 4 + 4y + y^2 \Rightarrow 4x^2 + 3y^2 - 4y = 4.$$

5. (a) The equation $r = \dfrac{1}{1 + 2\cos\theta}$ is in the standard form $r = \dfrac{ed}{1 + e\cos\theta}$,
where $e = 2$ and $ed = 1 \Rightarrow d = \tfrac{1}{2}$. Since $e > 1$, the equation describes a
hyperbola with axis along the x-axis and vertical directrix $x = \tfrac{1}{2}$.

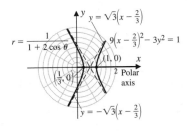

(b) Since $r^2 = x^2 + y^2$ and $x = r\cos\theta$, we have

$$r = \frac{1}{1 + 2\cos\theta} \Rightarrow r + 2r\cos\theta = 1 \Rightarrow$$

$$r = 1 - 2x \Rightarrow r^2 = 1 - 4x + 4x^2.$$

So

$$x^2 + y^2 = 1 - 4x + 4x^2 \Rightarrow 3x^2 - 4x - y^2 = -1.$$

7. **(a)** The equation $r = \dfrac{3}{1 - 2\sin\theta}$ is in the standard form $r = \dfrac{ed}{1 - e\sin\theta}$, where $e = 2$ and $ed = 3 \Rightarrow d = \frac{3}{2}$. Since $e > 1$ the equation describes a hyperbola with axis along the x-axis and horizontal directrix $y = -\frac{3}{2}$.

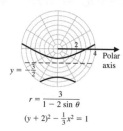

$$r = \frac{3}{1 - 2\sin\theta}$$
$$(y + 2)^2 - \tfrac{1}{3}x^2 = 1$$

(b) Since $r^2 = x^2 + y^2$ and $y = r\sin\theta$, we have

$$r = \frac{3}{1 - 2\sin\theta} \Rightarrow r - 2r\sin\theta = 3 \Rightarrow$$
$$r = 3 + 2y \Rightarrow r^2 = 9 + 12y + 4y^2.$$

So

$$x^2 + y^2 = 9 + 12y + 4y^2 \Rightarrow 3y^2 + 12y - x^2 = -9.$$

9. Since the directrix $x = -4$ is vertical and to the left of the pole, the conic has the form $r = \dfrac{ed}{1 - e\cos\theta}$. Since $e = 2$ and $d = 4$, we have

$$r = \frac{8}{1 - 2\cos\theta}.$$

11. Since the directrix $y = -\dfrac{1}{4}$ is horizontal and below the pole, the conic has the form $r = \dfrac{ed}{1 - e\sin\theta}$. Since $e = 1$ and $d = \dfrac{1}{4}$, we have

$$r = \frac{1/4}{1 - \sin\theta} = \frac{1}{4 - 4\sin\theta}.$$

13. Since the directrix $x = 2$, is vertical and to the right of the pole, the conic has the form $r = \dfrac{ed}{1 + e\cos\theta}$. Since $e = 3$ and $d = 2$, we have

$$r = \frac{6}{1 + 3\cos\theta}.$$

15. Since the directrix $y = 1$ is horizontal and above the pole, the conic has the form $r = \dfrac{ed}{1 + e\sin\theta}$. Since $e = \frac{1}{4}$ and $d = 1$, we have

$$r = \frac{1/4}{1 + \frac{1}{4}\sin\theta} = \frac{1}{4 + \sin\theta}.$$

17. The conic has equation in the form $r = \dfrac{ed}{1 + e\cos\theta}$, and since $(1,0)$ and $(3, \pi)$ are on the curve,

$$1 = \frac{ed}{1 + e\cos 0} = \frac{ed}{1 + e} \Rightarrow 1 + e = ed,$$

and

$$3 = \frac{ed}{1 + e\cos\pi} = \frac{ed}{1 - e} \Rightarrow 3 - 3e = ed \Rightarrow$$

$$1 + e = 3 - 3e \Rightarrow e = \frac{1}{2},$$

and

$$\frac{1}{2}d = 1 + \frac{1}{2} \Rightarrow d = 3.$$

So

$$r = \frac{ed}{1 + e\cos\theta} = \frac{3/2}{1 + 1/2\cos\theta} = \frac{3}{2 + \cos\theta}.$$

19. Since the orbit is elliptical if we assume the axis is horizontal, the form of the equation is

$$r = \frac{ed}{1 - e\cos\theta},$$

with $e < 1$. The satellite will be at its maximum height when $\theta = 0$, so

$$\frac{ed}{1-e} = 3960 + 560 = 4520,$$

and at its minimum height when $\theta = \pi$, so

$$\frac{ed}{1+e} = 3960 + 145 = 4105.$$

Then

$$4520(1-e) = ed = 4105(1+e) \Rightarrow 8625e = 415,$$

and

$$e = \frac{415}{8625} \approx 0.048 \Rightarrow ed = 4105\left(1 + \frac{415}{8625}\right) \approx 4303.$$

The equation of the orbit is

$$r = \frac{4303}{1 - 0.048\cos\theta}.$$

5.7 Parametric Equations

In this method for representing curves in the plane, the x and y coordinates of a point on the curve are each specified separately by functions of a third variable, called the *parameter*. If the functions are f and g, and the parameter t, then

$$x = f(t), \quad y = g(t)$$

are called the *parametric equations* for the curve. As t varies, the point $(x, y) = (f(t), g(t))$ traces out the curve.

EXAMPLE 5.7.1   Sketch the graph of the curve described by the parametric equations

$$x = 2t - 1, \; y = t^2 + 1,$$

and find a corresponding rectangular equation.

SOLUTION: First plot some representative points on the curve by making a table of values.

t	0	1	2	-1	-2
$x = 2t - 1$	-1	1	3	-3	-5
$y = t^2 + 1$	1	2	5	2	5

The points are plotted in the figure and a curve traced in between. The curve appears to be a parabola. Notice as t increases the curve is traced in the direction of the arrows.

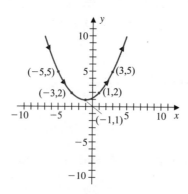

Figure 5.27

To find a corresponding rectangular equation for the curve, *eliminate the parameter*. That is, solve for t in either the equation for x or the equation for y, which ever is easiest, and then substitute the value for t in the other equation. So

$$x = 2t - 1$$
$$2t = x + 1$$
$$t = \frac{x + 1}{2},$$

and then substituting in the equation for y gives

$$y = t^2 + 1$$
$$= \left(\frac{x + 1}{2}\right)^2 + 1$$
$$= \frac{1}{4}(x + 1)^2 + 1.$$

The equation describes a parabola with vertex at $(-1, 1)$ and opening upward.

◇

EXAMPLE 5.7.2   For each set of parametric equations, find a corresponding rectangular equation, and sketch the graph of the curve, showing, if possible, when $t = 0$ and the direction of increasing values of t.

(a) $x = t - 1$, $y = \sqrt{t}$

(b) $x = e^{2t} + 2$, $y = e^t - 3$

(c) $x = 4 + 2\cos t$, $y = 6 + 2\sin t$

SOLUTION:

(a) Solve for t in the equation for x, and substitute it into the equation for y. This gives

$$t = x + 1,$$

so

$$y = \sqrt{x + 1},$$

or

$$y^2 = x + 1.$$

Be careful here that you have the *correct* curve. Is the curve the entire parabola given by the equation $y^2 = x + 1$? In this case the answer is no! The reason is that

$$\text{since } y = \sqrt{t}, \quad y \text{ must be greater than or equal to } 0.$$

The curve consists *only* of the portion of $y^2 = x + 1$ above the x-axis. As t increases, x and y both increase and the curve is traced in the direction of the arrows in the figure.

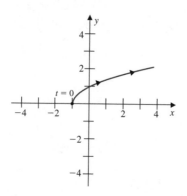

Figure 5.28

(b) In this example, we do not have to find t since finding an expression for e^t will do. So

$$y = e^t - 3$$

$$e^t = y + 3$$

$$e^{2t} = \left(e^t\right)^2 = (y+3)^2,$$

and

$$x = e^{2t} + 2 = (y+3)^2 + 2,$$

with the restriction that, since $e^t > 0$, we have $y = e^t - 3 > -3$.

The curve is then,

$$x = (y+3)^2 + 2, \quad \text{for} \quad y > -3.$$

Note as $t \to -\infty$, $x = e^{2t} + 2 \to 2$ and $y = e^t - 3 \to -3$. The point on the curve corresponding to $t = 0$ is $(3, -2)$.

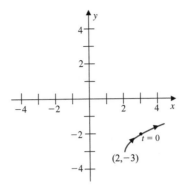

Figure 5.29

(c) In problems involving sine and cosine, always be on the look out for

$$(\sin t)^2 + (\cos t)^2 = 1.$$

If $x - 4$ and $y - 6$ are each squared and then added, the resulting expression involves this basic identity and can then be simplified. That is,

$$
\begin{aligned}
(x - 4)^2 + (y - 6)^2 &= (4 + 2\cos t - 4)^2 + (6 + 2\sin t - 6)^2 \\
&= (2\cos t)^2 + (2\sin t)^2 \\
&= 4(\cos t)^2 + 4(\sin t)^2 \\
&= 4((\cos t)^2 + (\sin t)^2) \\
&= 4,
\end{aligned}
$$

which is the equation of the circle with center $(4, 6)$ and radius 2.

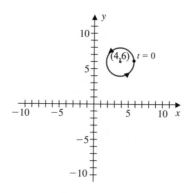

Figure 5.30

◇

EXAMPLE 5.7.3   Sketch the graph described by the parametric equations, if possible showing when $t = 0$ and the direction of increasing t.

(a) $x = t, y = t^3$ (b) $x = t^3, y = t$ (c) $x = t^2, y = t^6$ (d) $x = t^6, y = t^2$

SOLUTION:

(a) <u>Eliminating the parameter:</u> $y = x^3$

<u>When $t = 0$:</u> $(x, y) = (0, 0)$

<u>Increasing t:</u> As t increases from 0 toward ∞, both x and y increase, so the points trace the curve upward along the right half of $y = x^3$. As t goes toward $-\infty$, x goes to $-\infty$ and $y = x^3$ goes to $-\infty$. So as t increases from negative values towards 0, the left half of the curve is traced upward toward 0.

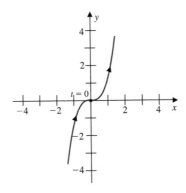

Figure 5.31

(b) <u>Eliminating the parameter:</u> $x = y^3$

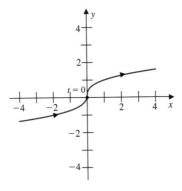

Figure 5.32

(c) <u>Eliminating the parameter:</u> $y = x^3$

But since $x = t^2 \geq 0$, the curve is restricted to $x \geq 0$.

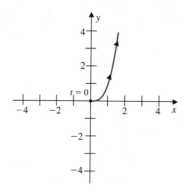

Figure 5.33

(d) <u>Eliminating the parameter:</u> $x = y^3$, $x \geq 0$

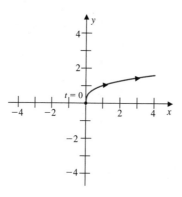

Figure 5.34

◇

EXAMPLE 5.7.4   Find parametric equations for the line passing through the points $(4, 5)$ and $(-1, 3)$.

SOLUTION: First find the rectangular equation of the line. We need the slope, which is

$$m = \frac{3 - 5}{-1 - 4} = \frac{-2}{-5} = \frac{2}{5}.$$

Using the point $(4, 5)$ and the point-slope equation of a line gives

$$y - 5 = \frac{2}{5}(x - 4).$$

One set of parametric equations for the line can be found by setting $t = x - 4$.

So

$$x = t + 4$$
$$y - 5 = \frac{2}{5}t$$
$$y = \frac{2}{5}t + 5.$$

A set of parametric equations is given by

$$x = t + 4, \quad y = \frac{2}{5}t + 5.$$

Note that $t = 0$ gives $(4, 5)$, and $t = -5$ give $(-1, 3)$.   ◇

EXAMPLE 5.7.5   Find parametric equations in the parameter t, $0 \le t \le 2\pi$, for the circle with equation $x^2 + y^2 = 16$, so that it is traced

(a) once around, counterclockwise, starting at $(4, 0)$;

(b) twice around, counterclockwise, starting at $(4, 0)$;

(c) three times around, counterclockwise, starting at $(-4, 0)$;

(d) twice around, clockwise, starting at $(0, 4)$;

(e) three times around, clockwise, starting at $(0, 4)$.

SOLUTION: The idea is to recognize that if

$$x = 4\cos t, \; y = 4\sin t,$$

then

$$x^2 + y^2 = (4\cos t)^2 + (4\sin t)^2$$
$$= 16((\cos t)^2 + (\sin t)^2)$$
$$= 16,$$

which is the equation of the circle. By slightly modifying these parametric equations, the manner in which the circle is traced can be changed.

(a) No modification is required. The parametric equations

$$x = 4\cos t \quad \text{and} \quad y = 4\sin t, \quad \text{for} \quad 0 \le t \le 2\pi,$$

trace the circle starting at $(4,0)$, when $t = 0$, and ending back at $(4,0)$, when $t = 2\pi$. To check which direction the curve is traced, select a few values of increasing t. For example,

$$t = 0, \ (x,y) = (4,0)$$
$$t = \frac{\pi}{2}, \ (x,y) = (0,4)$$
$$t = \pi, \ (x,y) = (-4,0)$$
$$t = \frac{3\pi}{2}, \ (x,y) = (0,-4),$$

and the curve is being traced counterclockwise.

(b) To trace the circle twice, cut the period of the cosine and sine functions in half so that as t varies from 0 to 2π, the points on the circle are traced twice. To halve the period, multiply the argument t by 2. Note the period of $y = \cos 2t$ or of $y = \sin 2t$ is $\frac{2\pi}{2} = \pi$. So the parametric equations are

$$x = 4\cos 2t \quad \text{and} \quad y = 4\sin 2t, \quad \text{for} \quad 0 \le t \le 2\pi.$$

Notice also that when,

$$t = 0, \ (x, y) = (4, 0)$$

$$t = \frac{\pi}{4}, \ (x, y) = (0, 4)$$

$$t = \frac{\pi}{2}, \ (x, y) = (-4, 0)$$

$$t = \frac{3\pi}{4}, \ (x, y) = (0, -4)$$

$$t = \pi, \ (x, y) = (4, 0).$$

(c) To go three times around starting at $(-4, 0)$, let

$$x = -4\cos 3t \quad \text{and} \quad y = -4\sin 3t, \quad \text{for} \quad 0 \le t \le 2\pi.$$

Then

$$t = 0, \ (x, y) = (-4, 0)$$

$$t = \frac{\pi}{6}, \ (x, y) = (0, -4)$$

$$t = \frac{\pi}{3}, \ (x, y) = (4, 0)$$

$$t = \frac{\pi}{2}, \ (x, y) = (0, 4)$$

$$t = \frac{2\pi}{3}, \ (x, y) = (-4, 0).$$

(d) To trace the curve clockwise, switch the roles of sine and cosine.

$$x = 4\sin 2t \quad \text{and} \quad y = 4\cos 2t, \quad \text{for} \quad 0 \le t \le 2\pi.$$

So for

$$t = 0, \ (x, y) = (0, 4)$$

$$t = \frac{\pi}{4}, \ (x, y) = (4, 0)$$

$$t = \frac{\pi}{2}, \ (x, y) = (0, -4)$$

$$t = \frac{3\pi}{4}, \ (x, y) = (-4, 0)$$

$$t = \pi, \ (x, y) = (0, 4).$$

(e) Finally, to go around three times clockwise, let

$$x = 4 \sin 3t \quad \text{and} \quad y = 4 \cos 3t, \quad \text{for} \quad 0 \le t \le 2\pi.$$

So for

$$t = 0, \ (x, y) = (0, 4)$$

$$t = \frac{\pi}{6}, \ (x, y) = (4, 0)$$

$$t = \frac{\pi}{3}, \ (x, y) = (0, -4)$$

$$t = \frac{\pi}{2}, \ (x, y) = (-4, 0)$$

$$t = \frac{2\pi}{3}, \ (x, y) = (0, 4).$$

◇

Solutions for Exercise Set 5.7

1. (a) $x = 3t, y = \frac{t}{2} \Rightarrow t = \frac{x}{3}$ and $y = \frac{1}{2}\frac{x}{3} = \frac{x}{6}$.

(b)

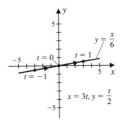

3. (a) $x = \sqrt{t}, y = t + 1 \Rightarrow t = x^2$ and $y = x^2 + 1$, only for $x \geq 0$.

(b)

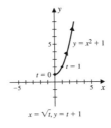

5. (a) $x = \sin t, y = (\cos t)^2 \Rightarrow y = 1 - (\sin t)^2 = 1 - x^2$ for $y \geq 0$.

(b)

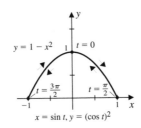

7. (a) $x = \sec t, y = \tan t$ and $(\tan t)^2 + 1 = (\sec t)^2$. So $y^2 + 1 = x^2 \Rightarrow$
$x^2 - y^2 = 1$. Since $x = \sec t, |x| \geq 1$.

(b)

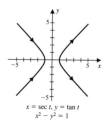

$x = \sec t, y = \tan t$
$x^2 - y^2 = 1$

9. (a) $x = 3\cos t, y = 2\sin t \Rightarrow \dfrac{x^2}{9} + \dfrac{y^2}{4} = \dfrac{9(\cos t)^2}{9} + \dfrac{4(\sin t)^2}{4}$
$\Rightarrow \dfrac{x^2}{9} + \dfrac{y^2}{4} = (\cos t)^2 + (\sin t)^2 = 1 \Rightarrow \dfrac{x^2}{9} + \dfrac{y^2}{4} = 1.$

(b)

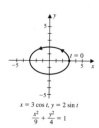

$x = 3\cos t, y = 2\sin t$
$\frac{x^2}{9} + \frac{y^2}{4} = 1$

11. (a) $x = e^t, y = e^{-t} \Rightarrow y = x^{-1} = \dfrac{1}{x}$, for $x > 0$.

(b)

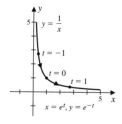

$y = \dfrac{1}{x}$

$t = -1$

$t = 0$

$t = 1$

$x = e^t, y = e^{-t}$

13. The graphs of the parametric equations are shown below.

(a) $x = t, y = t^2 \Rightarrow y = x^2$ (b) $x = t^2, y = t \Rightarrow x = y^2$

(c) $x = t^2, y = t^4 \Rightarrow y = x^2, x \geq 0$ (d) $x = t^4, y = t^2 \Rightarrow x = y^2, y \geq 0$

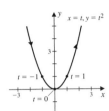 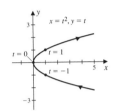 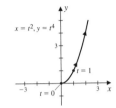 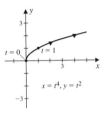

15. The graphs of the parametric equations are shown below.

(a) $x = \sin t, y = \cos t \Rightarrow x^2 + y^2 = (\sin t)^2 + (\cos t)^2 = 1$

(b) $x = \sin t, y = \cos t + 1 \Rightarrow x^2 + (y-1)^2 = (\sin t)^2 + (\cos t)^2 = 1$

(c) $x = \cos t + 1, y = \sin t \Rightarrow (x-1)^2 + y^2 = (\cos t)^2 + (\sin t)^2 = 1$

(d) $x = \cos t, y = \sin t + 1 \Rightarrow x^2 + (y-1)^2 = (\cos t)^2 + (\sin t)^2 = 1$

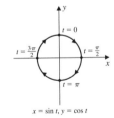 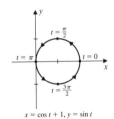 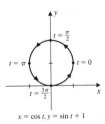

$x = \sin t, y = \cos t$ $x = \sin t, y = \cos t + 1$ $x = \cos t + 1, y = \sin t$ $x = \cos t, y = \sin t + 1$

17. The graphs of the parametric equations are shown below.

(a) $x = \cos t, y = \sin t \Rightarrow x^2 + y^2 = 1$

(b) $x = \sin t, y = \cos t \Rightarrow x^2 + y^2 = 1$

(c) $x = t, y = \sqrt{1 - t^2} \Rightarrow y = \sqrt{1 - x^2}$

(d) $x = -t, y = \sqrt{1 - t^2} \Rightarrow y = \sqrt{1 - x^2}$

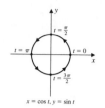

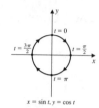

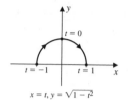

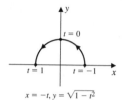

$x = \cos t, y = \sin t$

$x = \sin t, y = \cos t$

$x = t, y = \sqrt{1 - t^2}$

$x = -t, y = \sqrt{1 - t^2}$

19. The graphs of the parametric equations are shown below.

(a) $x = t, y = \ln t \Rightarrow y = \ln x$

(b) $x = e^t, y = t \Rightarrow \ln x = t, y = \ln x$

(c) $x = t^2, y = 2\ln t \Rightarrow y = \ln t^2 = \ln x$

(d) $x = \frac{1}{t}, y = -\ln t \Rightarrow y = \ln t^{-1} = \ln x$

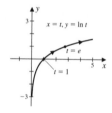

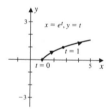

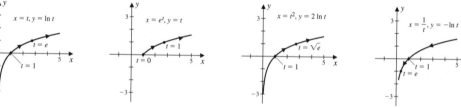

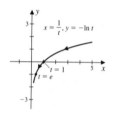

21. A rectangular equation of the line with slope $\frac{1}{3}$ and passing through $(2, -1)$ is

$$y + 1 = \frac{1}{3}(x - 2).$$

One set of parametric equations is obtained by setting

$$t = x - 2 \Rightarrow x = t + 2, y = \frac{1}{3}t - 1.$$

23. A parabola with vertex $(1, -2)$ and passing through the points $(0, 0)$ and $(2, 0)$ has the form $y = a(x - 1)^2 - 2$. Since $(0, 0)$ is on the curve,

$0 = a(-1)^2 - 2 \Rightarrow a = 2 \Rightarrow y = 2(x-1)^2 - 2$. One set of parametric equations is obtained by setting

$$t = x - 1 \Rightarrow x = t + 1, y = 2t^2 - 2.$$

25. (a) $x = r\cos t, \ y = r\sin t$; (b) $x = r\cos 2t, \ y = r\sin 2t$;

(c) $x = -r\cos 3t, \ y = -r\sin 3t$; (d) $x = r\sin 2t, \ y = r\cos 2t$;

(e) $x = r\sin 3t, \ y = r\cos 3t$

27. For $a = 1, b = 1$; $a = 1, b = 2$; and $a = 2, b = 1$ we have the following graphs.

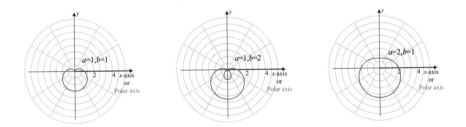

29. The parameter a determines the maximum height of the curve, $2a$, and the period of the curve, $2\pi a$, which equals the circumference of the rolling circle. The curve touches the x-axis for those values of t for which the y-coordinate is 0. That is,

$$a(1 - \cos t) = 0 \Rightarrow \cos t = 1 \Rightarrow t = \pm 2k\pi, \quad \text{for} \quad k = 0, 1, 2, \ldots$$

and

$$x = a(t - \sin t) = at = \pm 2ak\pi \quad \text{for} \quad k = 0, 1, 2\ldots$$

Solutions for Exercise Set 5 Review

1. $4y - x^2 = 0 \Leftrightarrow y = \dfrac{1}{4}x^2 \Rightarrow c = 1$

Vertex: $(0,0)$; Focus: $(0,1)$; Directrix: $y = -1$

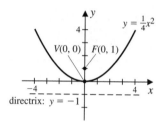

3. $4x - y^2 + 6y - 17 = 0 \Leftrightarrow -(y^2 - 6y + 9 - 9) = 17 - 4x$

$\Leftrightarrow -(y-3)^2 = 8 - 4x \Leftrightarrow x - 2 = \dfrac{1}{4}(y-3)^2 \Rightarrow c = 1;$

Vertex: $(2,3)$; Focus: $(3,3)$; Directrix: $x = 1$

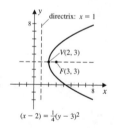

5. $x^2 + 4y^2 = 4 \Leftrightarrow \dfrac{x^2}{4} + y^2 = 1 \Rightarrow a = 2, b = 1 \Rightarrow c^2 = 3 \Rightarrow c = \sqrt{3}$

Foci: $\left(\sqrt{3},0\right), \left(-\sqrt{3},0\right)$; Vertices: $(2,0), (-2,0)$

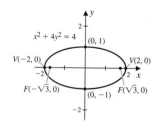

7. $4(x-1)^2 + 9(y+2)^2 = 36 \Leftrightarrow \dfrac{(x-1)^2}{9} + \dfrac{(y+2)^2}{4} = 1 \Rightarrow a = 3, b = 2$

$\Rightarrow c^2 = 5 \Rightarrow c = \sqrt{5}$

Foci: $\left(1 - \sqrt{5}, -2\right), \left(1 + \sqrt{5}, -2\right)$; Vertices: $(4, -2), (-2, -2)$

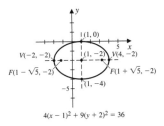

9. $x^2 - 2y^2 = 4 \Leftrightarrow \dfrac{x^2}{4} - \dfrac{y^2}{2} = 1 \Rightarrow a = 2, b = \sqrt{2} \Rightarrow c^2 = 6 \Rightarrow c = \sqrt{6}$

Foci: $\left(\sqrt{6}, 0\right), \left(-\sqrt{6}, 0\right)$; Vertices: $(2, 0), (-2, 0)$;

Asymptotes: $y = \pm\dfrac{\sqrt{2}}{2}x$

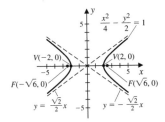

11. $2x^2 - 4x - 4y^2 + 1 = 0 \Leftrightarrow 2(x^2 - 2x + 1 - 1) - 4y^2 = -1$

$\Leftrightarrow 2(x-1)^2 - 4y^2 = 1 \Rightarrow a = \dfrac{\sqrt{2}}{2}, b = \dfrac{1}{2} \Rightarrow c^2 = \dfrac{1}{2} + \dfrac{1}{4} = \dfrac{3}{4} \Rightarrow c = \dfrac{\sqrt{3}}{2}$

Foci: $\left(1 - \sqrt{3}/2, 0\right), \left(1 + \sqrt{3}/2, 0\right)$; Vertices: $\left(1 + \sqrt{2}/2, 0\right), \left(1 - \sqrt{2}/2, 0\right)$;

Asymptotes: $y = \pm\dfrac{\sqrt{2}}{2}(x-1)$

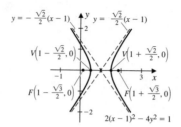

13. $x^2 = -2(y-5) \Leftrightarrow y - 5 = -\dfrac{1}{2}x^2$

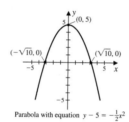

Parabola with equation $y - 5 = -\frac{1}{2}x^2$

15. $16x^2 + 25y^2 = 400 \Leftrightarrow \dfrac{x^2}{25} + \dfrac{y^2}{16} = 1$

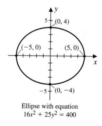

Ellipse with equation
$16x^2 + 25y^2 = 400$

17. $x^2 - 2x - 4y - 11 = 0 \Leftrightarrow x^2 - 2x + 1 - 1 = 4y + 11 \Leftrightarrow (x-1)^2 = 4(y+3)$
$\Leftrightarrow y + 3 = \frac{1}{4}(x-1)^2$

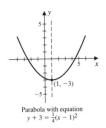

Parabola with equation
$y + 3 = \frac{1}{4}(x - 1)^2$

19. $9x^2 - 16y^2 = 144 \Leftrightarrow \dfrac{x^2}{16} - \dfrac{y^2}{9} = 1$

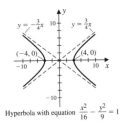

Hyperbola with equation $\dfrac{x^2}{16} - \dfrac{y^2}{9} = 1$

21. $16x^2 - 64x - 25y^2 + 150y = 561$

$\Leftrightarrow 16(x^2 - 4x + 4 - 4) - 25(y^2 - 6y + 9 - 9) = 561$

$\Leftrightarrow 16(x - 2)^2 - 25(y - 3)^2 = 400 \Leftrightarrow \dfrac{(x - 2)^2}{25} - \dfrac{(y - 3)^2}{16} = 1$

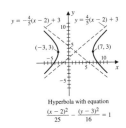

Hyperbola with equation
$\dfrac{(x - 2)^2}{25} - \dfrac{(y - 3)^2}{16} = 1$

23. The vertex is midway between the focus and the directrix so is $(0, 1)$ and the equation has the form $y - 1 = \dfrac{1}{4c}x^2$. Since the distance from the focus to the directrix is 1, and the directrix is above the focus, $c = -1$, so $y - 1 = -\frac{1}{4}x^2$.

25. Since the foci $(\pm 3, 0)$ are centered about the origin, the hyperbola is in standard position with axis on the x-axis, and $c = 3$. Since a vertex is $(1, 0)$, $a = 1$ and

$$c^2 = a^2 + b^2 \Rightarrow 9 = 1 + b^2 \Rightarrow b^2 = 8.$$

The equation is

$$x^2 - \frac{y^2}{8} = 1.$$

27. Since the foci of the ellipse are $(0, \pm 5)$, the equation has the form $\frac{y^2}{a^2} + \frac{x^2}{b^2} = 1$ and $c = 5$. Since $(4, 0)$ is on the curve,

$$\frac{16}{b^2} = 1 \Rightarrow b^2 = 16.$$

Then

$$c^2 = a^2 - b^2 \Rightarrow a^2 = 25 + 16 = 41.$$

The equation is

$$\frac{y^2}{41} + \frac{x^2}{16} = 1.$$

29. Since the directrix $y = -2$ is horizontal and below the pole the conic has the form $r = \dfrac{ed}{1 - e \sin \theta}$. Since $e = 3$, the conic is a hyperbola and

$$r = \frac{(3)2}{1 - 3 \sin \theta} = \frac{6}{1 - 3 \sin \theta}.$$

31. For $r = 4 + 4 \cos \theta$

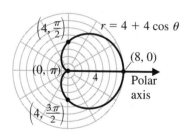

33. For $r = 1 + 3\sin\theta$

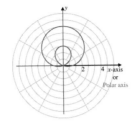

35. For $r = 2\sin\theta$

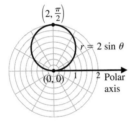

37. For $r = 2\cos 2\theta$

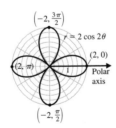

39. For $\theta = 1/2$

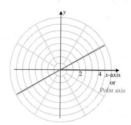

41. Since $x^2 + y^2 = r^2$ and $x = r \cos \theta$,

$$r = \frac{3}{1 + \cos \theta} \Leftrightarrow r + r \cos \theta = 3 \Leftrightarrow r = 3 - x \Leftrightarrow r^2 = 9 - 6x + x^2 \Leftrightarrow$$

$$x^2 + y^2 = 9 - 6x + x^2 \Leftrightarrow y^2 = 9 - 6x \Leftrightarrow y^2 = -6\left(x - \frac{3}{2}\right) \Leftrightarrow$$

$$x - \frac{3}{2} = -\frac{1}{6}y^2.$$

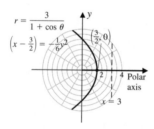

43. Since $x^2 + y^2 = r^2$ and $x = r \cos \theta$,

$$r = \frac{4}{2 - \cos \theta} \Leftrightarrow 2r - r \cos \theta = 4 \Leftrightarrow 2r = 4 + x \Leftrightarrow 4r^2 = 16 + 8x + x^2$$

$$\Leftrightarrow 4x^2 + 4y^2 = 16 + 8x + x^2 \Leftrightarrow 3x^2 - 8x + 4y^2 = 16,$$

and

$$2\left(x^2 - \frac{8}{3}x + \left(\frac{4}{3}\right)^2 - \left(\frac{4}{3}\right)^2\right) + 4y^2 = 16 \Leftrightarrow 3\left(x - \frac{4}{3}\right)^2 + 4y^2 = \frac{64}{3}.$$

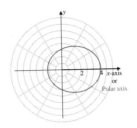

45. $x = t^2 - 1, y = t + 1 \Rightarrow t = y - 1, x = (y-1)^2 - 1 \Rightarrow x + 1 = (y-1)^2.$

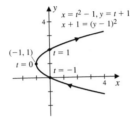

47. $x = e^t, y = 1 + e^{-t} \Rightarrow y = 1 + x^{-1} \Rightarrow y = 1 + \frac{1}{x}$ for $x > 0$, since

$x = e^t > 0.$

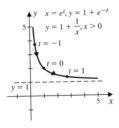

49. $x = (\sin t)^2 + 1, y = (\cos t)^2 \Rightarrow x + y = (\sin t)^2 + (\cos t)^2 + 1 = 2 \Rightarrow$

$y = -x + 2$, for $1 \le x \le 2.$

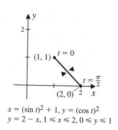

$$x = (\sin t)^2 + 1, y = (\cos t)^2$$
$$y = 2 - x, 1 \leqslant x \leqslant 2, 0 \leqslant y \leqslant 1$$

51. The number of leafs in the curves $r = 1 + \sin 2n\theta + (\cos n\theta)^2$ is $2n$.

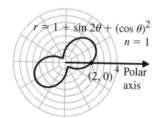

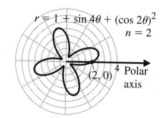

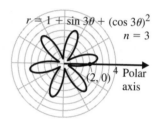

53. The curves in the figures are given by the parametric equations

$$x = (1 - a\sin t)\cos t, y = (1 - a\sin t)\sin t, \quad \text{for} \quad 0 \leq t \leq 2\pi.$$

If $a = 1$, the curve is a cardioid. If $a < 1$, the curve is a limacon without a loop.

If $a > 1$, the curve is a limacon with a loop.

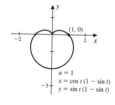

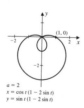

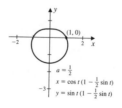

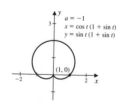

55. Epicycloids defined by the parametric equations

$$x = (a+b)\cos t - b\cos \frac{a+b}{b}t, \quad y = (a+b)\sin t - b\sin \frac{a+b}{b}t, \quad \text{for} \quad 0 \le t \le 2\pi.$$

are shown below.

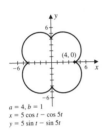

$a = 4, b = 1$
$x = 5\cos t - \cos 5t$
$y = 5\sin t - \sin 5t$

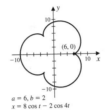

$a = 6, b = 2$
$x = 8\cos t - 2\cos 4t$
$y = 8\sin t - 2\sin 4t$

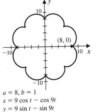

$a = 8, b = 1$
$x = 9\cos t - \cos 9t$
$y = 9\sin t - \sin 9t$

Solutions for Exercise Set 5 Calculus

1. (a) The equation has the general form $y = ax^2$. Since (x_1, y_1) is on the curve, $y_1 = ax_1^2 \Rightarrow a = \dfrac{y_1}{x_1^2}$ and

$$y = \frac{y_1}{x_1^2}x^2.$$

(b) The equation has the general form $y - k = a(x - h)^2$. Since (x_1, y_1) is on the curve, $y_1 - k = a(x_1 - h)^2 \Rightarrow a = \dfrac{y_1 - k}{(x_1 - h)^2}$ and

$$y - k = \frac{y_1 - k}{(x_1 - h)^2}(x - h)^2.$$

3. (a) The equation has the general form $y - k = a(x - h)^2$. Since $(h+1, k+1)$ is on the curve, $k + 1 - k = a(h + 1 - h)^2 \Rightarrow a = 1$ and $y - k = (x - h)^2$.

(b) The equation has the general form $x - h = a(y - k)^2$. Since $(h+1, k+1)$ is on the curve, $h + 1 - h = a(k + 1 - k)^2 \Rightarrow a = 1$ and

$$x - h = (y - k)^2.$$

5. The equation of the ellipse is $\dfrac{x^2}{4} + \dfrac{y^2}{9} = 1$, so $a^2 = 4, b^2 = 9$. The equation of the tangent at the point $\left(1, \frac{3\sqrt{3}}{2}\right)$ is

$$\frac{x(1)}{4} + \frac{y\left(3\sqrt{3}/2\right)}{9} = 1 \Rightarrow 9x + 6\sqrt{3}y = 36 \Rightarrow y = -\frac{\sqrt{3}}{2}x + 2\sqrt{3}.$$

7. The equation of the hyperbola is $\dfrac{x^2}{4} - \dfrac{y^2}{3} = 1$, so $a^2 = 4, b^2 = 3$. The equation of the tangent at the point $(2\sqrt{2}, \sqrt{3})$ is

$$\frac{x(2\sqrt{2})}{4} - \frac{y\left(\sqrt{3}\right)}{3} = 1 \Rightarrow 6\sqrt{2}x - 4\sqrt{3}y = 12 \Rightarrow y = \frac{\sqrt{6}}{2}x - \sqrt{3}.$$

9. Since the two triangles in the figure are similar, $\dfrac{\sqrt{36-y^2}}{6} = \dfrac{x}{2}$, so the equation for the point on the ladder is

$$\sqrt{36-y^2} = 3x \Rightarrow 9x^2 + y^2 = 36.$$

So the point moves along the portion of the ellipse $y = 3\sqrt{4-x^2}$ from the point $(0,6)$ to the point $(2,0)$.

11. The light source should be placed at the focal point of the parabola. Since $y = \dfrac{1}{4c}x^2 = \dfrac{1}{4}x^2$, the axis of the parabola is the y-axis and $c = 1$, so the focal point is $(0,1)$. Place the light source one unit along the axis from the vertex.

13. If the parabolic cross section of the mirror is a parabola of the form $x = \dfrac{1}{4c}y^2$, the information implies the point $(3.75, 100) = \left(\frac{15}{4}, 100\right)$ lies on the parabola, so

$$\frac{15}{4} = \frac{1}{4c}(100)^2 \Rightarrow 4c = \frac{4(100)^2}{15} = \frac{8000}{3} \Rightarrow c = \frac{2000}{3}.$$

So the focal point of the mirror is $\left(\dfrac{2000}{3}, 0\right)$ and the observers viewing area is $\dfrac{2000}{3}$ inches from the center of the mirror.

15. Since

$$r = a\cos\theta + b\sin\theta \Rightarrow r^2 = ar\cos\theta + br\sin\theta \Rightarrow x^2 + y^2 = ax + by,$$

we have

$$x^2 - ax + y^2 - by = 0 \Rightarrow x^2 - ax + \frac{a^2}{4} + y^2 - by + \frac{b^2}{4} = \frac{a^2 + b^2}{4},$$

and

$$\left(x - \frac{a}{2}\right)^2 + \left(y - \frac{b}{2}\right)^2 = \frac{a^2 + b^2}{4}.$$

17. $r = \dfrac{ed}{1 - e \cos \theta}$

(a) The aphelion, a, occurs when $\theta = 0$ and the perihelion, p, when $\theta = \pi$. So

$$a = \frac{ed}{1 - e \cos 0} = \frac{ed}{1 - e}, \qquad p = \frac{ed}{1 - e \cos \pi} = \frac{ed}{1 + e},$$

and

$$\frac{a(1 - e)}{e} = d = \frac{p(1 + e)}{e} \Rightarrow a(1 - e) = p(1 + e) \Rightarrow a - ae = p + pe,$$

so

$$a - p = ae + pe \Rightarrow a - p = e(a + p) \Rightarrow e = \frac{a - p}{a + p}.$$

(b) The length of the major axis, R, satisfies

$$2R = a + p = \frac{ed}{1 - e} + \frac{ed}{1 + e}.$$

From part (a),

$$d = \frac{a(1 - e)}{e} \Rightarrow 2R = \frac{e\left(\frac{a(1-e)}{e}\right)}{1 - e} + \frac{e\left(\frac{a(1-e)}{e}\right)}{1 + e} \Rightarrow 2R = a + \frac{a(1 - e)}{1 + e},$$

so

$$2R = \frac{a(1 + e) + a(1 - e)}{1 + e} = \frac{2a}{1 + e} \Rightarrow a = R(a + e).$$

Also from part (a),

$$d = \frac{p(1 + e)}{e} \Rightarrow 2R = \frac{e\left(\frac{p(1+e)}{e}\right)}{1 - e} + \frac{e\left(\frac{p(1+e)}{e}\right)}{1 + e} \Rightarrow 2R = \frac{p(1 + e)}{1 - e} + p,$$

so

$$2R = \frac{p(1 + e) + p(1 - e)}{1 - e} \Rightarrow 2R = \frac{2p}{1 - e} \Rightarrow p = R(1 - e).$$

19. We have

$$\sqrt{x^2 + (y-1)^2} + 1 = \sqrt{x^2 + (y+1)^2}$$

$$x^2 + y^2 - 2y + 1 + 2\sqrt{x^2 + (y-1)^2} + 1 = x^2 + y^2 + 2y + 1$$

$$2\sqrt{x^2 + (y-1)^2} = 4y - 1$$

$$4(x^2 + y^2 - 2y + 1) = 16y^2 - 8y + 1$$

$$12y^2 - 4x^2 = 3,$$

which is the equation of a hyperbola.

Appendix

Two copies of a placement examination have been included here so that you can assess your readiness for Precalculus and Calculus. These tests are similar to those used at Youngstown State University for a number of years, so we have confidence in the . We would expect you to be successful in a Precalculus using the Faires and DeFranza PreCalculus book if you score above about 16 before entering the course. A score of 28 or above indicates that you are likely to be well-prepared for a University Calculus sequence. All of this is, of course, made under the assumption that you will do a sufficient amount of homework in the respective courses. There is little substitute for diligent work when it comes to learning mathematics, or any other worthwhile skill.

We suggest that you take one of these examinations before you enter your Precalculus course and the other after you complete the course. This will show how much you have accomplished in Precalculus as give you a brief review of the topics that you will see regularly in your Calculus course.

The answers to the placement examinations given at the end of this Appendix, and both the examinations and the answers can be downloaded from the WWW site

http://www.as.ysu.edu/~faires/PreCalculus2/Placement

PRECALCULUS-CALCULUS
READINESS EXAMINATION
Test # 1

1. When $a = -2$, $b = 3$, and $c = 4$, the value of the expression $\dfrac{b^2 - c}{a^2 - b^2}$ is

    (a) 1   (b) -1 (c) $-\dfrac{5}{13}$ (d) $\dfrac{5}{13}$

    (e) None of the above.

2. The expression $(-3b)\left(-2a^2b\right)^2$ simplifies to

    (a) $-12a^4b^3$ (b) $36a^4b^4$ (c) $12a^4b^3$ (d) $6a^2b^2$

    (e) None of the above.

3. The expression $6x^2 - 7[-2x^3 - 6(3x^3 - 1)]$ simplifies to

    (a) $112x^3 - 6x^2 - 42$ (b) $140x^3 + 6x^2 - 42$ (c) $-140x^3 + 6x^2 + 42$

    (d) $-112x^3 + 6x^2 - 42$ (e) None of the above.

4. If $3(x + 2) + x = 4\left(1 - \dfrac{1}{2}x\right)$,   then x is

    (a) $-\dfrac{1}{2}$ (b) $-\dfrac{1}{3}$ (c) 0   (d) -1

    (e) None of the above.

5. The expression $\dfrac{\sqrt{x} \ \sqrt[3]{x}}{\sqrt[4]{x}}$ simplifies to

    (a) x (b) $x^{29/6}$ (c) x^{-1} (d) $x^{7/12}$

    (e) None of the above.

6. The quotient $\dfrac{\dfrac{1}{x+h}-\dfrac{1}{x}}{h}$ simplifies to

(a) $\dfrac{1}{h^2}$ (b) $\dfrac{-h^2}{(x+h)x}$ (c) $-\dfrac{1}{x(x+h)}$ (d) $\dfrac{1}{x(x+h)}$

(e) None of the above.

7. If $-6x+1 < 2$, then

(a) $x > -\dfrac{1}{6}$ (b) $x > -\dfrac{1}{2}$ (c) $x < -\dfrac{1}{6}$ (d) $x < -\dfrac{2}{5}$

(e) None of the above.

8. If $\dfrac{7}{5x-6} = \dfrac{1}{x-1}$, then x is

(a) $\dfrac{1}{12}$ (b) $\dfrac{1}{2}$ (c) $-\dfrac{5}{2}$ (d) $-\dfrac{13}{2}$

(e) None of the above.

9. The quadratic expression $x^2 - 19x + 84$ factors as

(a) $(x-12)(x-7)$ (b) $(x+12)(x+7)$ (c) $(x-21)(x-4)$

(d) $(x-28)(x-3)$ (e) None of the above.

10. The expression $\dfrac{4x-5}{x^2-3x+2} - \dfrac{3}{x-2}$ simplifies to

(a) $\dfrac{3}{x-1}$ (b) $\dfrac{x-8}{(x-2)(x+1)}$ (c) $\dfrac{1}{x-1}$ (d) $\dfrac{x-8}{(x-2)(x-1)}$

(e) None of the above.

11. The polynomial $24x - 14x^2 - 3x^3$ can be factored as

(a) $-x(x+6)(3x-4)$ (b) $-3x(x+4)(x-2)$ (c) $-x(3x-8)(x+3)$

(d) $-x(x-6)(3x+4)$ (e) None of the above.

12. The expression $\dfrac{x^2 - 9}{x + 3} \cdot \dfrac{2x - 6}{x^2 - 6x + 9}$ simplifies to

(a) 1   (b) $\dfrac{1}{3x}$ (c) $\dfrac{2}{x + 3}$ (d) 2

(e) None of the above.

13. The quadratic equation $x^2 - x - 7 = 0$ is satisfied when x is

(a) $1 \pm 2\sqrt{2}$ (b) $\dfrac{1}{2}(1 \pm 2\sqrt{2})$ (c) $\dfrac{1}{2}(1 \pm \sqrt{29})$ (d) $\dfrac{1}{2}(-1 \pm \sqrt{29})$

(e) None of the above.

14. If $|x - 3| \le 4$, then

(a) $1 \le x \le 7$ (b) $-7 \le x \le -1$ (c) $-1 \le x \le 7$ (d) $-7 \le x \le 1$

(e) None of the above.

15. One factor of $25y^{16} - 4x^2$ is

(a) $5y^8 + 4x$ (b) $5y^{12} - 2x$ (c) $5y^4 + 2x$ (d) $5y^8 - 2x$

(e) None of the above.

16. The inequality $(2x - 3)(x - 4) < 0$ is satisfied when

(a) $\dfrac{3}{2} < x < 4$ (b) $x > 4$ only   (c) $x < \dfrac{3}{2}$ only   (d) $x < \dfrac{3}{2}$ or $x > 4$

(e) None of the above.

17. The expression $\left(\dfrac{x^8 y^{-4}}{z^{-8}}\right)^{-\frac{3}{4}}$ is equivalent to

(a) $x^6 y^3 z^{-6}$ (b) $x^{-6} y^{-3} z^6$ (c) $x^{-6} y^3 z^6$ (d) $x^{-6} y^3 z^{-6}$

(e) None of the above.

18. The slope of the line with equation $4x + 3 = y - 2$ is

(a) $\dfrac{4}{5}$ (b) $\dfrac{4}{3}$ (c) $-\dfrac{4}{3}$ (d) 4

(e) None of the above.

19. A line with slope 2 and y-intercept 3 has the equation

(a) $y = 3x + 2$ (b) $x = 2y + 3$ (c) $y = 2x + 3$ (d) $x = 3y + 2$

(e) None of the above.

20. If (x,y) satisfies both of the equations $2x + 3y = 7$ and $2x - 3y = 4,$ then x is

(a) $\dfrac{3}{4}$ (b) 5   (c) $\dfrac{5}{4}$ (d) $\dfrac{11}{4}$

(e) None of the above.

21. If $f(x) = 2 - 3x,$ then $f(x-1) + f(x) - 1$ is

(a) $2 - 6x$ (b) $-6x$ (c) $6 - 6x$ (d) $8 - 6x$

(e) None of the above.

22. The largest set of real numbers in the domain of the function $f(x) = \dfrac{1}{\sqrt{3-x}}$ is

(a) $x < 3$ (b) $x \leq 3$ (c) $x < -3$ (d) $x \leq -3$

(e) None of the above.

23. If $f(x) = 3x - 5$ and $g(x) = 2x^2,$ then the composition $(f \circ g)(x) \equiv f(g(x))$ is

(a) $6x^2 - 5$ (b) $2x^2 - 5$ (c) $18x^2 - 60x + 50$ (d) $6x^3 - 10x^2$

(e) None of the above.

24.   The equation $\dfrac{x^2 - 3x + 2}{x - 1} + 2x = 4$ is satisfied when x is

(a) 1 or 2   (b) $\dfrac{2}{3}$ only   (c) 2 only   (d) -1 or $\dfrac{2}{3}$

(e) None of the above.

25.   The expression $\dfrac{(1 - x)(x + 2) + (x + 2)^2}{(1 - x)(x + 2)}$ simplifies to

(a) $x^2 + 4x + 4$ (b) $x^2 + 4x + 5$ (c) $x + 3$ (d) $\dfrac{3}{1 - x}$

(e) None of the above.

26.   The distance between the points $P(3, 8)$ and $Q(-5, 2)$ is

(a) 14   (b) 18   (c) $2\sqrt{10}$ (d) 10

(e) None of the above.

27.   For positive real numbers m, n, and r, which of the following are true?

$$\textbf{I.} \quad \log(mn) = \log m + \log n$$
$$\textbf{II.} \quad \log\left(\frac{m}{n}\right) = \frac{\log m}{\log n}$$
$$\textbf{III.} \quad \log(m^r) = r \log m$$

(a) I, II, and III (b) I and II   (c) I and III   (d) II and III

(e) None of the above.

28. Suppose that 3^8 is approximately $7,000$. Which of the following best approximates 3^{16}?

(a) $(7,000)^8$ (b) $490,000$ (c) $14,000$ (d) $49,000,000$

(e) None of the above.

29. A rectangle has a length that is 6 meters more than its width. What is the width of the rectangle if the perimeter of the rectangle is 156 meters?

(a) 36 meters   (b) 42 meters   (c) 75 meters   (d) $-3 + \sqrt{165}$ meters

(e) None of the above.

30. The volume of a sphere is proportional to the cube of its radius. Suppose the sphere S has a radius that is twice the radius of the sphere C and that the volume of C is V. Then the volume of S is

(a) $2V$ (b) $3V$ (c) $6V$ (d) $8V$

(e) None of the above.

31. The graph of $y = \cos \frac{1}{3}x$, for x in the interval $[0, 3\pi]$, crosses the x-axis at

(a) $0, \frac{3\pi}{2}$, and 3π (b) 0 and 3π (c) $\frac{3\pi}{2}$ only   (d) 0 only

(e) None of the above.

32. When the expression $\dfrac{1}{\tan t + \cot t}$ is defined, it is equivalent to

(a) 1 (b) $\dfrac{1}{\sin t \cos t}$ (c) $\sin t \cos t$ (d) $\dfrac{\sin t \cos t}{\sin t + \cos t}$

(e) None of the above.

33. The solutions of the equation $2 \cos x - \sqrt{3} = 0$ that lie in the interval $[0, 2\pi]$ are

(a) $\frac{\pi}{6}$ and $\frac{11\pi}{6}$ (b) $\frac{\pi}{6}$ and $\frac{5\pi}{6}$ (c) $\frac{\pi}{3}$ and $\frac{5\pi}{3}$ (d) $\frac{\pi}{3}$ and $\frac{2\pi}{3}$

(e) None of the above.

34. The value of $\sin(t - \pi)$ is the same as the value of

(a) $-\sin t$ (b) $\sin t$ (c) $-\cos t$ (d) $\cos t$

(e) None of the above.

35. An open rectangular box has height h and a square base with a perimeter $4x$. The surface area of the box is

(a) $x^2 h$ (b) $4x + h$ (c) $2x^2 + 4xh$ (d) $x^2 + 4xh$

(e) None of the above.

36. In an isosceles right triangle two sides have length 5. The length of the hypotenuse is

(a) $\sqrt{10}$ (b) $\dfrac{5\sqrt{3}}{2}$ (c) $5\sqrt{2}$ (d) $5\sqrt{3}$

(e) None of the above.

37. An angle of $\dfrac{7\pi}{3}$ radians is the same as an angle of

(a) $180°$ (b) $420°$ (c) $240°$ (d) $30°$

(e) None of the above.

38. A rectangle R has width x and length y. A new rectangle S is formed from R by multiplying all the sides of R by 6. How much more area does S have than R?

(a) $5xy$ (b) $6xy$ (c) $35xy$ (d) $36xy$

(e) None of the above.

39. The equation $x^2 + y^2 + 6x - 2y - 15 = 0$ describes a circle with

(a) center $(-3, 1)$ and radius 5   (b) center $(3, -1)$ and radius 5

(c) center $(-3, 1)$ and radius 25   (d) center $(3, -1)$ and radius 25

(e) None of the above.

40. The graph of the equation $y = \dfrac{2x}{x^2 - 2}$ has a vertical asymptote whose equation is

(a) $y = 2$ (b) $x = -\sqrt{2}$ (c) $y = \sqrt{2}$ (d) $x = 2$

(e) None of the above.

PRECALCULUS-CALCULUS
READINESS EXAMINATION
Test # 2

1. When $a = 2$, $b = -3$, and $c = 5$, the value of the expression $\dfrac{a^2 - b}{c^2 - b^2}$ is

   (a) $\dfrac{1}{3}$ (b) $\dfrac{7}{34}$ (c) $\dfrac{7}{16}$ (d) $\dfrac{1}{16}$

   (e) None of the above.

2. The expression $(-2a)\left(-4a^2b^3\right)^2$ simplifies to

   (a) $16a^5b^6$ (b) $-32a^5b^5$ (c) $64a^6b^6$ (d) $-32a^5b^6$

   (e) None of the above.

3. The expression $3x^2 + 5[4x^2 - 6(3x + 1)]$ simplifies to

   (a) $23x^2 - 90x + 30$ (b) $23x^2 - 90x + 5$ (c) $23x^2 - 90x - 30$ (d) $23x^2 - 18x + 1$

   (e) None of the above.

4. If $2(x + 3) + x = 6\left(1 - \dfrac{1}{3}x\right)$, then x is

   (a) $\dfrac{3}{5}$ (b) $-\dfrac{3}{8}$ (c) 0 (d) $\dfrac{1}{5}$

   (e) None of the above.

5. The expression $\dfrac{\sqrt[4]{x}\,\sqrt[3]{x}}{x}$ simplifies to

   (a) $x^{-5/12}$ (b) x^{-8} (c) $x^{5/12}$ (d) $x^{19/12}$

   (e) None of the above.

6. The quotient $\dfrac{(x+h)^2 - x^2}{h}$ simplifies to

(a) h (b) $2x + h$ (c) $x + h$ (d) $2xh + h$

(e) None of the above.

7. If $-5x + 1 < 5$,   then

(a) $x > -\dfrac{4}{5}$ (b) $x < -\dfrac{4}{5}$ (c) $x > -\dfrac{6}{5}$ (d) $x < -\dfrac{5}{4}$

(e) None of the above.

8. If $\dfrac{5}{3x - 4} = \dfrac{1}{x - 2}$,   then x is

(a) -1 (b) 3 (c) $\dfrac{3}{4}$ (d) 1

(e) None of the above.

9. The quadratic expression $x^2 - 7x + 10$ factors as

(a) $(x + 2)(x + 5)$ (b) $(x - 2)(x - 5)$ (c) $(x + 1)(x + 10)$ (d) $(x - 1)(x - 10)$

(e) None of the above.

10. The expression $\dfrac{3x - 4}{x^2 - 5x + 6} - \dfrac{1}{x - 3}$ simplifies to

(a) $\dfrac{2(x - 1)}{(x - 2)(x - 3)}$ (b) $\dfrac{2}{(x - 2)}$ (c) $\dfrac{2(x - 5)}{(x + 2)(x - 3)}$ (d) $\dfrac{2(2x - 5)}{(x - 2)(x - 3)}$

(e) None of the above.

11. The polynomial $x^3 + 7x^2 + 6x$ can be factored as

(a) $x(x - 1)(x - 6)$ (b) $x(x - 2)(x - 3)$ (c) $x(x + 2)(x + 3)$

(d) $x(x + 1)(x + 6)$ (e) None of the above.

12. The expression $\dfrac{x^2 - 16}{x + 4} \cdot \dfrac{3x - 12}{x^2 - 8x + 16}$ simplifies to

(a) 1   (b) 0   (c) 3   (d) $-\dfrac{3}{8x}$

(e) None of the above.

13. The quadratic equation $x^2 - x - 5 = 0$ is satisfied when x is

(a) $1 \pm \sqrt{6}$ (b) $\dfrac{1}{2}(1 \pm \sqrt{21})$ (c) $-\dfrac{1}{2}(1 \pm \sqrt{21})$ (d) $\dfrac{1}{2}(1 \pm \sqrt{6})$

(e) None of the above.

14. If $|x - 2| \le 5$, then

(a) $3 \le x \le 7$ (b) $-7 \le x \le -3$ (c) $-3 \le x \le 7$ (d) $-7 \le x \le 3$

(e) None of the above.

15. One factor of $9y^6 - 25x^2$ is

(a) $3y^3 + 5x$ (b) $9y^4 - 5x$ (c) $y^3 - 5x$ (d) $3y^4 + 25x$

(e) None of the above.

16. The inequality $\dfrac{x - 2}{x + 5} > 0$ is satisfied when

(a) $x < -5$ only   (b) $x < -5$ or $x > 2$ (c) $x < 2$ or $x > 5$ (d) $x > 2$ only

(e) None of the above.

17. The expression $\left(\dfrac{x^{12}y^{-3}}{z^{-3}}\right)^{-\frac{4}{3}}$ is equivalent to

(a) $x^{-16}y^{-4}z^4$ (b) $x^{-16}y^4z^4$ (c) $x^{16}y^4z^{-4}$ (d) $x^{-16}y^4z^{-4}$

(e) None of the above.

18. The slope of the line with equation $3x + 4 = y - 5$ is

(a) $\dfrac{3}{5}$ (b) 3   (c) $\dfrac{3}{4}$ (d) $-\dfrac{3}{5}$

(e) None of the above.

19. A line with slope 3 and y-intercept 5 has the equation

(a) $y = 5x + 3$ (b) $x = 5y + 3$ (c) $y = 3x + 5$ (d) $x = 3y + 5$

(e) None of the above.

20. If (x,y) satisfies both of the equations $3x + 4y = 9$ and $3x - 4y = 6$,   then x is

(a) $\dfrac{5}{2}$ (b) $\dfrac{1}{2}$ (c) $\dfrac{3}{2}$ (d) $\dfrac{9}{2}$

(e) None of the above.

21. If $f(x) = 2 - 4x$,   then $f(x - 2) + f(x) - 2$ is

(a) $14 - 8x$ (b) $10 - 8x$ (c) $-8x$ (d) $10 + 8x$

(e) None of the above.

22. The largest set of real numbers in the domain of the function $f(x) = \dfrac{1}{\sqrt{27 - x^3}}$ is

(a) $x < -3$ (b) $x \le -3$ (c) $x < 3$ (d) $x \le 3$

(e) None of the above.

23. If $f(x) = 2x - 3$ and $g(x) = 8x^2$,   then the composition $(f \circ g)(x) \equiv f(g(x))$ is

(a) $8x^2 - 3$ (b) $32x^2 - 96x - 72$ (c) $16x^2 - 3$ (d) $16x^3 - 24x^2$

(e) None of the above.

24. The equation $\dfrac{x^2 - 5x + 6}{x - 2} + x = 3$ is satisfied when x is

(a) 3 only (b) 2 or 3   (c) 0 only   (d) 2 or $\dfrac{3}{2}$

(e) None of the above.

25. The expression $\dfrac{(2-x)(x+3)+(x+3)(x-4)}{(x+3)(2-x)}$ simplifies to

(a) $x^2 - x - 11$ (b) $x^2 - x - 12$ (c) $\dfrac{2}{x-2}$ (d) $x - 3$

(e) None of the above.

26. The distance between the points $P(4,7)$ and $Q(-1,-5)$ is

(a) 5   (b) 17   (c) 13   (d) $\sqrt{13}$

(e) None of the above.

27. For positive real numbers m, n, and r, which of the following are true?

$$\textbf{I.} \quad \log(mn) = (\log m)(\log n)$$
$$\textbf{II.} \quad \log\left(\frac{m}{n}\right) = \log m - \log n$$
$$\textbf{III.} \quad \log(m^r) = r \log m$$

(a) I, II, and III (b) I and II   (c) I and III   (d) II and III

(e) None of the above.

28. Suppose that 2^{11} is approximately $2,000$, which of the following best approximates

(a) $(4,000)^{11}$ (b) $40,000$ (c) $4,000,000$ (d) $(2,000)^{11}$

(e) None of the above.

29. A rectangle has a length that is 2 meters more than its width. What is the width of the rectangle if the perimeter of the rectangle is 52 meters?

(a) $-1 + \sqrt{53}$ meters (b) 25 meters   (c) 14 meters   (d) 12 meters

(e) None of the above.

30. The volume of a sphere is proportional to the cube of its radius, and the surface area of a sphere is proportional to the square of the radius. Suppose the sphere S has a surface area that is 4 times the surface area of C. If the volume of C is 27, then the volume of S is

(a) 108   (b) 162   (c) 216   (d) 243

(e) None of the above.

31. The graph of $y = \sin \frac{1}{3}x$, for x in the interval $[0, 3\pi]$, crosses the x-axis at

(a) $0, \dfrac{3\pi}{2}$, and 3π (b) 0 and 3π (c) $\dfrac{3\pi}{2}$ only   (d) 0 only

(e) None of the above.

32. When the expression $\sin t(\tan t + \cot t)$ is defined, it is equivalent to

(a) 1   (b) $\dfrac{1}{\sin t}$ (c) $\dfrac{1}{\cos t}$ (d) $(\sin t)^2 \cos t$

(e) None of the above.

33. The solutions of the equation $2 \cos x - \sqrt{2} = 0$ that lie in the interval $[0, 2\pi]$ are

(a) $\dfrac{\pi}{4}$ and $\dfrac{3\pi}{4}$ (b) $\dfrac{\pi}{4}$ and $\dfrac{5\pi}{4}$ (c) $\dfrac{\pi}{4}$ and $\dfrac{7\pi}{4}$ (d) $\dfrac{3\pi}{4}$ and $\dfrac{7\pi}{4}$

(e) None of the above.

34. The value of $\tan(t - \pi)$ is the same as the value of

(a) $\tan t$ (b) $-\tan t$ (c) $\cot t$ (d) $-\cot t$

(e) None of the above.